FELIPE SABOYA DE SANTA

MKT

VOLUME 2

# INVASION AND MIND CONTROL

FELIPE SABOYA DE SANTA CRUZ ABREU

# MKTECH
## VOLUME 2
# INVASION AND MIND CONTROL

MKULTRA 2.0 experiments in the 21st century included

The technology of **invasion, control, reading and torture of the mind** that will change humanity forever.

2023 © Felipe Saboya de Santa Cruz Abreu

TITLE: MKTECH: Invasion and Mind Control - Volume 2
1st edition – October 2020

Author: **Felipe Saboya de Santa Cruz Abreu**
felipessca@gmail.com
Diagrams: **Felipe Saboya de Santa Cruz Abreu**
Text formatting: **Felipe Saboya de Santa Cruz Abreu**

Illustrations: **Eloy Rondon**
eloyartes@hotmail.com

Translation: **Luene Langhammer Alves**
luene.langhammer.alves@gmail.com

Book design: **Rubens Lima**
https://capista.com.br/
Book cover illustration: Freepik.com

ISBN: **9798862295993**

All rights reserved and protected under Brazilian Law number 9610 from 02/19/1998.

No part of this publication may be reproduced or transmitted in any form or by any means, electronic or mechanical, including photocopying, recording or any information storage and retrieval system, without permission in writing from the author.

The rights of this book belong to the author/editor FELIPE SABOYA DE SANTA CRUZ ABREU.

Official website:
www.invasionandmindcontrol.com

## CHAPTER 4 — 13

### MK-ULTRA 1950 - THE ORIGINS OF PSYCHO-ELECTRONIC/NEUROELECTRONIC/ELECTROMAGNETIC WEAPONS AND MIND CONTROL EXPERIMENTS — 13

4.1 - MK-ULTRA — 23
4.1.1 - The Experiments — 28
4.1.2 - Experiments conducted in Canada — 34
4.1.3 - Frank Olson — 34
4.2 - Evolution to "non-lethal" electromagnetic energy weapons — 37
4.2.1 - Delgado's Researches — 39
4.2.2 - Publications — 41
4.3 - MK-ULTRA and electromagnetic weapons — 43
4.4 - Projeto Montauk — 49
4.5 - Summary of Acronyms and Cryptonyms — 52
4.6 - Patents — 52
4.6.1 - Microwave Voice — 54
4.6.2 - Behavior change — 54
4.6.3 - Silent subliminal presentation system — 54
4.6.4 - Transmission of EEG Signals to the Brain — 54
4.6.5 - Radio Frequency Hearing Effect — 54
4.6.6 - VHF and electromagnetic radiation — 55

## CHAPTER 5 — 57

### MKULTRA 2.0 – Mind Control Experiments in the 21st Century — 57

5.1 - The 80s — 58
5.2 - The 90s — 58
5.3 - 2000, 21st century — 62
5.4 - Modern MKULTRA 2.0 - 2001 to 20XX — 66
5.5 - About the Targeted Individual — 71
5.6 - Operators — 73
5.7 - Organized Professional Stalkers (OPS) / Organized Professional Torturers (OPT), Gang-Stalking or "Cyberstalking" — 76
5.8 - Winter Soldier, "Programmed" Killer or "The Manchurian Candidate" — 84
5.9 - Full picture of the MKTECH universe — 87

**CHAPTER 5.10**                                                            **89**

**HOW ELECTRONIC TORTURE WORKS IN PRACTICE, "ELECTRONIC HARASSMENT" OR "CYBER TORTURE" INVOLVING ORGANIZED PROFESSIONAL STALKERS, MKTECH AND MKULTRA 2.0**    **89**

| | |
|---|---|
| 5.10.1.1 – Zero Hour | 90 |
| 5.10.1.2 - Proceeding with plan A… | 94 |
| 5.10.1.3 – EMR (Electronic Mind Reading), V2K | 95 |
| 5.10.1.4 - EMR, V2K, D2K (Synthetic Electronic Dream) | 96 |
| 5.10.1.5 – SYNTELE (Synthetic Electronic Telepathy) | 100 |
| 5.10.1.6 - EMR, V2K, D2K AND SYNTELE | 101 |
| 5.10.2 - My house is the worst place ever | 104 |
| 5.10.3 - Car | 107 |
| 5.10.4 - Bathroom | 108 |
| 5.10.5 - Personal Data | 110 |
| 5.10.6 - Trips | 111 |
| 5.10.7 - Paranoia | 113 |
| VI - Tips. Rule #1 for psychotronic warfare targets | 117 |
| 5.10.8 - Techniques employed by operators around the world | 119 |
| VII - Tips for the targets | 124 |
| VIII - Tips for the targets | 124 |
| IX - Tips for the targets | 125 |
| 5.10.9 - Side effects of the attacks | 126 |
| 5.10.10 - Peace and Quiet | 128 |
| X - Tips for the targets | 128 |

**CHAPTER 5.11**                                                         **131**

**ADVANCED TORTURE TECHNIQUES USING MKTECH**    **131**

| | |
|---|---|
| 5.11.1 - Advanced tactics used by OPS using MKTECH | 134 |
| 5.11.1.1 - Macabre and resilient companion | 135 |
| 5.11.1.2 - Controlling the environment | 136 |
| 5.11.1.3 - End of privacy | 136 |
| 5.11.1.4 – Bathroom: taking a bath without privacy | 137 |
| 5.11.1.5 - Your home becomes the most unbearable place in the world | 138 |
| 5.11.1.6 - Surveillance 24 hours a day, 7 days a week, 365 days a year | 139 |

| | |
|---|---|
| 5.11.1.7 - Cognitive deprivation | 140 |
| 5.11.1.8 - Decimate the will/Self-esteem/Motivation | 141 |
| 5.11.1.9 - Moral and sexual harassment | 142 |
| 5.11.1.10 - Fear | 145 |
| 5.11.1.11 - Tension | 146 |
| 5.11.1.12 - Maintaining emotional superiority over the victim | 147 |
| 5.11.1.13 - Anxiety | 147 |
| 5.11.1.14 - Chronic stress | 147 |
| 5.11.2 - Electronic Psychophysiological Warfare Protocols (Psy Warfare) | 148 |
| 5.11.2.1 - SYNTELE + D2K = a havoc combo | 148 |
| **5.11.2.2 – Reliving the past** | 151 |
| 5.11.2.3 - Night attacks, nightmares and screams | 153 |
| 5.11.2.4 - Sleep deprivation, the mainstay of modern electronic torture | 154 |
| XI - Tips for the Targeted Individual — Psychotronic Warfare Tactics | 157 |
| 5.11.2.5 - Psychic driving | 157 |
| 5.11.2.6 - Thought filter | 160 |
| 5.11.2.7 - Keeping the mind busy, "clogging" the primary reception systems with useless content | 160 |
| XII - Tips for the Targeted Individual — Psychotronic Warfare Tactics | 161 |
| 5.11.2.8 - Using hatred to destroy the target, to make them "implode" in their own rage | 161 |
| 5.11.2.9 - Ability to cease attention and divert the focus | 162 |
| 5.11.2.10 - Classic brainwashing techniques | 164 |
| 5.11.2.11 - Intensity of attacks during long periods of exposure to Microwave Voice (V2K), Synthetic Electronic Telepathy (SYNTELE) and Synthetic Electronic Dream (D2K) | 165 |
| 5.11.2.12 - Activity levels | 166 |
| 5.11.2.13 - Computerized Swarm Attack | 167 |
| 5.11.2.14 - Electronic narcosis | 169 |
| 5.11.3 - Advanced and complementary techniques employed in the use of SYNTELE | 171 |
| 5.11.3.1 - Mind bridge, electronic transfer of vocalized thought | 171 |
| 5.11.3.2 - Hidden interlocutor | 171 |
| 5.11.3.3 - Directed murmur | 172 |
| 5.11.3.4 - V2K audio for everyone | 173 |
| 5.11.3.5 - Crossed thoughts | 173 |
| 5.11.3.6 - Echo in thought | 174 |

| | |
|---|---|
| 5.11.3.7 - Electronic gaslighting | 175 |
| 5.11.3.8 - Artificially-induced suicidal ideation | 175 |
| 5.11.4 - Torture content | 176 |
| XIII - Tips for the Targeted Individual — Psychotronic Warfare Tactics | 178 |
| 5.11.5 - Altering raw brain waves to modify the victim's behavior | 178 |
| 5.11.6 - Consequences of persecution, torture and long-term exposure to the technology | 179 |

## CHAPTER 5.12   183

| | |
|---|---|
| Detailed techniques for extracting information from the human memory | 183 |
| 5.12.1 - What is memory and how does it work? — part 2 | 186 |
| 5.12.2 - Information extraction via Synthetic Electronic Telepathy and Intracranial Voice | 189 |
| 5.12.3 - Direct interference in memory | 192 |
| 5.12.4 - Evoking memories | 194 |
| 5.12.5 - Digging up memories | 197 |
| 5.12.6 - Complete the sentence | 199 |
| 5.12.7 - First impressions and associations | 200 |
| 5.12.8 - Disrupting the mental defense mechanism against associations | 202 |
| 5.12.9 - Electronic amnesia | 203 |
| 5.12.10 - Paramnesia | 206 |
| 5.12.11 - Access restriction | 206 |
| 5.12.12 - Electronic hypermnesia and memory triggers | 208 |
| 5.12.13 - Keeping secrets in the hacked mind under constant MKTECH attack | 210 |
| 5.12.14 - Mental defenses against mind reading and invasion | 211 |

## CHAPTER 6   215

| | |
|---|---|
| Danger in the use of the technology (part 6) - "Winter Soldier" Aaron Alexis | 215 |
| 6.1 - Attack on the naval base | 216 |

## CHAPTER 7   225

**USE OF THE TECHNOLOGY FOR FRAUD IN CIVIL SERVICE COMPETITIVE EXAMINATIONS**    225

7.1 - THE BIG FRAUD! THE FRAUD OF THE CENTURY: CIVIL SERVICE COMPETITIVE
EXAMINATIONS IN BRAZIL    228
7.1.1 - THE SCHEME    229
7.2 - A STEP-BY-STEP OF HOW THIS FRAUD WORKS    233
7.3 - RADIO FREQUENCIES AND ELECTRONIC/SPECTRAL TRACKING    240
7.3.1 - AMPLITUDE MODULATION (AM)    241
7.3.2 - FREQUENCY MODULATION (FM)    243
7.3.3 - FREQUENCY BANDS    245
7.4 - TECHNIQUES FOR SENDING QUESTIONS FROM INSIDE THE EXAM SITE TO THE HQ IN A
SECURE WAY    249
7.5 - GETTING QUESTIONS ANSWERED    250
7.6 - SIGNALS INTELLIGENCE    253
7.7 - THE STEPS TO TAKE TO IDENTIFY THE SIGNALS    255
7.8 - CONCLUSION    256

## CHAPTER 8    259

**DANGER IN THE USE OF THE TECHNOLOGY (PART 9) – TERRORIST ATTACK
ON THE U.S. EMBASSY IN CUBA AND CHINA (HAVANA SYNDROME)**    259
8.1 - EMBASSY WAR    262
8.2 - OFFICIAL PRELIMINARY MEDICAL/NEUROLOGICAL ASSESSMENT    268
8.3 - A SUMMARY OF THE ARTICLE    268
8.4 - CONCLUSION ABOUT ATTACKS AND KNOWN PROBLEMS WITH MICROWAVES    269

## CHAPTER 9    273

**COMPUTER SYSTEMS, AI AND SATELLITES USED IN MKTECH**    273
9.1.1 - ARTIFICIAL INTELLIGENCE OF THE MKTECH SYSTEM. THE DARK SIDE OF AI    273
9.1.2 - COMPUTATIONAL INTELLIGENCE    278
9.1.3 - AI - INFRASTRUCTURE    280
9.1.4 - AI - RAW BRAIN WAVES    280
9.1.5 - AI - MIND READING AND EMOTIONAL STATES    281
9.1.6 - EMOTIONS AND FEELINGS    286
9.1.7 - DISTORTING THE PERCEPTION OF REALITY    290
9.1.8 - ANSWERS AND SURVEYS    291

9.1.9 - Conclusion ........................................................................ 292

## CHAPTER 9.2 .......................................................................... 293

**TERRORIST SATELLITES, SATELLITE ELECTROMAGNETIC WARFARE, NEURONAL SATELLITES — SATELLITES, SATELLITES AND MORE SATELLITES**
................................................................................................. 293

9.2.1 - But what is a satellite? ..................................................... 295
9.2.1.1 - Types of orbits .............................................................. 298
9.2.1.2 - Frequencies ................................................................. 299
9.2.1.3 - Transmitters ................................................................ 301
9.2.1.4 - Terrestrial and space antennas .................................... 301
9.2.2 – Military/Intelligence/Neuronal satellites ..................... 302
9.2.3 - How many satellites currently orbit Earth? .................. 309
9.2.3.1 - How many satellites are currently in operation? ....... 309
9.2.3.2 - Whose satellites are these and what do they do? ..... 309
9.2.3.3 - The satellites in operation are divided into: ............... 310
9.2.3.4 - Who uses what? .......................................................... 310
9.2.3.5 - Earth observation - Intelligence data collection and environmental monitoring ........................................................ 310
9.2.3.6 - Some relevant data .................................................... 310
9.2.3.7 - Multiple uses .............................................................. 311
9.2.4 - Range of satélites ........................................................... 312
9.2.5 - Thoughts on Hertz (Hz) frequencies and data transmitted through MKTECH satellites ................................. 314
9.2.5.1 - Input ........................................................................... 314
9.2.5.2 - Output ........................................................................ 316
9.2.6 - Data transmission techniques ....................................... 322
9.2.7 - Conclusion ..................................................................... 325
9.2.8 - National sovereignty and terrorism using space, neuroelectronic weapons ........................................................ 328

## CHAPTER 12 ............................................................................ 333

**PRESENT, FUTURE AND CONCLUSION** .................................... 333
12.1 - Future of intimate relationships .................................... 340

| | |
|---|---|
| 12.2 - NEURAL COMPUTING, DIGITAL NEURAL MANIPULATION OR NEURAL PROGRAMMING | 343 |
| 12.3 - D2K – SYNTHETIC ELECTRONIC DREAM AND ITS THOUSAND FACETS | 347 |
| 12.3.1 - ONLINE (NEURAL) MULTIPLAYER MATCHES AND VIRTUAL RELAXATION ROOM | 347 |
| 12.3.2 - R.E.M GAMES | 349 |
| 12.3.3 - OPERATING ROBOTS AND MACHINES UNCONSCIOUSLY VIA D2K | 351 |
| 12.3.4 - CRYPTOCURRENCY MINING AND MENTAL DATA PROCESSING | 355 |
| 12.4 – A DECADE AND CENTURY OF THE BRAIN AND OF THE MIND | 358 |
| 12.4.1 - REGULATIONS AND HUMAN ENHANCEMENT | 360 |
| 12.5 - DARK MIND WEB, BRAIN NET, DEEP BRAIN WEB, BRAIN/NEURAL SATELLITES | 361 |
| 12.6 – KIDNAPPING MINDS | 364 |
| 12.7 - STATE MONITORING THOUGHTS. COGNITIVE MORALITY POLICE | 366 |
| 12.8 - SECRET POLICE (AS PER NKVD) | 371 |
| 12.9 - HAARP | 372 |
| 12.10 - THE FUTURE OF WAR | 376 |
| 12.11 - ARTIFICIAL INTELLIGENCE USING MKTECH IN THE FUTURE | 381 |
| 12.12 - FINAL CONCLUSIONS AND CONSIDERATIONS | 383 |
| **GLOSSARY I** | **390** |
| **GLOSSARY II** | **393** |
| **ABOUT THE AUTHOR** | **395** |
| **REFERENCES** | **396** |

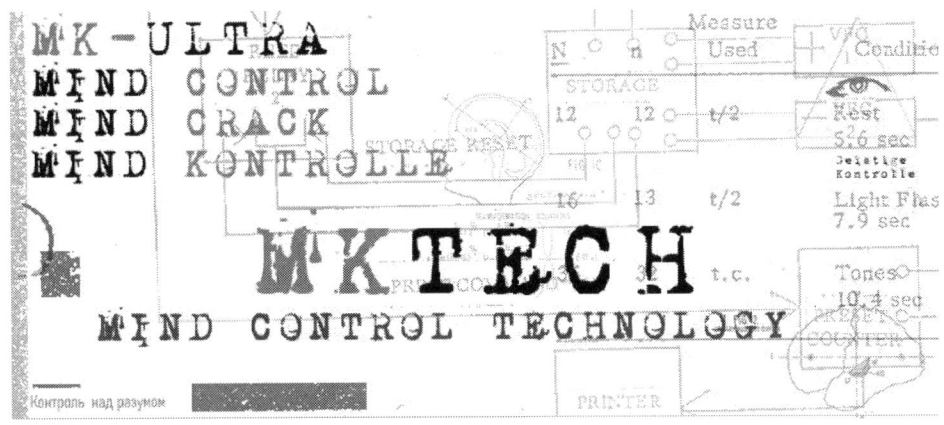

# CHAPTER 4

## MK-ULTRA 1950 - THE ORIGINS OF PSYCHO-ELECTRONIC/NEUROELECTRONIC/ELECTROMAGNETIC WEAPONS AND MIND CONTROL EXPERIMENTS

What is left of the documents that weren't destroyed in 1973 gives us a small taste of what the ultimate race for control of the human mind was like. A state project that lasted more than 28 years officially mobilized virtually all government branches of the largest countries. It involved doctors, universities, scientists, companies, hospitals, penitentiaries, military agencies, intelligence agencies, among others.

With the imminent creation of the second greatest weapon ever conceived by mankind, billions of dollars were spent and tests and experiments of all kinds were conducted directly on human beings. The development of the dreadful electromagnetic weapons of today responsible for all kinds of human evils began: deaths, suicides, thefts of thought and intellectual property, massacres perpetrated by people affected by electronic torture — "Winter Soldiers" who are nothing more than the modern denomination of the 21st century Manchurian Candidates.

In this scenario a new type of war began to emerge: the **Electronic Psy Warfare** together with the creation of potential agents for various purposes, such as drugs and new weapons for unconventional warfare. At this stage of the book, the origins of the weapons we've seen throughout the previous volume will be revealed, with all the horrors committed during the process of conducting these experiments directly on humans. Since this theme is extremely vast, the subject has been summarized as far as possible, delimiting the course of experiments that are connected with modern psychotronic weapons.

The construction of scientific thinking makes use of the accumulation of knowledge and information passed down over generations. MKTECH followed this same precept, in which it was trapped like a time capsule in the form of documents that were hidden from everyone until they were disclosed to the public years ago. I came across these documents in my research while looking at the patterns of current experiments and the way they're conducted. Such documents show exactly why the experiments never stopped and where the torture with a clear bias of scientific experiment with several purposes comes from. That way, I got a glimpse of how it all started.

Until recently I'd never heard of such experiments. I only began to understand when I had direct contact with modern weapons and observed the negative power of interaction with a person's nervous system to a degree never imagined and the extensive experiments that are being conducted on people remotely these days. Keep in mind that this is an extremely extensive subject. The released documents alone add up to tens of thousands of pages that resemble the Nazi and Russian concentration camps in the mid-twentieth century, but torture is applied for scientific purposes, not political or racial ones.

**Welcome to the origins of today's psychotronic weapons and the story behind it all.**

Tests to try to directly control people's minds and understand how humans react to certain types of torture and violence began in Nazi concentration camps during World War II, where acts of atrocity — targeting data on how the prisoners would respond physically and psychologically to torture — were committed by sadistic scientists such as Josef Mengele. There, torture and interrogation techniques actually began with the scientific study showing that the work of doctors and scientists involved in mind control didn't spontaneously come out of nowhere.

The importation of Nazi doctors in the United States through secret programs like PAPERCLIP is part of the context. After the end of World War II, German scientists and technical experts were being held in detention camps. One of these Nazi doctors had previously conducted experiments using mescaline — a plant that contains high-potency psychedelic alkaloids — on prisoners in the Dachau concentration camps, as he was looking for an effective method of extracting confessions through a truth serum. His name was **Kurt Friedrich Plötner.**

The British, French, Americans and Russians engaged in highly competitive recruiting efforts to secure the services of such experts. The prospect of losing the scientists' scientific knowledge led to the creation of projects for the extraction and expatriation of German dissidents and their technology. More than 1,000 German scientists were secretly brought to the U. S. with the State Department's approval. The most famous individual was Werner von Braun, the rocket scientist.

**Wernher Magnus Maximilian von Braun** was best known for heading up the German V2 rocket program, Vergeltungswaffen, the "retaliatory weapons" — which terrorized Europe during the conflict period — and later for leading the development of rockets for the U. S. government. The NASA rocket that took Neil Armstrong to the moon was built by von Braun and his colleagues.

Doctors were also brought through the Paperclip Project as von Braun so they would be able to continue the experiments they had been carrying out in concentration camps, but under the tutelage of the United States Department of Defense (DOD). With the onset of the Cold War between Russian communism and American capitalism shortly after World War II, in the mid-1950s, a secret and innovative battle for mind control began. Even then the Russians were well advanced in research into prisoner brainwashing, torture and mind control using invasive methods. This is because the Soviets have been working and developing experiments on mind control as of 1911. Since 1867, the famous psychiatrist Y. L. Ohotrovich (Ю.Л.Охотрович) developed hypnotic magnetic therapy based on the theory of animal magnetism. According to Ohotrovich, all living animals emit a magnetic field of organic origin. Various theories and primary experiments with mental suggestions and mind reading, magnetic fields and their effects on the mind have been studied.

In the post-Russian revolution, between 1917 and 1937, control programs with electricity and electromagnetic waves, involving theories of distance telepathy, thought transmission and the study of the brain, together with mental control techniques, were put into practice to test a series of theories relating to the field of bioelectricity.

Some of the various experiments documented between 1931 and 1937 that were under the tutelage of the USSR (the Soviet Union) stood out from the rest. One of them theorized on how to send information to the brain of a human (telepathy) using rustic equipment, however, it sought concretely the feasibility of transferring biological information. In order to measure the magnitude of this particular experiment and its importance to the government, the best Soviet centers were made available to conduct the tests, such as in Leningrad at the Bechterev's Brain Institute and in Moscow at the Laboratory of Biophysics, Academy of Science.

During this period of research, in an attempt to find "telepathic radiation" and thus transfer information to the brain, several phenomena were found in the search process between the emission of electromagnetic waves and some nerve activities in the brain, in particular interactions with microwaves that were discovered by the scientist B. G. Michaylovskiy (**Б.Г.Михайловский**). Nevertheless, telepathic radiation itself has never been found as theorized.

This is how these tests directed Soviet pre-war programs. In this sense, the interaction between electromagnetic waves and their effects on specific areas in the brain and on biological tissues became the goal. Variation of results based on how each brain area absorbed this radiation and its subsequent effects on cognition and emotions were analyzed. In the same experiment, they tried to answer questions related to how each band of the electromagnetic spectrum was able to affect different living organisms, turning into subtopics to be explored within the tests.

All the important research and results were confidential. Only S. J. Turlygn — a research scientist — published a summary of the results of these researches that theorized advanced interactions in the field of microwave radiation with humans in the 1940s. If someone theorizes that this sort of thing is possible, someone will inevitably test whether or not it is actually feasible, right?

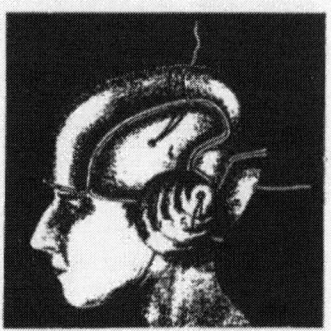

Figure 4.1 One of A. Barchenko's (А.Барченко) articles on mental suggestion and thought transmission published in 1911. "ПЕРЕДАЧА МЫСЛЕЙ НА РАССТОЯНИИ" (DISTANCE THOUGHT TRANSMISSION).

There were two scientists directly involved in the research that greatly helped in directing investigations in this field. The first was Dr. Leonid Vaseliev, Head of the Department at the University of Leningrad, who was active in research in the field of parapsychology, leading to many experiments in thought transference through electronic telepathy. The search for this type of phenomenon revealed valuable data in the discovery of electrical potentials of the brain (EEG), or that gave rise to the electroencephalogram. The continuation of his studies by the fellow Russian psychiatrist Hans Berger brought good results. A 1976 publication on his work entitled Remote Influence showed how it was possible to remotely influence the human brain in several different areas, such as motor and visual functions, affecting sleep/wakefulness stages and changes in electrodermal activities.

Even with classified Russian files, we can now confirm that the Soviets pioneered mind control tests. These studies constitute the source of information for the execution of torture techniques by the secret police from that period, such as the NKVD (People's Commissariat for Internal Affairs — **Народный комиссариат внутренних дел**), conducted in detainees of the most diverse nationalities in its numerous prison camps. These secret police are the same that gave rise to the famous KGB in 1954.

One of the first results of the new Russian brainwashing and manipulation techniques occurred at the beginning of World War II in trials of prisoners who were political opponents of the regime. The event became known as the Moscow Trials. In 1938, Josef Stalin's opponents had confessions extracted from them under torture, coercion and blackmail in prisons and facilities where they were located. During the trials they confessed that they'd committed the crime of conspiring against the October Revolution of which they were accused. However, what caught the attention of the whole world was the way in which the convicted individuals confessed their participation in the crime: in an extremely voluntary, calm and unnatural way, in addition to the acceptance in knowing that they would be executed. Then, the question arose: did the Russians discover the technique to manipulate the human mind?

Another case that drew attention was that of Cardinal Jozsef Mindszenty, who was fiercely opposed to communism and its ideology in Hungary. His case has been considered one of the most successful dezinformatsiya[43] operations in Russian history. Mindszenty was arrested by the Hungary's State Security Agency, the AVO (Allamvedelmi Osztaly), on December 26, 1948. He was imprisoned for 39 days and was forced to wear a clown outfit. The guards spent night and day laughing, talking loudly, telling sick jokes and smoking in an unventilated place on the outskirts of his cell. Every night Mindszenty was beaten until he passed out, but the guards wouldn't let him sleep after the acts. Drugs were administered in his food on a daily basis and he also underwent systematic interrogations. He was slandered, framed[44] with fake documents, judged in a show trial and convicted of being an American spy (under confession), of conspiring against the Soviet regime and attempting to start a Third World War. Advanced brainwashing torture techniques were used by the secret police. Mindszenty said he was so exhausted physically and mentally that he barely knew what he was saying or doing. Once again absurd confessions of crimes that the Cardinal clearly couldn't have committed caught the attention of intelligence agencies. The documents pointed out that the Soviets — in addition to being at the forefront of an advanced program in interrogation and mind control

---

[43] The Soviet and Russian term for disinformation operations. Disinformation is a secret tool for Intelligence, aiming to obtain a western, non-governmental seal of approval for the lies of the Soviet government. It's the use of communication and information techniques to mislead, deceive or give a false image of reality, by suppressing or hiding information, minimizing its importance or modifying its meaning. Its purpose is to deceive public opinion in order to protect private or political interests.

[44] Framed/Framing is a specialty of Soviet disinformation. It consists of changing a person's past by inverting their characteristics and the way the public perceives them, transforming the villain into a hero or vice versa. In other words, you must slander your enemies as having been pro-Nazis or pro-Fascists whenever possible. A core of truth is used to create slanderous and defamatory stories. For this, the target's entire contact network, intimacy and routine are studied. Original documents are used to forge fake ones with true characteristics.

techniques — had already managed to create false and confusing memories in certain prisoners.

The last case that finally woke up American intelligence to what was happening was during the Korean War, in which American soldiers trapped in combat became human subjects for brainwashing experiments. The episode took place during a confession made in front of the camera in which the prisoners admitted the use of biological weapons to attack Korea. The confession, however, wasn't true as soldiers never bombed Korea with biological weapons. It was then that the Americans tried to find out how the Russians managed to generate such a result in order to replicate it later in American subjects, that is, in their own citizens. Many questions remained in the air: how was it done? How to coerce people into confessing crimes they didn't commit? How to make prisoners believe that they really committed the crime?

The military and the CIA assumed that the people named in these events had their minds manipulated directly through some hitherto unknown technique of mental persuasion, leading to suspicion that Russia had found mechanisms that turned a human into a kind of robot that obeys orders without question and would be capable of committing crimes without knowing or remembering what they were doing at the time. Nevertheless, confirmation and detailed study of the results of these advanced techniques that these methods were being implemented came only when other Americans taken prisoner during the Korean War returned. Many of them had symptoms of brainwashing, temporary lapse in memory, multiple personalities, among other disorders. Another similar event occurred in 1960, when the spy pilot Gary Powers was captured and confessed that he worked for the CIA — one more fact that referred to the efficiency of the Soviet project at the time.

The CIA released 800,000 files with 13 million pages of secret documents from various operations over the years until the 1990s. From that date on, documents linked to intelligence remained confidential. The documents make it clear that Russia was always ahead in the matter of Intelligence. The CIA, for example, didn't know about its main enemy, the USSR. So, the agency was taken by surprise due to Stalin's death in 1953. They had no information about his successor Nikita Khrushchev.

The CIA has declassified around 18,000 pages of documents about the extensive mind control program started in the 1950s, which can be searched by anyone. One of the documents from **Subproject 61 MK-ULTRA** shows the CIA's interest in the subject of mind control along with military funds to execute experiments with implants in the brains of human subjects. In the beginning, the use of electrodes and implants was the only way to control the victim's mind.

*"The CIA was interested in this phenomenon after the return of some American POWs from the Korean War who were taken into the custody of the communist police and had their minds destroyed with the use of torture and brainwashing".*

Given the new current scenario on the development of something capable of altering the minds of prisoners with the new geopolitical situation, in order not to be left behind in this new field of study — which seemed to be the most promising of recent times —, the project BLUE BIRD was started. It was approved by the CIA director on April 20, 1950 with the intention of kick-starting mind-control experiments in humans. This would be the embryo of the greatest mind study project of all time, the **MK-ULTRA**. Some questions about the program were promptly raised by the directors and those involved, for example:

* Can we induce people — in a matter of hours or days — by using advanced hypnosis to take actions against their will and for our own benefit?

* Can we use people to commit acts such as attacks on aircraft, sabotage of public transportation (subways, trains, etc.) and cars?

* Is it possible to use temporal amnesia methods and hypnosis techniques to force the target to travel long distances, commit specific acts, and return to us with sensitive documents and materials?

* Can we guarantee total amnesia under any conditions?

* Can we change a person's personality?

* Is it possible to hide sleep-inducing chemicals in items for human consumption (e.g., cigarettes, teas, beers, pills)?

* Can we conduct experiments that make an individual act against their principles and their moral archetypes?

With this set of inquiries in hand, the CIA began its testing journey toward the goal of fully understanding and mastering the field of mind control.

In August 1951, the project was renamed ARTICHOKE. Bluebird and Artichoke have included a great deal of work on amnesia through torture, trying to use hypnosis as a way to generate personalities in order to create a programmed killer — The Manchurian Candidate — who would be an easy-to-manipulate agent and able to carry out any orders given to them. They would turn anyone into a programmed assassin so they could work as a CIA agent without knowing it.

Another objective of the Artichoke Project would be to develop more efficient interrogation techniques to the point of brainwashing the tortured person and getting as much information as possible. In order to achieve these primary goals, it'd be necessary to study the human brain and human behavior in detail; to formulate drugs such as the "truth serum" to persuade interrogation prisoners to tell the whole truth by taking people to a state of absolute suggestion; to create temporal amnesia; to conduct all kinds of experiments and test substances in order to fully understand their role in the central nervous system and find out which of these new chemical agents could be useful for the program.

The development and monitoring of all segments linked to human cognition in this process needed to be scientifically cataloged with the intention of achieving the primary goals, such as implanting false memories and creating multiple personalities. For this they should make use of various substances, a combination between them, in addition to cataloging its effects and consequences on humans. The details of the programs were kept secret even from other professionals inside the CIA.

More than 3,000 documents were released on just a few incidents of MK-ULTRA due to the Freedom of Information Act, a detailed demonstration of how thorough the experiments were. These documents described the horrors suffered by human subjects, equipment requests and contracts with large companies to supply substances for the tests, cataloging tortures and experiments with humans of all kinds, and the results obtained in the process. During the course of this chapter, I'm going to present a summary about this time period. Although I know that it'd take a separate book to cover all the information contained in it, I've summarized the data as much as possible. If you're interested in going deeper in the subject, you can check the documents in the public domain. They're available to everyone.

Operation Artichoke involved the detailed, systematic creation of specific techniques to develop amnesia, new identities, hypnotically implanted codes, and activity triggers to perform a particular action remotely. An Artichoke document dated January 7, 1953 presents how this whole scheme worked. In this particular case, it shows ways to create programmed killers, as Dr. Coling Ross found:

*"Some techniques included doping someone with Phenobarbital, turning on the lie detector, injecting intravenous drugs so that the individual would stay awake but in a state of suggestion. Then, false memories were introduced through interrogation. This was repeated over and over again until a story which was unconsciously interpreted as true was created in the subject's mind. Later, the false story/memory was described by the subject and they easily passed the lie detector test. Other techniques consisted of chemical brainwashing with sensory deprivation, causing the victim to be activated by a radio signal that only they would hear. It'd be a trigger to activate another personality, thus performing certain tasks without remembering them later. This creates the dissociative amnesia. Several tests with these assumptions were carried out for three years in a row."*

* Implementation of new identities activated by some secret code (such as a radio signal);

* Fabrication of memory lapses and amnesia, so the victim doesn't remember the acts performed;

\* Use of subjects in tests with controlled conditions and in the field.

Hypnosis wasn't the only method devised by doctors to create electroshock-controlled amnesia in the brain; drugs, **magnetic fields**, sound waves, sleep deprivation, isolation and many other methods have been studied.

At this stage, the use of hallucinogenic substances such as Amanita muscaria, Amanita pantherina and Amanita phalloides was exhaustively tested, as well as other alkaloids and fungi to search for techniques that included hypnosis to implement controlled temporal amnesia (which would be invaluable for operations), and combinations of various substances such as **Scopolamine** together with **Amobarbital** or **Sodium Amytal** administered in different doses for numerous purposes, including epilepsy induction to create advanced interrogation techniques and subdue human guinea pigs.

## 4.1 - MK-ULTRA

Artichoke and Bluebird were officially renamed MK-ULTRA by the CIA on April 13, 1953. The experiments became broader in scope, as they diversified the activities to various fields linked to mind control. It maintained the original goals, but with more in-depth knowledge of the topic.

MK-ULTRA is divided into 150 different subprojects, each with a particular objective and sharing data and results with each other. One group, for example, was responsible for testing and developing drugs that would directly control the minds of human subjects. Another group was responsible for developing similar techniques through electromagnetism. One of the goals was to produce components that had no color or smell and created partial or total amnesia and completely altered the human psyche. It was then that LSD began to be introduced into the studies.

LSD, or more precisely LSD25, is a pure white, crystalline compound that occurs naturally as a result of the metabolic reactions of the fungus Claviceps purpurea, which results in the alkaloid. In 1943, Swiss chemist Albert Hofmann, while working at Sandoz, "accidentally" discovered the complex and varied effects of the synthetic chemical which he became an enthusiast until his death at age 102.

LSD produces a very strong psychoactive reaction. It alters the cognitive mechanics of the brain, causing external and internal sensory stimuli to be interpreted in an abnormal way. In this context, depending on various external and internal aspects, it can lead a person to happiness and euphoria with a deep contemplative thought about the world and its mysteries, or it can trigger a terrible bad trip, in which the substance enhances every bad thought, leading the person to despair, depression, mental confusion, persecutory paranoia, etc. It was precisely this ability to profoundly alter the functioning of the mind that the CIA was seeking as a component, which would be used as a potentiating agent for "bad trips" in interrogations and in mind control experiments.

An emblematic feature of LSD is that it can make a person analyze memories without the scrutiny of emotional modulation, that is, to analyze such memories separately and in a timeless way; or it could be the other way around: memories can acquire an intense emotional charge, generating the opposite effect. Another striking feature of LSD is its ability to alter auditory perception, modifying the way sound is interpreted by the auditory cortex.

Some Top-Secret classification level documents showed how companies like Eli Lilly received $400,000 in 1954 to manufacture a large amount of LSD for the CIA. Others show the connection of various psychiatric institutions across the United States and other countries linked to MK-ULTRA and the CIA, as if they were a kind of agreement, further expanding the range of human guinea pigs and increasing the amount of data collected.

The program was gaining momentum around the world. It quickly obtained positive results in several fields of studies. Pioneer Russia was very advanced in the field and implemented its own MK-ULTRA, as well as China, Cuba and Korea. England, Canada and other countries also had their own MK-ULTRA projects or a connection with the American program.

Several subprojects have specialized in certain areas of mind control knowledge. All torture techniques could be tested in the field in order to observe the direct consequences on humans, thus qualifying which techniques were most effective. Everything was impeccably cataloged and stored in an old database: in sheets of paper. Amongst the techniques were some of the horrors black people, prostitutes, the mentally ill, beggars, inmates, immigrants and minorities in general — that were the main subjects of the entire experiment — went through. Many victims were tested under the influence of drugs and were never identified or compensated for the damage caused to them.

The drugs used in MK-ULTRA were aimed at altering the functions of the human brain and manipulating its mental state. Such drugs were used without the knowledge or consent of those to whom they were administered. In fact, one of the objectives of the project was to develop ways to administer such substances without the victim's knowledge.

Evidence published through the release of only part of the MK-ULTRA Project documents indicates that the research involved the use of animals and various types of drugs. However, it has led the industry to develop and test drugs on human beings without having to go through the normal human testing process.

Below are some data on the scale obtained by the war effort at the time for the torture experiments that recruited around 80 renowned institutions, such as 44 universities working directly/indirectly (knowingly or not, and for what purpose they were working), 12 psychiatric hospitals — such hospitals did know what was happening —, and 185 private research institutes.

MK-ULTRA continued with its official schedule until June 7, 1964. During this period, several experiments were conducted, as we're going to see in the next pages.

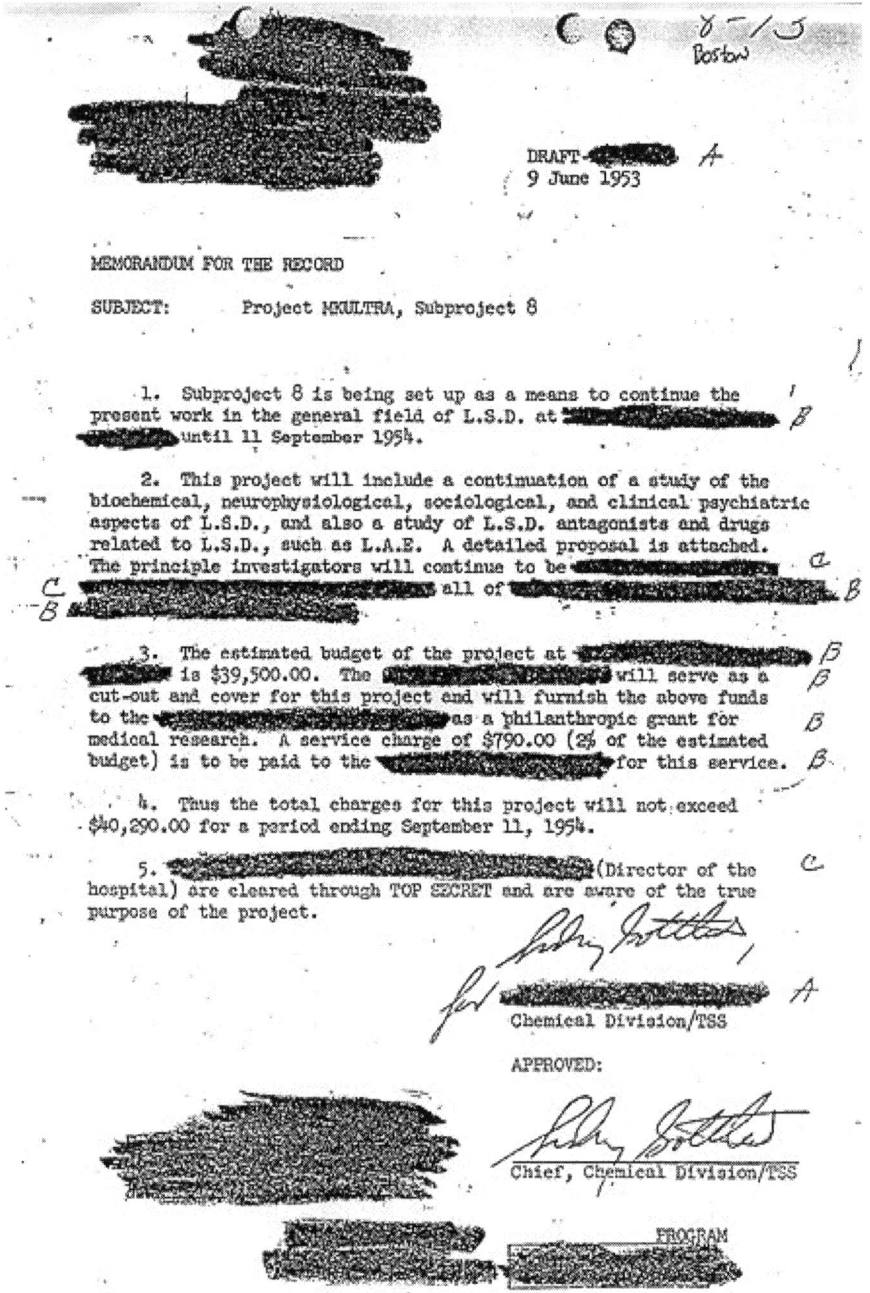

**Figure 4.2** Contract with the pharmaceutical company Eli Lilly that guaranteed the supply of large quantities of LSD for the CIA and MK-ULTRA/Subproject 8.

**Figure 4.3** The abbreviation "LSD" is from the German "Lysergsäurediethylamid" — lysergic acid diethylamide —, which is one of the most potent hallucinogenic substances known.

## 4.1.1 - The Experiments

In order to brainwash (completely or temporary) the countless human subjects chosen for the experiments, several invasive techniques were used to bring people to a state of complete disorientation, to remove any logical thinking, to make them weak, subdued and capable of being mentally modified. The acts of torture perpetrated by the tormentors of the time had a purpose: to answer the questions mentioned before and to improve what already existed in the context of the total knowledge of the human brain. The largest experiment in human history based on torture for scientific purposes had begun. The intention was to create suicides or programmed killers through the ingestion of drugs and hypnosis techniques. The experiments had the ultimate aim of studying how to control human behavior, thinking and emotions, using all available means to do so. And this maxim extends to the present day.

Electric shock to sensitive parts of the body, sense deprivation, sleep deprivation, confinement in boxes or coffins with small openings, all kinds of psychological and physical abuse. Extremes of heat and cold, including submersion in ice water and burning chemicals on the skin and eyes (e.g., Agent Orange); blinding light near the eyes; forced ingestion of bodily fluids; hanging them in painful or upside-down positions; keeping the victim hungry or thirsty for days, weeks or months; isolation of perception – making them not feel their senses (sight, hearing, touch, taste and smell); pulling or dislocating the upper and lower limbs; fainting tests involving carotid pressure, and so son.

Drugs were used to create illusion, confusion and amnesia, often given by intravenous injection; ingestion or use of intravenous toxic chemicals to cause pain or illness, including chemotherapeutic agents. They also used snakes, spiders, larvae, rodents and other animals to provoke fear, disgust and horror, and tested the reaction to substances secreted by those animals in the bloodstream and central nervous system.

Near-death experiences, commonly suffocation or drowning with immediate resuscitation; victims were forced to witness abuse, torture of people and animals. Threats to family, friends and loved ones, to force the achievement of certain goals were also exhaustively tested. Besides this, they changed the victim's personality by using LSD and all kinds of psychoactive drugs to see which left them more susceptible to hypnosis and thus "remove" the person's personality to insert a new one.

Taking this into consideration, some subprojects have specialized in researching new chemical agents with potential for military purposes in order to create weapons, including the "truth serum" and programmed killers. **Truth serum** is the term used to refer to a drug or a set of drugs that given to a suspect or patient who has something important to hide would force them to reveal such hidden information through suppression and even elimination of their will power. Several drugs were used for this purpose. The most common and known are scopolamine and sodium thiopental.

There was also the search for chemical agents efficient in causing confusion and amnesia, the insertion of memories based on a "fictitious reality" that was created from several torture sessions to later examine whether these realities were instilled in the subject's mind using the polygraph as a test. Also, use of substances and torture techniques such as:

* Belladonna (one of the most toxic plants);
* Morphine;
* Nerve gases;
* Atropine;
* Mescaline;
* Anesthetic gases;
* Nitrous oxide - commonly known as laughing gas;
* Animal toxin tests;
* Tests with paralyzing fish toxins that affect the nervous system;
* Attempt to reduce chronic pain and later reactivate it in order to cause sudden intense pain;
* Activation and inactivation of epileptic seizures that momentarily cloud the memory and cause extreme feelings of intimidation;

* Antihistamine (Anti-allergy) to cause emotional sensitivity in children and adults;

* Pyridoxine (Vitamin B6);

* Use of carbon dioxide indoors at various levels (30%, 40% and 80%), provoking changes in consciousness levels;

* Amphetamines and other substances used together to create the perfect brainwashing;

* More efficient drugs to induce deep sleep and alter the properties of the nucleus of the amygdala;

* Stimulation of certain brain regions with electrical current, ultrasound and other means by using radiant energy and interspersing it with x-rays;

* Surgical removals of parts of the brain like the cerebral amygdala responsible for emotions;

* Analgesic drugs were synthesized with the intention of taking away the pain without altering consciousness, as morphine did;

* Bulbocapnine is an alkaloid found in *Corydalis* (Papaveraceae) used to inhibit the reflex of motor activities in order to cause catatonic states — such as catatonic schizophrenia in some respects —, but temporarily. Small doses of bulbocapnine produce a state of tranquility and stillness that includes a certain degree of suggestibility and systematic torture;

* Combination of various brainwashing techniques;

* Degradation of mental states to weaken the will to live and the motivation that moves human beings in their daily lives;

* Chlorpromazine (an antipsychotic medication, a prototype in the treatment of schizophrenic patients);

* Use of carbon dioxide ($CO_2$) to produce a cataleptic state — in which the patient's limbs remain in whatever position was assigned to them by others;

* Injection of cocaine derivatives directly into the frontal lobe producing free and spontaneous conversations;

* Use of botulinum toxin in several cases. This is a neurotoxin produced by the bacterium *Clostridium botulinum* that causes decreased muscle action when injected;

* Use of Cohoba, Yopo or DMT (highly hallucinogenic drugs);

* Classical conditioning (Pavlovian or respondent conditioning), and constant reconditioning;

* Integrated tests for response at all levels, conscious and unconscious;

* Tests were combined with deprivation techniques and most of them would aim to improve the efficiency of the interrogation and collection of cause-and-effect information, food deprivation, sleep and sensory deprivation, water deprivation, and so on;

* Tests were always followed by a medical and scientific board that indicated the mental state produced by each deprivation or the set of them and the physical and psychological consequences observed;

* Creation and use of Amobarbital (sodium amytal), Sodium Pentothal (thiopental) and Secobarbital (Seconal);

* Use of hypnosis and violent electroshocks to the brain;

* Tests with hospitalized patients in order to change the personality of an enemy agent before they return to their original personality with no recollection of what they've said;

* Alcohol;

* Barbiturates;

* Amphetamines;

* Chemical lobotomy to replace usual lobotomy;

* Combined tests and hardships to the majority in order to improve interrogation techniques;

* Plant toxins with Pyridoxine;

* Ultrasound tests ("*sleep ray*") – electroshock to reach the sleep control center;

* They considered the use of atomic particles for this purpose;

* Tests with *Rauvolfia* with LSD-25 and electroshock;

* Polygraph used to detect physiological changes during testimonies with the use of sedatives that lead to induction of seizures;

* **Psychic driving** – torture technique in which the human guinea pig is subjected to an audio containing a message repeated in a loop 24/7 until their behavior completely changes, generating erratic and anomalous actions, modifying their personality and creating an emotional trigger that can drag on for years;

* **Ultrasonic Techniques** – to destroy the physical balance of the bulb and cerebellum of the brainstem using sonic weapons with frequencies of 2,000 Hz for 30 minutes at 140 to 150 db;

* **Vibration** – to cause fluctuations in the environment with low-frequency sound waves below what the human hearing is capable of detecting in order to make the person feel lost — to make them "drift", potentiating emotional problems and facilitating the acquisition of information; creating the false illusion of superiority on the part of the torturers over the victim;

* **Twilight Sleep** – an amnesic state that is characterized by insensitivity to pain without loss of consciousness, induced by an injection of morphine and scopolamine;

* **Brainwashing** – total isolation for long periods that produces a state of apathy, lack of purpose;

* **Flicker** – stroboscopic oscillation to create induced epilepsy or tests with induced seizures for various purposes. One of them is to

create a physical and psychological persuasion agent for anyone, which could be used later as a severe threat to the individual being questioned and thus convince the subject of the seriousness of the event. Reports say that tests combined with electroshock at various voltages to induce epilepsy caused some human subjects to bite so hard due to the pain that they broke their jaws in the process. All kinds of electroshock tests were conducted and various reactions tested in order to create a temporary (controlled) amnesia.

The documents show in detail how the experiments were conducted and the purpose of each test with their respective substances. There are thousands of pages. If you want to go deeper, I suggest that you access these documents. Here I just mentioned some torture techniques employed in the nearly 25 years of MK-ULTRA's existence. There is no doubt that the data collection results for a new war to come has been achieved.

More study seems worthwhile.

See: Chlorpromazine
Barbiturates

COPRINUS ATRAMENTARIUS — See Fungus Toxins

CORTISONE & A C T H

Cortisone may sensitize to convulsion, or convulse, persons who take it over a considerable period of time; it is given by physicians for a number of chronic illnesses — chiefly chronic arthritis, and chronic asthma — so that if a Subject has been taking it, his convulsive threshold may be appreciably lower than normal.

ACTH (adrenocorticotropic hormone) stimulates the body's production of Cortisone, along with some other effects. The Cortisone-effect, as above, might interest us.

Neither of these would be used actually to sensitize or to convulse, for our purposes. They would be too slow, and would involve a number of complications very undesirable.

CURRENT, galvanic — See Iontophoresis

Current, high frequency — See Iontophoresis
High Frequency and Radar Fields

Current, induced — See Electric Shocks
Electromagnetic Field

Current, Alternating — See Electromagnetic Field
Electric Shocks

DECONDITIONING — See Conditioning

DEPRIVATION-CONVULSIONS — See Deprivations
Sedative-Deprivations

DEPRIVATIONS

(1

See Brainwashing. See also SEDATIVE DEPRIVATIONS; &, DEPRIVATION OF FOOD, QUANTITATIVE.

Deprivations of dietary components, and Alterations of physical environments, can also be applied over a period of time (without other brainwashing stresses) to weaken a Subject non-specifically (#14), i.e., to make him more easily fatigued and to lower his resistance generally.

Specifically desirable mental states should be sought, in addition to a non-specific (#14) weakening of the Subject which would lower his resistance and make him more easily fatigued.

Mentioned under Brainwashing, is the drowsiness-tendency (#10) produced by a mild degree of starvation.

Under Sensitizing to Convulsion (page 2) is a list of biochemical and dietary means that may so act, viz.

DEPRIVATIONS

(2

| Dietary Means that probably Sensitize to Convulsion (not listed in order of probable effectiveness, which is unknown) | MENTAL STATES produced (other than generally lowered resistance, raised fatiguability, & the Sensitizing-to-Convulsion) |
|---|---|
| **Bio-Chemical** | |
| Low.tot.bl.electrolytes) Lowered blood-magnesium) Raised blood-potassium ) | other effects unknown |
| **Vitamins** | |
| Niacin (B2) deficiency | (if severe deficiency, (irritability & confusion |
| Pyridoxin (B2) " | other effects unknown |
| Riboflavin (B2) " | " " " |
| **Amino-Acids** | |
| Glutamin deficiency | other effects uncertain |
| **Antimetabolites** | |
| (producing deficiency of a specific Vit. or Am.Ac.) | effects, as above |

Questions:

Is it practical to alter the proportions of blood-electrolytes?

by feeding as a part of the regular meals:

a) certain salts?

b) certain ion-exchange resins?

c) vegetables of ordinary species, but grown in especially altered soils?

d) "mutant" vegetables, i.e., new species selected from freaks grown under influence of radio-activity?

Or to alter the vitamin-content of edible plants, by (c) or (d) above?

**Figure 4.4** MK-ULTRA document related to experiments with fungus toxins.

## 4.1.2 - Experiments conducted in Canada

The experiments carried out in Canada have ruined many people's lives. One of those directly responsible for running the program, Dr. D. Ewen Cameron, conducted a series of mind control experiments under the direct authority of the CIA on more than 53 people — all patients seeking treatment at the Allan Memorial Institute of McGill University between 1957 and 1961.

Selected patients were treated with high doses of LSD for 63 consecutive days and with electroshock therapy 75 times greater than that commonly used in patients. In order to mold new personalities, Cameron forced his patients to listen to the same message for 16 hours straight. This form of torture is known as psychic driving. Cameron and the CIA were interested in testing brainwashing and the individual's ability to redirect thoughts and actions.

Patients never authorized this type of procedure and were never informed that they were part of a large research study. A son of a victim who went through the program in Canada told a newspaper that his father, who was an independent, intelligent and caring person, became a completely different individual. Every time he was visited at the hospital, his father didn't speak much. When he did, he talked gibberish. If he wasn't sleeping, he was drowsy. The son talked to the father and noticed that he had no more memories — they seemed to have been erased. He also acted strangely, performing actions without remembering what he did or where he was. Later, in the 1980s, some victims went to Court and received compensation from the government of Canada.

The author and psychiatrist Harvey Weinstein established the direct relationship of mind control research done in England by the British psychiatrist William Sargant, involved in MK-ULTRA research in England, with Ewen Cameron's experiments in Canada, also for MK-ULTRA and with methods currently used as means of torture (for example, the use of hallucinogenic drugs, chemical agents, and sleep deprivation). Ewen Cameron often had the collaboration of William Sargant, both of whom were linked to CIA experiments.

## 4.1.3 - Frank Olson

During the early years of MK-ULTRA, a big problem arose for the CIA, and it involved a distinguished person: Frank Olson. He was an expert in biological and chemical weapons, highly respected in the American military's highest rank.

Frank Olson ended up "committing suicide" nine days after attending a meeting with CIA agents in November 1953. Olson threw himself from the tenth floor of a New York hotel. It was an unlikely death for someone with his prestige and service to the nation. However, he had been showing a sudden change in behavior. His striking and always good-humored personality presented signs of depression and psychotic symptoms.

In the week before his death, Olson and other scientists had attended a meeting with Sidney Gottlieb, hitherto the director of the Central Intelligence Agency and head of MK-ULTRA. At the time, Gottlieb would have slipped a considerable amount of LSD into Olson's drink. As a consequence, the scientist was placed in a sanitarium to be treated by the doctors involved in the program due to the result of the dosages. Olson completely freaked out days before he died. Why was he chosen? Because he would have discovered that the CIA was using American citizens in mind control experiments. Outraged, he would have threatened to leave the MK-ULTRA project. The problem was that Olson knew too much. Therefore, Gottlieb decided not only to test the power of LSD, but to put into action the plan to eliminate a prospective witness. Since Olson was doped, it would have been easy for the CIA agents to throw the scientist out of the hotel window and fake a suicide.

The program was "officially" shut down in the 1970s, and almost all of its material was destroyed, but some documents were recovered and a series of lawsuits against the CIA and the U. S. government commenced. Reports have been documented in several countries, including Sweden, where victims like Robert Naslund claimed that SABU — the secret police at the time — turned him into a human guinea pig to secretly monitor his movements using non-invasive radio waves. He claimed that words that trigger actions in the killers, or radio signals that would trigger such a state, were used as a posthypnotic trigger test.

# 25 Years Of Nightmares

## Victims of CIA-Funded Mind Experiments Seek Damages From the Agency

By David Remnick
Washington Post Staff Writer

Harvey Weinstein, a quiet, bearded man who practices psychiatry at Stanford University, says there are days when he is "ashamed" of his profession, nights when he cannot stop thinking about the Canadian psychiatrist who "ruined my father's life . . . Left him with nothing. It's a nightmare that never ends."

With funding from the CIA, the late Dr. D. Ewen Cameron did a series of mind-control experiments on 53 people, including Harvey Weinstein's father, Louis, a prosperous Montreal businessman. All had come to the Allan Memorial Institute of McGill University in Montreal between 1957 and 1961 for treatment of various psychological ailments.

The experiments, Weinstein says, left his father "a human guinea pig, a poor pathetic man with no memory, no life. He lost his business, he lost everything." Weinstein is one of nine plaintiffs in a lawsuit, seeking damages from the CIA.

To erase or "de-pattern" personality traits, Cameron gave his subjects megadoses of LSD, subjected them to drug-induced "sleep therapy" for up to 65 consecutive days and applied electroshock therapy at 75 times the usual intensity. To shape new behavior, Cameron forced them to listen to repeated recorded messages for 16-hour intervals, a technique known as "psychic driving." Cameron and the CIA were interested in brainwashing and the ability to redirect thought and action. The patients did not consent to the treatment and were never told they were being used for research.

"When you're 13 years old and you see your father—an independent, kind, smart person—become a different man before your eyes, it's impossible to accommodate that," Weinstein says. "I remember one of his first visits home from the hospital. He didn't talk much, and when he did talk it made no sense. When he wasn't sleeping he was drowsy. He asked us things about his parents, even though they'd been dead for years. His memory was gone. At night once, when I was in bed, I saw him come into my room and urinate on the floor. He didn't know where he was.

"My father has ended up feeling guilty that he had done something to deserve this punishment. He is convinced the CIA listens to his telephone. He's ashamed, embarrassed. My mother died without seeing the end of this. It will be a tragedy if my father dies without restoring some sense of dignity to his life."

Today Louis Weinstein lives alone in Montreal, cared for by his two grown daughters.

No one knows the whereabouts of all the subjects, some of whom may be dead. But Louis Weinstein and eight others, including Velma Orlikow, the wife of a New Democratic Party member of the Canadian parliament, claim they have been injured irreparably by the experiments. "I'd say Velma operates at about 20 percent of capacity," David Orlikow says. "It's horrific."

The CIA's involvement in mind control experiments has been coming to light for years. The suit filed by the group against the U.S. government has been pending here in U.S. District Court since December 1980 before Judge John Garrett Penn. The plaintiffs originally asked for $1 million each in damages but have cut that to $175,000. The government has offered to pay $25,000. The group's attorney, Joseph Rauh Jr., calls the settlement offer "demeaning" and contends that the CIA has managed to delay the proceedings by "stonewalling."

The CIA's counsel, Lee Strickland, declined to comment on the case. Agency spokeswoman Kathy Pherson said, "We don't comment on cases under litigation. It's inappropriate to try cases in the press."

In Cameron's defense, Brian Robertson, the present director of the Allan Institute, and James Farquhar, a psychiatrist there, wrote in the Montreal Gazette that "we have not been able to uncover a single shred of evidence that Dr. Cameron knew of the CIA connection with his research funding." They said Cameron's work "must be placed in its historical context" and that "in Cameron's day [researchers] were not expected to inform their patients of the nature of their research in the way that they are today."

The CIA has asked Judge Penn to block Rauh from taking depositions from two key agency figures—Stacey Hulse and John Knaus, who have been publicly identified as former CIA station chiefs in Ottawa. They are both retired.

Cameron, who died of a heart attack while mountain climbing in 1967, had been one of the most prominent psychiatrists in North America. A former president of both the Canadian and American psychiatric associations, he was selected to diagnose Nazi

*The Washington Post* 28 July 1985

**Figure 4.5** The Washington Post's publication on LSD, brainwashing, and mind control experiments.

28 November 1953

MEMORANDUM FOR THE RECORD: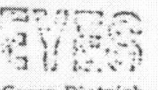

SUBJECT: Suicide of Frank OLSEN,
Army Civilian Employee, Camp Dietrich.

1. I was called by telephone at 5:00 A.M. this date by a CIA Officer and asked to meet with him, Dr. Gibbons and Dr. Sidney Gottlieb in the latter's office in Quarters Eye. I arrived there about 5:40 and was told the following story:

2. Mr. Robert V. Lashbrook, a TSS employee, had been in New York City on two occasions during the last week accompanying Subject, who was taking psychiatric treatment from a Dr. Abramson. Olsen is a civilian employee of the Chemical Corps at Camp Dietrich. The arrangements for Olsen to proceed to New York were made by Colonel Vincent Ruette (sp.), the officer in charge of Special Operations at Camp Dietrich, and at the suggestion of Dr. Gottlieb. The latter stated that he suggested Dr. Abramson due to the fact that the latter is a cleared consultant of both this Agency and the Chemical Corps, and that the sensitive nature of Olsen's work, part of which he was performing for TSS, made this appear desirable.

3. Dr. Gibbons stated that Olsen had been treated five or six times during the last week by Dr. Abramson. Olsen and Lashbrook returned to the Washington area for Thanksgiving but went back to New York City for further consultation with Abramson. Yesterday, Abramson decided that Olsen should be placed in a sanitarium for treatment for a period and apparently arrangements were made with a sanitarium near Rockville, called Chestnut Hill. Gottlieb reported that Subject had stated he was willing to take this treatment.

4. Last night, Lashbrook and Olsen had a room at the Statler Hotel. At 2:30 A.M. Lashbrook was awakened by a crash, awoke and found that

**Figure 4.6** Original CIA document on the investigation of the death of Frank Olson in 1953.

## 4.2 - Evolution to "non-lethal" electromagnetic energy weapons

Research would have continued and evolved naturally. In 1964, MK-ULTRA was renamed MK-SEARCH, however, the program will always be known as MK-ULTRA. After exhausting all the possibilities and tests related to physical and psychological torture during the 20 years in which the experiments were conducted as detailed above, MK-ULTRA already covered numerous different subprojects in 1968, each of them focused on a specific area that used the extraordinary amount ranging from 6 to 10% of the entire U. S. defense and intelligence budget.

The persistence and dirty work over the years were rewarded with the creation of various new technologies, chemicals, drugs, pathological agents, intelligence data and a profound and unprecedented knowledge of the human psyche, its reaction to maximum stress and the disclosure of hidden mental and behavioral secrets. Thus, advances in many areas of science were made on a scale of evolution and knowledge that can only be measured in the face of a war effort, such as the technological advance that was achieved during and after World War II (the computer, radar, rocket and atomic energy, for example).

So, the experiment began to take another direction — they followed another path. An area emerged as the consolidation of all the achievements, and it would be the newest and most promising area of study: mind control using electromagnetic waves, radio frequency — HF (High Frequency), SHF (Super High Frequency), VHF (Very High Frequency), UHF (Ultra High Frequency), and EHF (Extremely High Frequency) — more efficient to erase memories, implant false memories and encourage multiple personalities. All that was researched so far was the culmination of experiments with radio waves!

One of the people directly responsible for this direction was José Manuel Rodríguez Delgado. He claimed that he had never worked directly with MK-ULTRA, but he received financial incentives from the Department of Defense in order to carry out his tests, more specifically from the Psychiatric Research Fund, the United States Public Health Service, the Office of Naval Research and the U. S. Army Aeromedical Research Laboratory. Therefore, his work contributed to this funneling of projects towards mind control by electromagnetism. It was the milestone in the program that started the foundation of today's contemporary model as we know it, the Mind Control Technology, which is treated as a non-lethal weapon and secret of state.

In the historical context, this leap was also possible thanks to the development of increasingly sophisticated and improved equipment in space projects. Everyone benefited from this evolution provided by NASA and its programs, which were evolving several strands of science at the time due to the space race against the Soviet Union. During this period, the Cold War addressed several disputes in the most varied fields, from the best known, such as the development of increasingly powerful atomic bombs, to the lesser known, but no less important, such as mind control projects.

José Delgado was the author of more than 134 scientific publications between the 1950s and 1970s on electrical stimulation in cats, monkeys, mammals and humans. In 1963, the New York Times reported his experiments on its front page: he had implanted a "stimoceiver" in the caudate nucleus of a fighting bull.

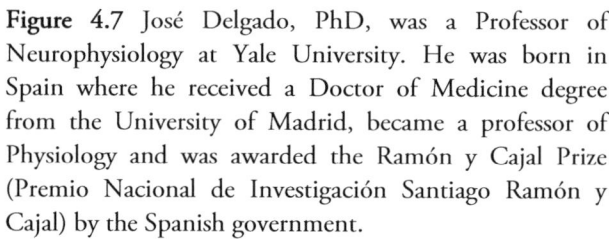

Figure 4.7 José Delgado, PhD, was a Professor of Neurophysiology at Yale University. He was born in Spain where he received a Doctor of Medicine degree from the University of Madrid, became a professor of Physiology and was awarded the Ramón y Cajal Prize (Premio Nacional de Investigación Santiago Ramón y Cajal) by the Spanish government.

Delgado received funding from the Office of Naval Research (Research Laboratory - Grant number F29600-67-C-0058) and social welfare for brain electrode research on children and adults.

In 1954, he received $7,950.00 for the Neurological mechanisms in epilepsy; in 1955, it was $9,610.00 for the Neurological behavior in epilepsy and in group behavior. It was $9,610.00 in 1956 and $10,000.00 in 1960 for the most diverse experiments with radio waves and the mind.

Dr. Delgado did similar research in monkeys and cats. Some documents describe these cats as "electronic toys". He was able to control the movements of his animal and human subjects by pushing buttons on a remote transmitter box coupled to their brains. Another case involving an 11-year-old boy who was a patient of Dr. Delgado shows how he was successful in partially altering human identity upon remote stimulation of the brain electrode.

In his book Physical Control of the Mind: Toward a Psychocivilized Society, Delgado believed that the control of people's thoughts through remote stimulation by electrodes offered man another step up the evolutionary ladder. With this technology, we could directly control our own minds by modifying our psychic states and altering our moods, as well as acquiring complete knowledge of all mental processes without the use of chemicals (such as medicines)!

## 4.2.1 - Delgado's Researches

Delgado's research interests centered on using electrical signals to evoke responses in the brain. He was fascinated by the possibility of finding a way to communicate directly with the brain. His work began with cats, but later he experimented with monkeys and humans, including psychiatric patients. Delgado's experiments aimed to explore the use of intracranial stimulation to provide knowledge about the brain.

Much of his work was based on an invention of his own called the Stimoceiver, a device that combined a radio antenna and an electrical stimulator of brainwaves with a receiver that monitored the EEG waves and sent a signal in separate channels. Some of these devices were as small as a coin. This allowed the object of the experiment, the subject, to have complete freedom of movement, generating greater dynamism for scientists to control the experiments together with the reduction of equipment and radio signals. Thus, several different brain regions could be stimulated simultaneously. This was a big improvement of his equipment which in the beginning used large electrodes implanted in the brain and worked with the use of wires and accessories.

The stimoceiver could be used to stimulate emotions and control behavior. According to Delgado, "Radiostimulation of different points of the amygdala and hippocampus in the four patients elicited a variety of responses, including pleasant sensations, elation, deep concentration, thoughtful state, odd feelings, relaxation, colored visions and so on." Delgado stated that *"brain transmitters can remain in a person's head for life. The energy to activate the brain transmitter is transmitted by way of radio frequencies."*

Using the stimoceiver, Delgado found out that he could not only provoke emotions, but he could also elicit specific physical reactions. These reactions, such as the movement of a limb or the clenching of a fist, were achieved when Delgado stimulated the motor cortex. A man who had one of these implants inserted in his brain was stimulated to produce a specific physical and mental reaction and he reported that he was unable to resist it. The patient said, "*I guess, doctor, that your electricity is stronger than my will*". Some consider one of Delgado's most promising finds is that of an area called the septum within the limbic region. This area, when stimulated, produced feelings of strong euphoria. These feelings were sometimes strong enough to overcome physical pain and depression.

The apparatus was capable of transmitting amygdala signals to a computer using a program called Amygdala Spindles, which results in the synchronous activation of significant populations of amygdala neurons, reducing amygdala spindles and modifying the behavior of Paddy, a monkey (subject) in the experiments.

Delgado created many inventions. Other than the stimoceiver, he created a "chemitrode": an implantable device that released controlled amounts of a drug into specific areas of the brain. Furthermore, Delgado invented an early version of what is now a cardiac pacemaker.

In Rhode Island, Delgado did some work in what is now a closed mental hospital. He chose patients who were desperately ill whose disorders had resisted all previous treatments. So, electrodes were implanted in about 25 of them who suffered from neurological problems. The devices induced the blocking of various behaviors, such as complex motor actions, aggression and sexual desires. Most of these patients suffered from schizophrenia or epilepsy. To determine the best placement of electrodes within human patients, Delgado initially researched the work of Wilder Penfield, who studied the brains of epileptics in the 1930s, as well as earlier animal experiments and studies of people with brain damage. He also used the technique to control aggressive behavior in monkeys that was later replicated by the CIA in humans.

## 4.2.2 - Publications

Delgado was invited to write his book *Physical Control of the Mind: Toward a Psychocivilized Society* as the forty-first volume of a series entitled *World Perspectives*, edited by Ruth Nanda Anshen. In the book, Delgado discussed how we have managed to tame and civilize our surrounding nature, arguing that now it was time to civilize our inner being. The book has been a center of controversy since its release. The tone of it was a challenge, and the philosophical speculations went beyond the data. Its intention was to encourage less cruelty, more happiness and greater benevolence, as well as to improve mankind, however it came into conflict with religious beliefs.

José continued to publish his research and philosophical ideas through articles and books for the next quarter century. He wrote over 500 articles and six books. His final book in 1989 was named *Happiness* and had 14 editions.

Another important event: Delgado started to conduct the same experiments remotely, without electrodes, using only radio waves — that is, without equipment physically implanted in the subjects' heads. In one of his researches, he managed to make a monkey sleep and dream using only radio waves.

Even in 1960, Delgado already knew that his researches would one day culminate in a profound human-machine interaction and brain-computer interfaces, using radio waves as a communication vector:

*"We are advancing rapidly in the pattern recognition of electrical correlates of behavior and in the methodology for two-way radio communication between brain and computers. Electronic knowledge and microminiaturization have progressed so much that the limits appear biological rather than technological".*

Delgado was a true exponent of his time. He was always looking for results that could benefit the human being, focusing on the well-being and improvement in the quality of life of people with severe neurological and psychopathological disorders. In his book *Physical Control of the Mind: Toward a Psychocivilized Society*, he states:

*"Fears have been expressed that this new technology brings with it the threat of possible unwanted and unethical remote control of the cerebral activities of man by other men, but this danger is quite improbable and is outweighed by the expected clinical and scientific benefits".*

Unfortunately, in 2020 — 51 years after publishing his predictions — I can categorically state that Delgado was wrong regarding such predictions made in 1969. Maybe it was because he believed too much in human goodness and in the peaceful purposes that his experiments always aimed at. Perhaps it never crossed his mind that they would use the basis of his research to help develop the most cruel and inhumane weapon ever produced by mankind.

It's undeniable the technological innovation and its less invasive methods for complete knowledge of the functioning of the mind that Delgado has exquisitely created, indirectly contributing to the making of today's psychotronic weapons, mainly due to the continuation and constant improvement of his researches by other scientists, using only electromagnetic waves without the need for any implant. Later, within the project, such researches were classified as top-secret electromagnetic weapons.

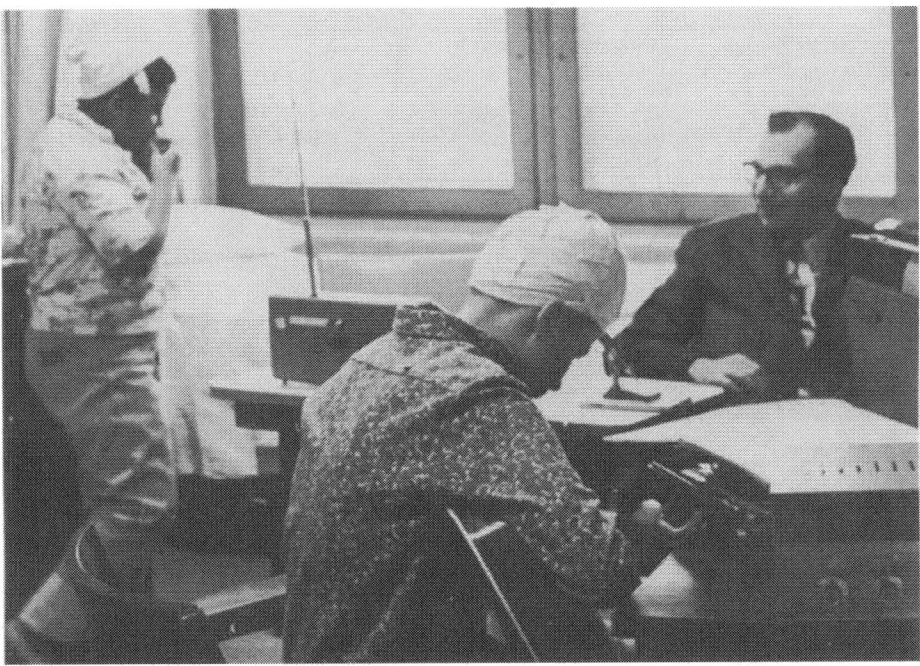

**Figure 4.8** Application of the stimoceiver by radio stimulation and telemetric electroencephalography of his patients' brains.

## 4.3 - MK-ULTRA and electromagnetic weapons

New advances introduced microwaves and stimuli via radio frequencies into the experiments. At that point, important changes in field tests relating to techniques of mental persuasion, torture and personality change began to take place. Gradually, the use of chemical agents gave way to electromagnetic waves as the main source of research. In this new stage, scientists were able to obtain significant results with the electronic stimulation of the brain through the implantation of a small probe into the subjects' heads. When triggered by radio waves, it would control emotions such as anger, sexual desire and tiredness.

They then realized that they could combine the strongest primitive human emotions with artificial external stimuli at any time. Thus was born the exploration of fear as a weapon that would later be of great value to communist totalitarian regimes and secret police. They could induce fear through electromagnetic waves without hurting people.

The tests also combined radio waves with MK-ULTRA's well-known studies of hypnosis. The hypnotist's words would reach the brain through electromagnetism, with no receiving device implanted. So, the intelligence service would intervene in the mind from a distance and the person being controlled wouldn't even notice it. That would put an end to the use of material evidence such as chips and probes.

Extensive research using radar waves concluded that the continuous use of this type of frequency could cause confusion in the subjects' minds. These results could also be obtained by using alternating current in a coil around the subject's head. In addition to causing mental confusion, it'd provoke a sensation of a flash of light and the electronarcosis, which is characterized by reversible and non-specific decreases in the excitability of neurons produced by various physical or chemical agents. This would lead the victim to a unique state, culminating in a "spiritual" emptiness and a type of surrender of the will to live, together with the loss of positive emotions.

Due to these unique characteristics, a project specializing in such effects within the mind control project was later codenamed Soul Catcher. It faithfully reflects the target's feeling when their mind is invaded via radio by unauthorized people that captures everything that goes on inside, imprisons and acquires the human essence, stealing their soul until it collapses. A combination of barbiturates with electromagnetic waves and electroshock was used to achieve such state.

Further studies performed later by the DOD and the CIA called **STARGATE** began to explore the communication of sound and images directly to the individual's brain using silent conversation (**the Silent Talk**). They were also tasked with analyzing supposed psychic powers and extrasensory perceptions abroad and at home, which automatically camouflaged themselves in the popular imagination and covered up the real nature of the experiments. On the Soviet side, they managed to reach these same levels in experiments as they created a new type of distraction to cover up their true effects: the famous remote viewing. That is, they took advantage of the innate quality in using these pseudosciences, elaborating theories that ended up leading to serious studies with concrete results, as was the case of the transfer of consciousness proposed in 1911 that culminated in all these studies and in the current neural weapons we're discussing at the moment.

The infamous remote viewers were the subjects and targets at the time. They received demodulated images and sounds in their brains, and they passed on information during these transmissions with details of specific geographic areas along with other various environmental characteristics. They also narrated certain events that occurred in the places indicated by the transmissions. So, remote viewers claimed to have paranormal powers capable of seeing crime scenes, explaining murder dynamics, describing remote structures, even deceiving some military segments that recruited such "remote viewers" to describe enemy bases and the artifacts that were found there. Needless to say, this wasn't a paranormal event. They just transmitted the information raised by the Soviet intelligence to be described by the supposed psychic, hiding details of how the technology worked and its purpose. These details were kept from the staff themselves, as they were to be used and tested, but without much fanfare, as it was a discovery that would completely change the course of wars and society. In this way, other military sectors didn't suspect what was behind the remote viewing; they just took advantage of the disclosed information, which for the most part was reliable and accurate.

The Soviet researcher and electrical engineer Professor I. M. Kogan postulated that the phenomenon of remote viewing was just the transfer of information via **ELF** electromagnetic waves, which would be able to penetrate even cages equipped to block different frequencies. This shows us why such attacks today are very difficult to stop.

Another subproject, number 129, was successful in its proposal to understand how the cerebral hemispheres worked together with the effects of weapons considered non-lethal, in a vast category of equipment with different purposes. It was aimed at using directed energy on humans and animals in order to momentarily disable or change the target's behavior.

Some experiments were conducted through universities that had contracts with the CIA.

Documents found from subproject number 62 dealt with how certain types of radio frequencies reversed neurological problems in chimpanzees. Project 54, on the other hand, investigated a way to cause trauma and concussions at a distance using bursts of mechanical waves (sound) that would propagate through the air and last for 10 seconds, causing a transient state of amnesia during impact, which would be used in brainwashing techniques. Last but not least, MK-ULTRA subproject 119 dealt with techniques for activating human beings by remote electronic means.

Research into the electromagnetic field's ability to insert false memories into altered states of consciousness or unconsciousness has been found in CIA files under the codename **Sleeping Beauty**. It referred to the Department of Defense studies on the influence of microwaves on the human mind during sleep, modifying the content of dreams and taking "complete" control of brain processes. This subproject is directly responsible for the creation of the current D2K - **Synthetic Electronic Dream** (Volume 1). The goal was to fully manipulate the thinking of the human guinea pig during a state of unconsciousness and to insert false memories — strong memories that would remain embedded in the brain for the rest of their lives. This objective was finally achieved in the mid-1980s. The technology, in addition to being fully operational, has been evolving satisfactorily with the decentralized experiments being carried out in the world today.

Stargate and **GRILL FLAME** started the project considered the most private and secret together with Sleeping Beauty in which the Department of Defense studied techniques to influence the human brain in a remote way by using electromagnetic waves, in this case microwaves. Once the study became feasible and the brain was completely hacked, Sleeping Beauty became the most powerful weapon in the mind control program (MKTECH) and was immediately classified as Top Secret. That's why nobody has a full patent or more information about it, just the result of the creation of a weapon that is being used all over the world nowadays.

Stargate was an interesting project aimed at improving the techniques for transmitting voice and images over radio waves. It's curious: the target who has their mind connected to the system actually assimilates all the data like a stargate. That is, an open portal in their mind representing point A; point B would be the place where sounds and images are generated and sent remotely — where any type of information travels without control and is absorbed by the mind with no restrictions. It takes advantage of the properties of the brain to send data, "upload" — thoughts, images, memories and raw EEG information — and receive data via "download". However, the phenomenon is perceived by the Targeted Individual as something that stimulates all cortical areas, including the sensation of being observed combined with a feeling that the atmosphere of a completely hostile environment is heavy, demodulating various cortical stimuli at once.

This mental phenomenon applies to the interpretation described by Rachel, who knows nothing about psychotronic weapons. Her story is at the beginning of Volume 1, chapter 2.3.2 – SYNTELE and Electronic Schizophrenia, do you recall? It says:

*"Sometimes she notices the presence of the voices, even though they say nothing. It'd be almost like a sensory experience, in addition to the voice".*

Stargate is exactly that. All current victims of psychotronic weapons feel the same. However, most of them don't understand what it is or can't put the feeling into words.

Subproject 129 of MK-ULTRA reported the success in the improvement of general understanding of the functioning of the cerebral hemispheres using the Computer Analysis of Bioelectric Response Patterns, which was a secret project conducted at George Washington University and University of Georgia. It involved implanting electrodes in the brains of animals in order to control their behavior by using a remote transmitter. They could then direct the animal's movements through radio signals, causing it to be used for attacks with biological and chemical weapons. The result was clear: controlling the mind and behavior and creating a dissociation through a combination of psychoactive drugs, sensory isolation, hypnosis, brain electrode implants, electric shock and a blast of different types of electromagnetic energy directly into the mind. The ability to create limited temporal amnesia was one of the first results achieved by a variety of different methods. The CIA was in favor of allowing direct hands-on experiments on humans, as there were several safety issues involving animals and the results would never be the same.

It's known that on the Russian side, some intelligence memorandums indicated that in 1961 the so-called unconventional research received a new boost; several projects were taken up again by the government.

Brain electrode research was also conducted independently by the co-authors of Delgado's research at Harvard, Drs. Vernon Mark, Frank Ervin and William Sweet, trained at Tulane University by brain electrode implant specialists. Drs. Vernon Mark and Frank Ervin were the authors of the book Violence and the Brain that describe the potential use of brain electrode implants to control urban violence. In the book they suggest that the use of this technology could be useful in controlling racial riots and crowds in urban disturbances of any kind. These experiments were initiated during the MK-ULTRA Project and formed one of the numerous foundations for the beginning of all technology in the development and conception of weapons considered non-lethal.

Dr. Adey was a member of the MIT Neurosciences Research Program and performed experiments with electromagnetic fields in the brains of human subjects. He edited a book entitled Brain Interactions with Weak Electric and Magnetic Fields, reported the document MK-ULTRA Subproject 8.

In 1956, in Subproject 62, Maitland Baldwind supervised the project involving radiofrequency stimulation in monkeys through the National Institute of Health (NIH), which was responsible for advanced research in the area of biomedicine. In 1957, in Subproject 68, an extensive experiment was conducted to study the effects of repeated verbal signals (sent in a loop) on the central nervous system and on human behavior. At that time, they were studying the famous Psychic Driving, and their results were even published in the American psychiatric journal of the time.

Subproject 138 was in charge of experiments involving the creation of biomedical sensors, including brain implants and radio frequency tests. Do you remember chapter 3, Volume 1, whose theme is focused on the incredible telemetric EEG capable of capturing countless internal electrical details of the target? Under the auspices of MK-ULTRA, several projects like these have led to today's advanced EEG monitoring tool.

Dr. Persinger was also an important figure. He published a research article entitled Psychophysiological Effects of Extremely Low Frequency Electromagnetic Fields: A Review. Another work involving brain studies was conducted in 1972 at UCLA. It was called Violence Project and was headed by Dr. Louis Jolyon West. It consisted of inserting electrodes into the brains of prisoners at Vacaville State Prison located in California, a place often used for mind control experiments with psychoactive drugs. The prisoners there were monitored remotely by radio waves. If they entered restricted places or exhibited sexual arousal that was detected by the electrodes via Telemetric EEG (Electroencephalography), a signal would be sent to the electrodes implanted in their brain and the prisoner would be immediately immobilized by a discharge. It was similar to the experiment conducted by Dr. José Delgado, in which he made a fighting bull stop its attack with the use of radio waves.

An immense network of connections with MK-ULTRA illustrated how everything was maintained: not through a central agency, but an interconnection of academic relations that guaranteed conferences and military commitments. Some people in this network had no direct relationship with the CIA or military agencies, but their work was directly related to the development of non-lethal mind control weapons for the creation of Manchurian Candidates.

In the years that followed, Russian scientists tested their new inventions based on their experiments with low-frequency wave radiation. They bombed the American Embassy in Moscow, resulting in several people seriously injured and dead (including people working at the embassy).

Some intelligence messages already addressed how the microwave voice could be extremely useful in various fields and devastating for those who suddenly started hearing voices out of nowhere! They still considered applying several configurations to manipulate human muscular systems and induce limb paralysis, similar to epilepsy, using synchronous configurations from 50 to 100 kV to work with the voltage of 2V in cell membranes through electromagnetic pulse attacks, which would cause the brain to synchronize certain regions at frequencies called alpha waves.

## 4.4 - Projeto Montauk

Montauk is located in Suffolk County, New York. It had a population of 3,851 as of 2000, but it became an active center in testing electromagnetic waves for mind control experiments. It's possible to immediately see a huge radar antenna that sets the tone for the type of experiments taking place there. Now that they could directly access the central nervous system and interact with its electromagnetic fields without the need for implants, they performed virtually every torture test based on that premise, but on another level. Montauk grouped a number of experiments with electromagnetism in a single location for years.

The testimonies of people tortured in this project are appalling and emotional. Some were tested with chemical agents, such as Agent Orange, which burned to the bone. Others who worked on the project lost their minds. Some of them walked with a pot over their heads, unable to explain the tests with psychotronic weapons that were being developed at the time. They actually confused them with time travel experiments due to the enormous influence of this weapon on our thoughts and an immense cognitive dissonance. Some subjects reported what happened there in a few words: *"It was like waking up inside a fiction film with no possibility of getting out alive"*. Individuals were forced to think what they didn't want to think, using combinations of frequencies that ranged from 435 MHz to 2 GHz. They even programmed children in a similar way to the Manchurian Candidates. These programs were carried out until the mid-1970s and were part of the MK-ULTRA apparatus whose code name would be **Phoenix Project.**

The place was terrifying. The facility was buried so deeply in order to insulate the premises from other electromagnetic waves that sunlight never reached it. Electrical equipment was visible everywhere like a funeral decoration. In this place, the main modern psychotronic weapons emerged and evolved. For years people served as mind-control guinea pigs with as much terror as possible. Some victims working at the site described a chair in which the subject sat surrounded by coils; there they received all psychotronic radiation directly into the brain, spent days listening to voices, having their dreams changed and memories altered. These are exactly the prototypes that preceded the weapons used remotely in satellites and terrestrial antennas capable of attacking the brain of anyone on the planet, at any time today. Montauk is a separate topic. There are numerous documents detailing what actually happened there with experiments between radio signals and the human mind. Some of them were called Mind Amplifiers — now known as the Electronic Mind Reading (EMR).

They improved all aspects of electromagnetic transmission and reception system and created the Montauk Chair. There were two versions of the Chair. In them, the coils provided three outputs of the signals that carried the white noise that correlated with the thoughts of the target in the chair — white noise is one of the carriers that produce the most accurate effects in order to insert microwaves in the human mind. The resulting signal was picked up by three receivers. From there, signals were processed by computers of the time, such as Cray-1, which analyzed and displayed the audio and video of the human guinea pig's thoughts.

The experiments conducted there were of paramount importance for the diversification of tests, including accurate knowledge of the triangulation in communication (between antennas), the creation of an infrastructure that would work properly with little or no signal loss, highly advanced tests with TV transmissions and satellites, radio signals tests, complex analysis techniques and diversified processes in transmission and the technique of hiding signals within signals, improvements in receivers, oscillators, transmitters, modulators, modulations, radar pulses, reception control and improvement in antennas and in the reception of radio signals. Montauk was the birth of psychotronic weapons and an electronic architecture capable of working with human consciousness as we know it today.

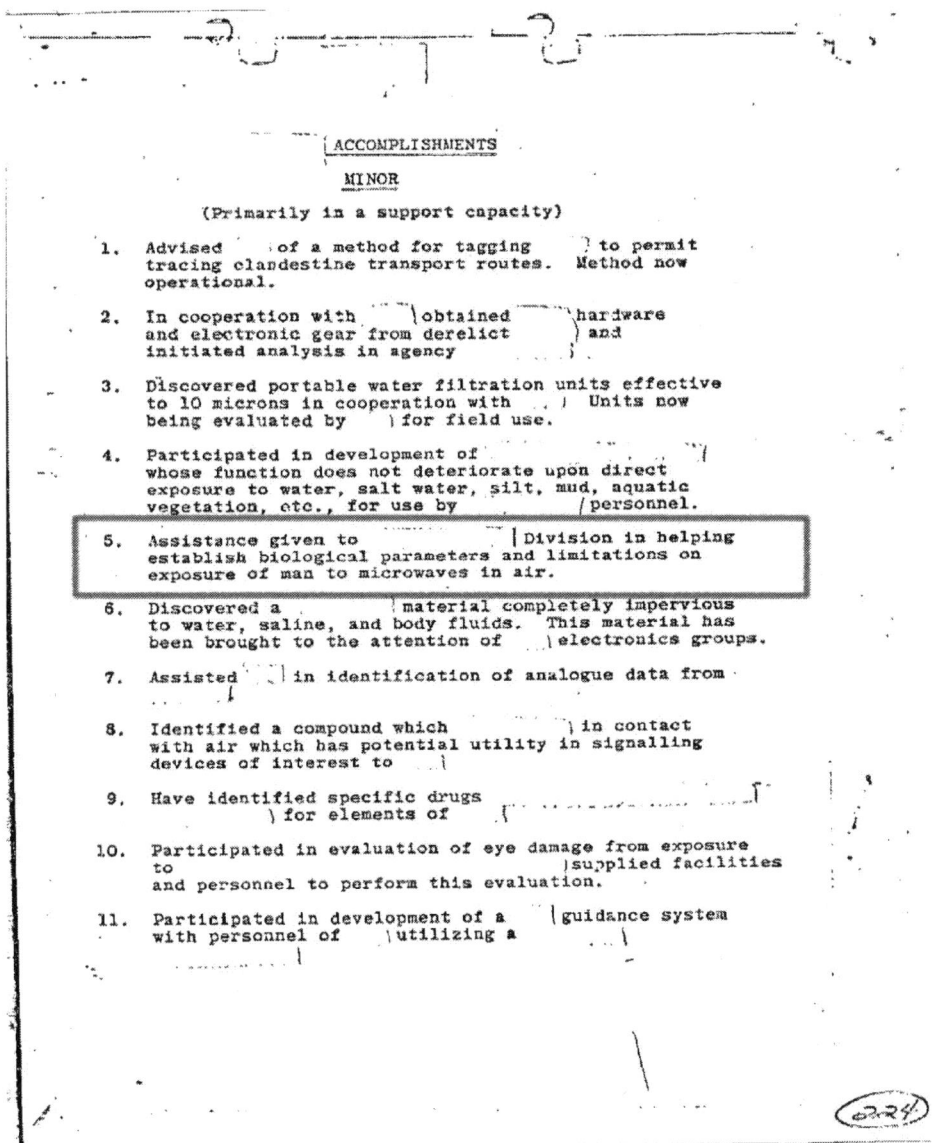

**Figure 4.9** One of the MK-ULTRA documents reporting the use of microwaves in order to set limits on both physical and psychological parameters in humans. There are several others with similar experiments.

Little information was documented, as Richard Helms, then director of the CIA, had all MK-ULTRA records destroyed in 1972. Documents recovered from the bonfire in the same year showed the horrors of 25 years of horrific experiences, perhaps as cruel as those conducted in concentration camps — in terms of scientific torture. MK-ULTRA officially ended in 1973, but the results of all experiments ushered in a new era for humanity. Needless to say, the experiments continued. They were now identified as "non-lethal" weapons. Then, they became a state project in 1975 with the compilation of data from more than 25 years of research, but without huge facilities and the use of detainees, beggars and minorities as live human subjects.

The truth is that mind control testing programs never really ended. They continue to this day. The modern MKULTRA 2.0 conducts the same types of experiments as its predecessor, the MK-ULTRA 1950. Now, however, remotely, non-invasively and performing other more advanced and in-depth tests linked to physical and psychological torture. They use adapted electronic means to perpetuate the techniques of torture and information theft derived from the gradual evolution from this dark period.

## 4.5 - Summary of Acronyms and Cryptonyms

CIA cryptonyms are code names or code words. They contain a two-character prefix called a digraph, which designates a functional or geographic area.

- **MK-CHICKWIT** – identification of new drug developments in Europe and Asia and acquisition of samples;

- **MK-DELTA** – accumulation of as much lethal biological and chemical agents as possible; it subsequently became MK-NAOMI;

- **MK-OFTEN** – testing of the effects of biological agents and their consequences in the field;

- **MK-SEARCH** – it became the mind control program after 1964;

- **MK-ULTRA** – mind-control research program.

## 4.6 - Patents

A patent registration protects an invention or creation from competitors. A patent is considered to be a formal document issued by a public agency through which ownership and exclusive use rights are granted and recognized for an invention described broadly. This is a privilege granted by the State to inventors (individual or legal entities) who hold the right to a product or a process, or the improvement of an existing one.

Keep in mind that **the patent itself doesn't verify that what is specified there actually works**. However, it serves as a guide to know a little about the technical part of the equipment and the idea conceived for the experiments we've seen throughout the chapters of this book. There are thousands of patents. Just browse and search for them. The equipment used in the most modern targets won't certainly be found in this public patent records, as they are considered sensitive and defense equipment, or military patents, intelligence equipment or national security apparatus.

The Invention Secrecy Act prevents patents on Mind Control Technology from being fully exposed. We only have fragments of some of them that are used in the technology or that were used as a basis for the conception of the projects.

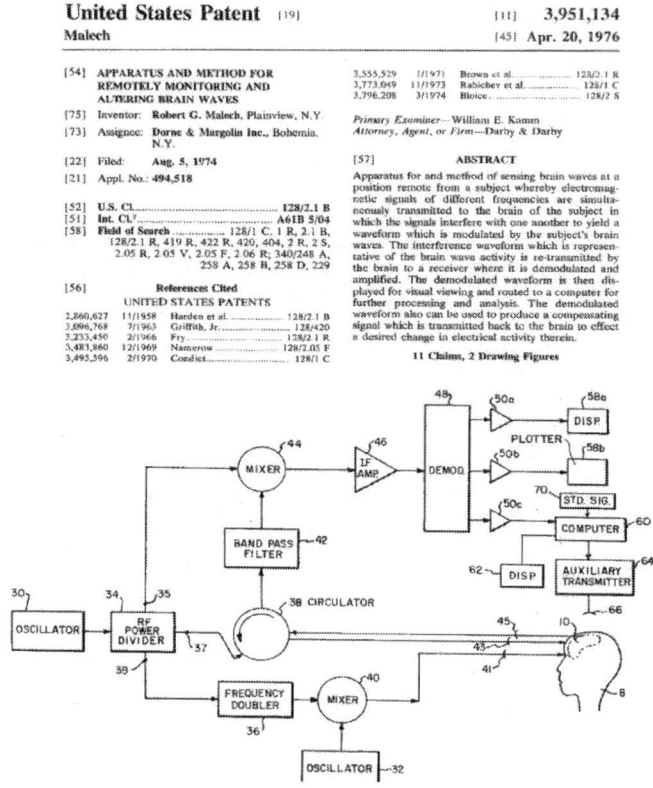

**Figure 4.10** One of the patents on remote EEG monitoring.

## 4.6.1 - Microwave Voice

* https://patentimages.storage.googleapis.com/68/e6/cc/1742b6ac0a d3d0/US6587729.pdf

* https://patentimages.storage.googleapis.com/0d/ad/ad/24b28667e 47641/US4858612.pdf

* https://patents.google.com/patent/US6587729B2/en

* https://patentimages.storage.googleapis.com/9b/5d/e1/5e51efb79c 085d/US4877027.pdf

* https://patentimages.storage.googleapis.com/60/23/f1/dcd0fc6cf0 de01/US3629521.pdf

* https://patentimages.storage.googleapis.com/66/a2/dc/3b6db8d01343e7/US3766331.pdf
* https://patentimages.storage.googleapis.com/00/24/7c/4cf02f4210343e/US6470214.pdf

## 4.6.2 - Behavior change

* https://patentimages.storage.googleapis.com/9d/c3/78/cf12e7f9c6b22b/US4717343.pdf
* https://patentimages.storage.googleapis.com/41/42/b5/258bc077289658/US5784124.pdf

## 4.6.3 - Silent subliminal presentation system

* https://patentimages.storage.googleapis.com/57/83/8a/a7-ba63d45ea3f4/US5159703.pdf

## 4.6.4 - Transmission of EEG Signals to the Brain

* https://patentimages.storage.googleapis.com/80/46/32/d4239f3ec5eb79/US3951134.pdf

## 4.6.5 - Radio Frequency Hearing Effect

* https://patentimages.storage.googleapis.com/68/e6/cc/1742b6ac0ad3d0/US6587729.pdf

## 4.6.6 - VHF and electromagnetic radiation

* http://www.freepatentsonline.com/3773049.html

Some patents have certain similarities to the way the scheme works. The excerpt below was taken from one of the patents on electromagnetic interaction with the human mind:

*"(A frequency) of 100 MHz and the other of 210 MHz are transmitted simultaneously and combine in the brain to form a resulting wave of frequency equal to the difference in frequencies of the incident signals, i.e., 110 MHz. The sum of the two incident frequencies is also available, but is discarded in*

*subsequent filtering. The 100 MHz signal is obtained at the output of an RF power divider into which a 100 MHz signal generated by an oscillator is injected.*

*The oscillator is of a conventional type employing either crystals for fixed frequency circuits or a tunable circuit set to oscillate at 100 MHz. It can be a pulse generator, square wave generator or sinusoidal wave generator. The RF power divider can be any conventional VHF, UHF or SHF frequency range device constructed to provide, at each of three outputs, a signal identical in frequency to that applied to its input. The 210 MHz signal is derived from the same 100 MHz oscillator and RF power divider as the 100 MHz signal, operating in concert with a frequency doubler and 10 MHz oscillator.*

*The frequency doubler can be any conventional device which provides at its output a signal with frequency equal to twice the frequency of a signal applied at its input. (...) When the signals are exactly 180 out of phase the combination produces a resultant waveform of minimum amplitude. If the amplitudes of the two signals transmitted to the subject are maintained at identical levels, the resultant interference waveform, absent influences of external radiation, may be expected to assume zero intensity when maximum interference occurs, the number of such points being equal to the difference in frequencies of the incident signals. However, interference by radiation from electrical activity within the brain causes the waveform resulting from interference of the two transmitted signals to vary from the expected result, i.e., the interference waveform is modulated by the brain waves."*

# CHAPTER 5

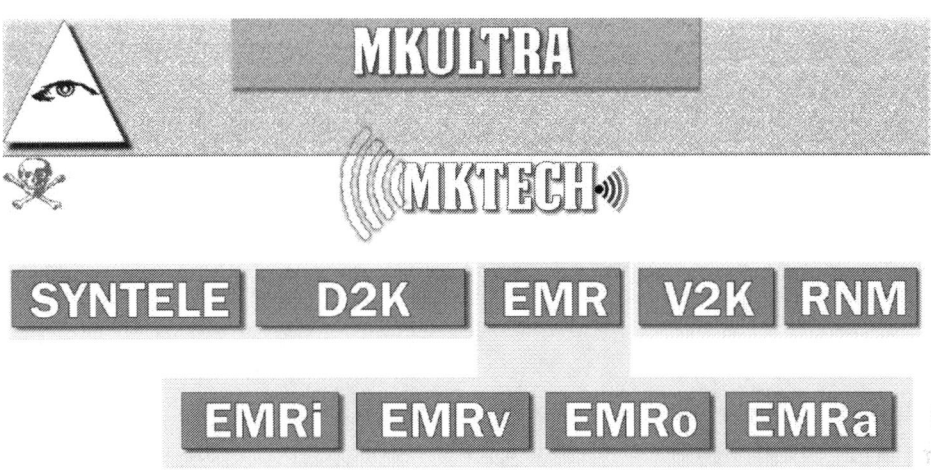

## MKULTRA 2.0 – MIND CONTROL EXPERIMENTS IN THE 21ST CENTURY

*"If I had access to a firearm during the height of MKTECH attacks – torture, paranoia, stress, cognitive deprivation, sleep deprivation and a suspicion that my neighbors were involved in some kind of bizarre scheme, spying on me 24 hours a day and having fun at my expense – then I would have shot everyone, without exception. I would have become a 'Winter Soldier', totally oblivious to it. I would be just another 'fun toy' for MKULTRA and its operators. In the end, I would find out that my neighbors had nothing to do with the torture!"*

**– Report of a Targeted Individual tormented by neuroelectronic weapons.**

After the scandals and investigations from 1973 to 1977, as MK-ULTRA officially ended and an investigation took place, all information from 25 years of experiments was being compiled behind the scenes in order to continue the mind control tests in another format, aiming at building weapons and the creation of interrogation protocols based on the lessons learned from that time. It was then that a new decade began and carried on everything that we've seen in the previous chapter.

## 5.1 - The 80s

In the mid-1980s, the decade considered the beginning of the information age, Diretas Já [Translator's note: "Direct (Elections) Now" was a civil unrest movement which, in 1984, demanded direct presidential elections in Brazil], weird hair styles, shoulder pads, Plano Cruzado [TN: "Cruzado Plan" was an economic plan that was introduced during José Sarney's government (1985-1990) in Brazil with the objective to achieve a rate of zero inflation], the beginning of free software, the development of the IBM PC and Apple Macintosh and the first graphical interfaces: XFree86, Windows and MacOS, the popularization of the CD and the beginning of extensive testing with electromagnetic weapons as we know it today, reports of people all over the world claiming they were being spied on, persecuted and that they were hearing voices that interacted with their mind began to outbreak, along with the use of a range of highly advanced weapons dedicated to remotely monitoring hundreds of thousands of citizens across the United States, Russia and other countries.

In March 1983, the President of the United States, Ronald Reagan, announced the Strategic Defense Initiative known as "Star Wars program". In this project, satellite technology in general had a substantial advance. Anti-ballistic equipment capable of tracking intercontinental missiles from space — which have a higher speed than conventional missiles —, using beams of atomic particles or laser beams were later modified with microwaves to track small objects and people.

Its capabilities include accessing vocalized thought and visual images from the cortex linked to vision and thoughts as well as the sounds that are picked up by the target's ear. Emotional re-engineering on victims and the beginning of the implementation of the modern torture protocol came to light, based on the set of best practice guidelines inherited from the MK-ULTRA of the 50's! Some 1989 patents that transmitted highly intelligible voices to the target's auditory system, similar to the weapons we have in use today, were discovered.

Behold...

## 5.2 - The 90s

The 90s was the decade in which the technology began to be tested on a worldwide scale and was constantly being improved. It happened in Brazil when the antenna era took place, in which massive tests started to be carried out. Brazil was definitely included in the MKULTRA 2.0 experiments. The 90s also brought: the collapse of the Soviet Union, Intel Pentium processors, PC 486, first cell phones, Rock, Grunge on the rise in music, Rap beginning to gain notoriety, Super Nintendo, Mega Drive, Macross, Rede Manchete [TN: "Headline Network" was a television network from Brazil], and so on. During this period, the United States disclosed classified documents. President Clinton publicly apologized to victims of experiments conducted in the 50's, 60's and 70's. Due to presidential and popular pressure, he asked the CIA to declassify documents over 30 years old, but the CIA claimed that they should pass the documents through a kind of sieve so as not to compromise groups of agents or national security. Therefore, the documents were gradually becoming available. However, only files that were released by the CIA's own analysis committee under the cloak of the country's security were revealed. What has been disclosed is in fact an infinitesimal number of the actual content from 30 years of MK-ULTRA and another 10 years of modern weapons. Access to Information Laws revealed some authorized documents of the most varied levels of restriction.

In 1992, several data were disclosed on this matter by former NSA employee **John St. Claire Akwei**. He sued his former employer, the NSA, due to the use of electronic weapons to monitor citizens within their privacy and the use of torture — remotely, if necessary —, exposing all the evils perpetrated by the agency. He was a kind of John Snowden[45], which makes sense from what could be observed in Brazil from the late 90's and

---

[45] Edward Joseph Snowden (Elizabeth City, born June 21, 1983) is a systems analyst, former CIA system administrator, and former NSA contractor who leaked highly classified information from the National Security Agency. He leaked the evidence to reporters from The Guardian and The Washington Post, giving details of the Global Surveillance carried out through many programs, including PRISM (the United States' surveillance program).

early 2000's regarding the use of neural weapons in civilians. This generated an unprecedented number of reports and their evil consequences on everyone who survives the attacks.

Formal apologies by President Bill Clinton were actually mere formality; his tears were just political theater. Experiments are still being conducted; their effectiveness advances, their results are increasingly accurate and their effects devastating. The network of operations spreads all over the world. From the 1990s onwards, the projects were completely classified and virtually leak-proof, as the requested mind control weapon was finally in full working order. Its patents are therefore inaccessible to people outside of the experiments.

The demystification of the use of the chip for mind reading occurred at that time — when the use of the technology was being observed. Allegations concerning the use of chips in humans were great distractions, as everyone who noticed the voices would look for a microchip implanted in their skull. It clearly didn't work, and the person would be taken to the nearest psychiatric hospital. This willful misinformation helps us determine that microchipping is completely unnecessary. The brain is the chip itself. All functions that could be done with the supposed chip are done using brainwave amplification techniques and monitoring.

Although the Ground-penetrating radar (GPR)[46] was used in 1929 to measure ice thickness in Austria, its functionality was only improved and distributed as a tool in 1970. In the 1990s, space and military experiments using technology similar to GPR (the Synthetic-aperture radar – SAR) began to be mounted on aircrafts, satellites and spacecrafts to map in 3D the ground and enemy bases from space. In 1994, Adrok's technology began to be applied using several patents, based on GPR and SAR, but with great capacity for penetration of the ground using different radio and microwave frequencies.

---

[46] It's a high frequency electromagnetic method (from 50 to 1600 MHz). This system generates images of the subsoil using as a transmitting source an electromagnetic antenna that emits a signal at a fixed frequency that can penetrate sediments, rock, ice or other types of natural or artificial materials.

Russian non-lethal weapons expert Igor Viktorovich Smirnov was invited to the US for a series of meetings that were held in Virginia on March 17, 1993. Several personalities from the most varied branches of government, representing the CIA, the Department of Defense (DOD), the FBI and the Defense Advanced Research Projects Agency (DARPA), attended the meeting. The meeting was also attended by civilians such as Dr. Richard Nakamura of the National Institute of Mental Health and Dr. Christopher Green, Director of Biomedical Research at General Motors Corp.

According to the article, the electromagnetic mind control equipment was not used due to software incompatibility issues, and thus cannot guarantee its safe use. A company called Psychotechnologies (Technologies linked to the brain and mind) has entered into an agreement with the Russians to share, adapt and develop the technology for use in the United States.

Every kind of weapon known as neuro-weapons, satellite terrorism, behavior modification through radio frequencies, manipulation of the nervous system via electromagnetism, remote mind manipulation, directed energy weapons, remote mind monitoring, advanced electronic surveillance, psychotronic warfare and cognitive isolation were now being gradually implemented in the field.

After deliberating for hours, the Parliament of the European Union decided to ban non-lethal weapons that alter human behavior and the functioning of the mind in February 1998. Parliament Resolution A4-005/99 on the Environment, Security and Foreign Policy passed on January 29, 1999. However, during the UN hearings on the topic, American representatives declined to participate in any resolution. They claimed total ignorance of the subject and that no congressman of the US knew what it was about.

The 1990s were also marked by the Oklahoma City Bombing, a terrorist attack carried out by American Timothy McVeigh on April 19, 1995, in Oklahoma, which targeted the Alfred P. Murrah Federal Building. The result was 168 dead and more than 500 injured. It was the worst attack in the United States since the explosion of a car bomb in

1993 at the World Trade Center in New York. Since then, it has been considered the most significant one, until the September 11 attacks, the worst that happened on American soil.

In addition to several reasons, the use of a pre-SYNTELE technology is considered: the Neural Phone. The classified intelligence apparatus consisted of an electronic device that used radio waves. It activated memory triggers on terrorists through which they transmitted direct orders to Timothy McVeigh, sentenced to death for the murder of 165 people during the attacks. While on death row, he claimed that microchips had been implanted in his body and it was through them that he heard the voices. This is a usual misconception by targets that have their minds "bugged" by MKTECH technology due to the surreal nature of the microwave voices together with the Synthetic Electronic Telepathy (SYNTELE) input. This is another extremely serious incident in which there is a strong suspicion of using MKULTRA to create terrorists, "Manchurian Candidates", involving government agencies and psychotronic weapons.

The renowned Russian psychologist Doyan wrote an article on advanced hypnosis methods. He detailed the experiments conducted by Dr. I.F. Tomashevsky who described the remote action of brain control by radio waves. The participants of this experiment didn't fully know what was going to happen. They were 2-3 quarters away from the radiation source. In this experiment and others conducted in another setting, participants fell asleep minutes after being affected by waves coming from the device.

On October 27, 1992, inventor Dr. Oliver M. Lowery was interested in the positive emotions caused by his U.S. Patent #5,159,703 — Silent Subliminal Presentation System. However, neither the military nor the Department of Defense were interested in this limitation. The president of the company that created the invention stated: "The system was successfully used and tested on battlefields in Iraq during Operation Desert Storm", an episode in which, according to documents, an entire platoon of Saddam Hussein's Republican Guard (the Iraqi Republican Guard) voluntarily surrendered after hearing "the voice of God" in their minds asking them to surrender without a fight. He claimed that using it on civilians was dangerous as it is virtually undetectable when employed. The weapon sent sounds to the minds of soldiers in the field and altered their emotional state. It was then submitted to the Department of Defense, where it joined the myriad of non-lethal mind-altering weapons.

Murray Gell-Mann, the 1969 Nobel Prize Winner for Physics, wrote in 1994:

*"Some day, for better or for worse, a human being could be wired directly to an advanced computer, and by means of that computer to one or more human beings. Thoughts and feelings would be completely shared, with none of the selectivity or deception that language permits. I am not sure that I would recommend such a procedure at all".*

The Nobel prize-winner was right in every word!

And so we come to the year:

## 5.3 - 2000, 21st century

There has been a boom in reports of harassment and cognitive and psychological violation in the world in a way never imagined: programmed killer attacks and suicide of people who were affected by the weapon, for example. In this way, MKULTRA 2.0 matured into what we have today, a modern format. Music was lost, culture was slowly withering around the world; social networks, Y2k Bug, Napster, eMule, China's economic growth, iPod, YouTube, Google... In the 2000s, the internet became all the rage. All kinds of services and possibilities were arising, grabbing people's attention and becoming part of their social dynamics.

It wouldn't be possible to conduct experiments along the same lines as those carried out in 1950 (even in criminals, people with mental illness or beggars) due to the abundance of easily accessible information — the ease of communication provided by the internet—, since people are today aware of their rights. The internet opened the door for reports of abuse by state agents. Atrocities like these would quickly "go viral".

9/11 put everything back to previous levels. Experiments conducted with psychotronic weapons were reclassified and debates about the significant decrease in privacy intensified. Taking advantage of the situation, tests with these weapons were accelerated, and their use on citizens in their own territory was no longer an obstacle. The budget for the DOD and DARPA has increased substantially to continue the development of increasingly invasive and advanced surveillance technologies. Under the guise of national security, weapons gained more and more power in this decade — in a wake of terror.

A DARPA experiment in San Antonio, Texas, tested new paradigms of electrical brain stimulation. A microelectrode for stimulation of the cortical region called the primary somesthetic cortex and a microelectrode in the medial bundle (a place where pleasurable sensations would be generated when stimulated) were implanted. Generally speaking, the scientists were able to control the mouse and train it to go through mazes and overcome difficulties and obstacles along the way.

We know the speed with which technology advances and how it can progress in a decade. The technological leap is immense; sometimes people don't even notice it. When they open their eyes, a certain technology is already part of their daily lives or coexists among them, as it happened before: cell phones, cars, computers, the internet itself, which is not even the shadow of what it was at the beginning of the century. Today, the commercial bias has taken over the entire network, and it is controlled in the West by four giant corporations, whose product is, well, you. In this way, they collect all your consumption habits and ideas, monitoring and tracking all users using their services. In general, the exponential advancement of technology that occurs from time to time has also boosted the development of neural weapons in all directions, both in terms of hardware (equipment) and software (programs). A state-of-the-art international infrastructure was created to support the peculiar use of electromagnetism in the shape of a weapon and to improve the accuracy of physical interaction with the brains of animals and humans. The equipment that makes up this entire network also followed the natural evolutionary flow, gaining strength and stabilizing the network around the world, as it enables it to achieve its goals wherever the target is physically located on the planet.

During this period, they implemented for testing a network called **SATAN** – **S**ilent **A**ssassination **T**hrough **A**daptive **N**etworks, which does exactly what it suggests: it kills silently. The network adapts to its targets, adjusts attacks with the aid of advanced AI, and improves its interaction tactics and violence according to the situation. Its adaptive principle is the model used up to the present in the most modern networks all over the world.

Among the equipment that make up the collection that represents MKTECH, there is an emphasis on the use of increasingly advanced satellites, modified radars and more efficient and longer-range radio and microwave antennas. Such equipment made possible the massively organized persecution of victims, propelling the modern MKULTRA experiment by turning its gears and reaching a level never theorized in terms of amplitude, magnitude and intensity. So, it was possible to open fronts around the world, continuing experiments of all kinds, including those of a commercial and financial nature behind this international organization with ramifications in practically all countries.

In order to corroborate these statements about the new arms race, it is important to point out that more than half of the enormous research budget of the U.S. Armed Forces is used to finance the development of weapons that use electromagnetism as their main object of research. Research that ranges from shooting with lasers capable of destroying cars, drones and small planes to psychotronic weapons capable of sending signals into the minds of targets and causing irreparable neurological damage. Directed energy weapons as they are known have become the war product with the biggest investment in research in the last few decades.

The program — in the same way as its namesake that began in 1953 using all the already established techniques of torture and persecution and the same international infrastructure resources that keep the victim 24 hours a day, 7 days a week, months or years on end under intense psychotronic torture and illegal surveillance with no privacy, everywhere they go — uses equipment that is the result of 50 years of development of previous experiments on the human mind honed by nature and evolution shaped by time.

At the beginning the intention was: a) to develop the best torture techniques for confessions in interrogations; b) to create remote assassins; c) to eliminate all threats to the system of the respective countries; d) political assassinations; and e) to subjugate people who had great social influence, among others, and who could engage in acts against the democracy in the United States and the communism in Russia. Today, the program has become quite diversified. It has other objectives in addition to those we've already seen. The experiments also entered the age of digital information with a commercial bias and for the sale of information derived from the results collected across the globe. It can also include the improvement of artificial intelligence research and of brain-machine interaction, behavior pattern recognition, commercial control of embedded technology, EEG telemetry and satellite neural tracking — psychotronic weaponry mounted on satellites capable of reaching anyone anywhere in the world, at any time, not respecting borders, cultures, nationalities, laws, privacy or human rights. Neural engineering is also a booming field. Modification and insertion of memories directly into people's minds is already a reality.

The transformation of the technology into a weapon for military use is also one of the key factors for program continuity. The continuous search for the most efficient weapon to destroy, control or disable the human mind — without the need of direct physical contact and being thousands of miles away from the target — has become the main objective of the agencies.

This is not science fiction or some futuristic daydream. This is the present we live in. We're facing an increasingly advanced technology that is merging with the human mind and generally with unorthodox goals. It turned into a great commercial experiment based on torture. A paid service, for fraud and death, destruction of adversaries, persecution and remote killing of the ordinary citizen affected by the attacks and cowardly subjugated in a short time.

Brazil, Russia, Japan, several countries in Europe, Asia, Africa, Oceania, Central America: virtually all countries and continents are using the "services" of the Mind Control Technology (MKTECH). They are actively building results for the modern MKULTRA or are using their citizens as guinea pigs. Reports come from all over the world.

Even though they know this is a weapon, targets are often battered. They wonder if maybe this is some kind of otherworldly technology, an illusion, or if it involves mental disorders. Well, no. As Arthur C. Clarke said: *"Any sufficiently advanced technology is indistinguishable from magic"*. This technology grows and evolves with each human being tortured; each person violated remotely. It develops in an increasingly powerful and optimized way to interact with human cognition. Every conflict generates additional advanced technology that later ends up serving peaceful purposes.

The war for the human mind is taking place today all over the planet. Many are falling; others are fighting bravely, each in their own way. In the end, every gesture, reaction and additional information is crucial for the next person being attacked by the MKULTRA experiments. They will be coming into this fight better prepared and in possession of information compiled, synthesized and captured with great difficulty from other victims until this type of attack becomes unfeasible, forcing its regression or changing the course of its evolution to another ramification in terms of psychotronic attack.

If in the past torture and experiments were conducted face-to-face and interpersonally, in the new MKULTRA experiments are conducted remotely and impersonally. This allows anyone to be able to perform such tests, regardless of age, physical strength or intellectual capacity. For example, a child using this weapon is capable of subjugating an adult target coldly, and of exquisitely conducting this barbarism, as complete anonymity, time and every advantage achieved by the scheme are on their side in an internal war for control of the targets' brains, minds and souls.

## 5.4 - Modern MKULTRA 2.0 - 2001 to 20XX

Imagine being compelled to live with strangers 24 hours a day. They're always there: when you wake up, eat lunch, work and go to sleep in your own bed. Imagine being forced to share all your intimacy, including your rational thoughts and memories; forced to follow an imposed agenda and to be attacked by anything that could violate democratic principles with the intention of torturing you, destroying your livelihood and completely deteriorating your quality of life, as they try to make you commit suicide, and, in the process, make you become a remote killer, or a total lunatic in a psychiatric clinic, defeated, humiliated and harmless.

At one end we have all the equipment capable of completely destroying society as we know it, as they collect data and samples for the improvement of the technology. At another, we find the dreadful operators that are responsible for the torture content and the persecution to keep the target under illegal surveillance wherever they are.

So, the modern MKULTRA 2.0 performs the same practices of the past, now remotely, non-invasively, decentralized and using a set of extremely advanced systems. Such systems range from electromagnetic neuro weapons, satellites, cutting-edge artificial intelligence, biophysiological interpretation systems of the human state to direct energy weapons that cause burns on people's skin — this is not the object of research of this book — among other tests related to physical and psychological pain, using electronic means adapted to improve torture techniques, theft of information and all kinds of experiences that involve the brain and the mind.

Thus, the most diverse return data of the attacks are captured, such as the prolonged consequences of the use of these weapons. Such consequences are devastating in humans and range from complete behavior modification that leads to serious self-mutilation, suicide, acts of violence to others, schizophrenia, temporal amnesia, psychotic episodes to problems in the brain area linked to language, speech and hearing. Technology operators can stay hidden and get information from thoughts, as they torture people and drive them to death. And all of these are used to study the technology of control and invasion of the mind and the cognitive limits of the human being.

Decentralized field experiments scattered across the globe, following a strict protocol of torture techniques, were a big leap for MKULTRA 2.0 and the MKTECH program. This took the program to a level never seen in history, thus adapting to new times. In the past, it was necessary to maintain huge private and government facilities — or government-associated facilities —, such as psychiatric institutes and federal and private universities. In addition, victims would have to be physically brought to these locations, where invasive experiments would be performed. Conducting experiments from a remote and decentralized location was the natural way found by government agencies in several countries around the world for the continuity of MK-ULTRA. The advantage was the access to a vast and abundant human material from different cultures with wide varieties of behavioral, religious and social traits.

Based on the immense variety of human nature, it enables a wide range of results regarding torture and its consequences through the direct invasion of the mind via electromagnetic weapons, but without bureaucracy — just a push of a button! It improves the knowledge of brain mechanics and mastery over it, along with the ultimate goal of total control of the human mind, as well as its transformation into an ultimate weapon of mass destruction capable of annihilating entire populations, completely weakening the target's brain, but keeping the environment and the infrastructure intact, and being able to be used with no concerns related to radiation exposure or anything of that nature. They would only have to deal with the putrefaction of thousands of corpses simultaneously. Anyone anywhere in the world is a potential target regardless of race, age or religion.

Taking advantage of this huge variety of human guinea pigs, a more sophisticated level of the experiment began in this century: an experiment that works with the deep mechanisms of the human psyche. It acts as a "brain parasite", as it integrates with cortical systems by interfering, deflecting and inserting information in the processing areas of primary reception such as vision and hearing that capture external stimuli, decoding them into an internal model. It acts directly in the processing between the reception of external stimulus, its transduction and its internal interpretation of reality, joining together as integral parts of the processes, causing behavior modification and substantial changes in the set of representations that define the essence of a person. This set of cognitive skills also includes memory, language, attention, learning and intellectual potential.

As if that wasn't enough, they conduct experiments with modification of REM dream content, mapping "raw" EEG waves — collecting data of pre-defined configurations of mental and emotional states —, appropriating private neural biometrics, assembling a complete picture without the victim's knowledge and, thus, physically locating him/her as one tracks down a device via GPS.

The aggravating factor is that *everything* is done remotely, without physical contact by anyone who has access to the technology or capital to pay for the services of torture for fun, Voyeurism and various anomalies and obscurities inherent to human nature that always finds a way to express such evils. They also work on experiments of a social nature, removing privacy and then neutralizing the target always on the basis of a constant and agonizing pain.

There are still many interests between the target and MKULTRA 2.0. On a deeper layer, the objective is to verify how different human guinea pigs evolve during the long process of torture; how targets deal with torturers and how they manage to hide information inside their minds, despite having all cognitive processes hacked and watched, and all the information that travels there being constantly stolen.

It's a whole new field for science and for the military and intelligence agencies around the world. Consequently, the results sought by the experiments are as follows:

* How the human being reacts, both in the short and long term, as all their thoughts become public;

* How to defend yourself against the attack without using electromagnetic shielding;

* How to change the processes of reception of external stimuli and the processes that govern brain dynamics between functional units.

With the data from experiments in progress around the globe operators will apply the results, and then condense them in order to update the "manual" of techniques on how to hide information in the mind and thus train their own personnel against an inevitable attack of such weapons. They need to discover in practice the most effective way of not revealing thoughts, memories and intentions; a way to keep secrets inside their own minds without equipment that blocks electromagnetic waves, and thus to instruct field agents or people in key roles in industries and governments who work with sensitive and confidential information during this psychotronic war fought in silence worldwide.

On another layer we have the interest in understanding all cognitive processes and the emotional reactions linked to such processes for industrial purposes — mainly in games, movies and series. Programs that control equipment such as drones, robots, unmanned vehicles through mind control without the use of any chip implanted in the person, in addition to using other systems that can interact with data from social networks for commercial purposes, and have access to the greatest commodities of the century: the sale of target customer profiles as advertising and direct sales.

The principle is similar to data tracking that occurs with your profile while browsing the internet. Now, however, it is based on thoughts, behavioral intentions, and occasional needs much more accurately and effectively than is currently implemented on the network. The compilation of this data will be extremely relevant for the entertainment industry. It can for example check which action on the screen generates a certain emotional reaction and reaches more people. Thus, it can improve scenes that touch and captivate a larger audience, verifying what pleases and what doesn't. One can even use this technology to produce well-being in the future, transmitting certain patterns of electromagnetic waves that stimulate an electrical configuration and positively affect the brain as a whole.

These tortures have a clear commercial intention. So, everyone benefits from the way criminal experiments are conducted. Such experiments completely violate human rights and the freedom of the intellect — which are hallmarks of our Western civilization — without submitting to the bureaucracy and without following the regulation of tests and experiments. For the most part, it requires years of tests using animals to later be performed in humans under strict and clear rules and with the due consent of the human guinea pig and also compensation to volunteers who undergo the tests. Experiments are being carried out across the globe hidden under the most diverse interests. Each country has a peculiarity in the use of this technology directly on humans, as well as different interests such as political, ideological and even religious ones.

One of the most worrying features is the Brain Net/Web, the internet 3.0. The network of thoughts is also improved and operates in total secrecy. It's possible to connect "minds" remotely, to transfer information and to hijack cognitive processes, creative thoughts and everyone's secrets. Everything is recorded, "logged in" and stored in the cloud, database and networks like the Dark Brain Net. This is the internet as we know it, however, composed of cognitive data and brain and mental processes. That is, an AI that interacts with the data and feeds on it, monitoring the reactions of the human body.

Another implementation would be the creation of more accurate equipment to control machines with no microchip at all. To centralize thoughts; to make the machine reason with the data collected through the detailed processes that the mind uses in order to generate reasoning, ideas and conclusions.

On the military side, the technology is extremely advanced. It follows the same logic as the other equipment: its outdated version is found in the civilian sector as the latest one (generations ahead) is already in their power. The money made by the first generation of the weapon makes up one of the financing arms of its most modern version. A demonstration of the advanced level of this weapon is present in **Chapter 8**, in which I described the last terrorist attack on the Embassy of the United States of America in Cuba using this stealthy, perverse weapon. The attack left many seriously injured. Another visible evolution was the microwave (hearing) weapons or intracranial voices from their original V2K version to the modern V2K that took a leap in quality and destruction.

Our society is heading to a point where thoughts, lines of reasoning and physiological reactions to certain stimuli will be virtual commodities traded between legal and illegal companies — similar to what happens with credit card information and profiles containing your personal preferences in the digital world.

So, today we have a version of MKTECH that is used for espionage, surveillance or theft of thoughts and the feature that uses the same technology to carry out modern MKULTRA 2.0 tests. The aim is to torture victims to death and at the same time obtain as much relevant data as possible from the results of the event in order to improve the weapon and update software and hardware; to improve "Psy Warfare" techniques (psychophysiological electronic warfare tactics), to improve military use of the technology against enemies and to develop more efficient tactics to cause deprivation of the senses and to inflict the greatest possible damage.

The continued use of electronic torture 24/07 in all cases shows that trials with human subjects continue to this day, but in a different format: there's always some motive to be used against people. It can range from land disputes, frequent attacks until the person is defeated and abandons the property, to revenge; one can also use MKTECH to cause excruciating pain to the opponent, whether for commercial disputes, family or political fights, wars between nations, among others.

## 5.5 - About the Targeted Individual

**TARGETED INDIVIDUAL**

Target # Victim/Human guinea pig\Lab rat! Mental prisoner # Electronic Holocaust Winter Soldier/!!* Loneliness ^^Madness $%Hatred(Psychological pain)*\\Fear: Anger -family--- Human Rights Violated$:::Buried democracy! ```I'm there, but not really # # # :&^%$[[P]]q]p_Did they take my mind??_-Who controls me-?-End of freedom? Struggle*****-Perception #$@!-Wake up-dream [Hacked Dream] fight- sleep- wake up&^^%$]

Who is the target anyway? Well, it can be anyone: me, you, a member of your family, friends or co-workers. Targets are people who unwittingly — overnight — woke up with their brain hijacked, that is, bugged with an "upload and download link" and with no way to control the flow. The target turns into a kind of wild animal in captivity, a human Tamagotchi[47],

---

[47] The Tamagotchi (たまごっち) is a toy that creates a virtual pet. It was released in 1996 and re-released in 2017 by Bandai in Japan. The objective is to take care of the

a jester, a clown, a sex toy, an animal bred for torture, a pet, or a lab test subject within their own life in which operators and torturers seize all private aspects of daily life, alluding to ancient and medieval punishments. However, all this is done without physical contact, leaving no visible traces that can be used as evidence in a trial.

This weapon has the peculiar characteristic of leaving no evidence of damage on the target and few traces in the electronic equipment in which the data circulates. There may be evidence on specific servers and devices. In them, metadata and thoughts travel and they're located where the attack is produced. They're used as a backdrop for the morbid theater that is a fundamental part for the good progress of the experiments and torture. But if traffic is not analyzed at the time of the attack — or captured at the point in time it occurs — it can never be recovered. The target will live with the permanent scars of the experiments and the terrorists will be unharmed. That's what happens today.

If they have to give a reason to carry out the torture process (that is, the torture and persecution of the victim), operators will use any story, from family issues, problems related to work or spouses, revenge, political dirty war to any other motive, as an excuse (a backdrop). Some victims report that they complained about a certain government body, criticized some groups publicly and the next day they began to feel the common indications of the beginning of the attacks. Others report that after waking up they simply started hearing maddening 3D voices and sounds.

Targets wake up and their lives are turned upside down. In other words, MKULTRA experiments as it used to, but in a hidden and remote way, using a symbiotic relationship between victims and torturers. The latter are responsible for maintaining a fierce persecution motivated by their interests while providing the means for it to be carried out. At the same time, they store the results of experiments. In other words, everyone wins: torturers satisfy their unhealthy desires by torturing the victim and the scheme that doesn't need to get down to business directly — that

---

virtual pet as if it were real, giving it virtual affection, food, bath, care, and so on. In the case of MKTECH, the target becomes an animal fed by hatred, torture and malice. Victims are deprived of all the care mentioned above.

doesn't need to maintain expensive facilities and specialized personnel — uses the enthusiastic torturers to get results. Only the victim will lose, as their brain will be used as an invasive data processing. Wherever they are, they will be the subject of torture; their mental, physical and financial resources will be drained, depleted to their immediate end by acute stress, stroke, cardiac arrest, insanity or even worse: by not being able to bear the torture and being driven to commit heinous crimes. But do not worry: you'll step into the shoes of a target in the next few pages to fully understand the whole scenario.

ILLEGAL ESPIONAGE

ILLEGAL REMOTE SURVEILLANCE

IMPLEMENTATION OF TORTURE PROTOCOL

BRIDGE BETWEEN THE TECHNOLOGY AND THE GANG-STALKING

IMPLEMENTATION OF IMPROVEMENTS IN TECHNOLOGY

TECHNOLOGY SUPPLY AND ITS MAINTENANCE

## 5.6 - Operators

Virtually all intelligence agencies, military and defense-related specialized departments have shared this MKTECH technology. Such technology (which uses only electromagnetic waves with no need for electrodes or implants) was able to develop software that manage technological resources such as antennas that sent sine waves or pulse trains, and they're primarily responsible for the viability of the entire scheme. We now know that this range of waves interacting with each other and directly with the human mind at a given set of frequencies and varying power is capable of mixing with the weak electrical currents of the brain, amplifying them and altering the natural functioning of an area of interest in any part of the cortex. In this way they measure, capture and change the frequencies, for example, in the visual cortex.

Modern operators were free to implement a new wave of experiments unprecedented in human history that covered a much larger and more diverse and lethal field than the original MK-ULTRA. The perfect weapon was created in its predecessor. Silent, invisible and capable of remotely accessing virtually all cognitive functions and the central nervous system — the central processor of the human being —, it gradually ended the lives of targets in the process. In the past, it was necessary to take the victim to a physical location to carry out the torture. Today, you just press a button and everyone on the planet is a potential human subject. Currently the order is to test everything possible and on a greater number of people until the current scheme falls and another one emerges.

With the infrastructure in place and the know-how of years of torture method, the scheme was thought about how to conduct the experiments without a physical location. Thus, the concept similar to that of cloud computing was adopted by the centers that hold the technology. MKTECH was distributed to various key people, military, defense and intelligence agencies, companies interested in the data collected, etc. People with high purchasing power, or who have great influence on some markets, can access the network, pay for the service and connect to the brain of someone of interest. Private and governmental agencies responsible for the apparatus only collect the data that goes through their neuro-satellites and processing servers, whether metadata or raw and analytical data about all aspects of the ongoing attack.

Among the new information being collected, one of the goals includes the improvement of the technology to its most advanced version, as it naturally occurs with computer systems. Another constant research aims at fostering improvements in the paths that lead the human being to commit suicide in an involuntary way or to be persuaded to become a "programmed" killer without even realizing it. Another scope of attacks includes how to neutralize a particular individual more quickly and efficiently.

Behind this technology, the objectives to be achieved are as follows:

* To improve methods of extracting information from the target's mind;

* To enhance the entire MKTECH - Mind Control Technology scheme;

* To improve torture methods and protocols with the use of the weapon and techniques to neutralize targets;

* To process and analyze the set of data collected during long-term exposure to torture and cognitive modifications by electromagnetic and sound stimulation along with physical and psychological changes;

* To modify the brain in order to negatively or positively alter some cortical field of interest;

* To improve or weaken human primary cognitive and perceptual skills;

* To insert false memories;

* Behavior and personality modification;

* Techniques that improve performance in certain areas of the brain and inhibit or reduce others (commercial value);

* Human Enhancement;

* To create remote killers who perform tasks for the operators and with no recollection of the events, or who commit acts through intense torture. This leads perpetrators to erroneously think that they're right in their actions in order to free themselves from electronic captivity;

* The target's brain becomes a kind of mobile lab. Wherever they go, experiments of all kinds take place and are processed by the mind, which affect their mental health, body and spirit.

Another example is that of the crooked paths that the sale of services of this infrastructure has taken. Billionaires and bored multi-millionaires use this technological apparatus for sadistic fun and morbid pleasure. Another way to get valid data for modern MKULTRA 2.0 research is to use the infrastructure for revenge, undermining the victim competing for work, quarrels over land, among other disputes or attacks on enemies of any kind.

So let's say a family wants to evict a person from the property in a land dispute or wants to get revenge on someone for some mysterious reason. This tool will be used and will easily destroy the target's quality of life, forcing them to give in to the attackers' demands. This directs the victim's actions towards the final goals that were planned by the attackers, even if such plans aren't clear or explicit, such as abandoning the disputed land, committing acts of violence, etc.

All this, reader, is already happening in Brazil. The tendency is for the attacks to escalate until the structure and secrets that make the technology possible — such as the equations, the nuances of the transmitted signals — become available to the general population.

No one is safe anymore. Anyone with access to this technology can simply connect to a brain anywhere in the world from the comfort of their own home. This person will have access to the victim's entire cognitive process, their intimacy, private thoughts, conversations and the content of their vision, with the ability to insert primary sensory data, "clogging" the victim's brain with a lot of useless information, diverting their attention and completely violating their privacy and fundamental rights using all MKTECH technological resources. This interference will cause extreme physical and psychological suffering with high levels of stress, leading to a slow painful death by mental breakdown.

And that's how a new category of gangs emerged, or evolved, coinciding with the advancement of the technology in the mid-90's. It revealed a class of "humans" who saw in this technology a way to externalize all their lack of character and their tendency to spread evil, to profit from it and to entertain themselves in a very safe way. People who enjoy watching and interacting with the torture process think of themselves as new century sugar mill lords. They enjoy torturing the targets systematically and the power they feel over their mental slaves. Meet the new type of gang: the Organized Gang-Stalking (OPS – Organized Professional Stalkers or POS – Professional Organized Stalking).

## 5.7 - Organized Professional Stalkers (OPS) / Organized Professional Torturers (OPT), Gang-Stalking or "Cyberstalking"

**GANG-STALKING - OPT/OPS**

DREAM CONTENT PRODUCTION

ILLEGAL 24-HOUR SURVEILLANCE

THEATRICAL PERFORMANCE/TORTURE CONTENT

NOISE IN THE TARGET'S CORTEX 24/7

HARASSMENT, STALKING AND TORTURE UNTIL THE TARGET DIES

Now let's get to know the people behind the torture in more detail and depth: the **operators**. I used this term to facilitate the understanding since we would only be able to comprehend the dynamics of all this absurdity as we began to scrutinize the most diverse characters and organizational levels of this diabolical scheme.

Organized Professional Stalkers (OPS) or Organized Professional Torturers (OPT) are a group or groups of people responsible for the content of torture that will be imposed on the victim in order to carry out the "stalking" and still maintain constant surveillance. They're responsible for creating a negative scenario. In these scenarios, fights, screams, verbal aggressions and threats are transmitted directly from the place where the performance is staged to the target's brain 24 hours, 7 days a week, without interruption. In addition to driving them mad, the intention is to make the victim feel uncomfortable, shaken, persecuted, tired, ridiculed, humiliated and mentally, physically and socially violated. It is to make them feel watched, observed, with their intimacy and privacy destroyed, wherever they go in their daily life — in the workplace, at their friends or family's house, and especially in their own home.

They're professionals in the art of remotely stalking (in absolute anonymity), and in creating a chaotic scenario, as well as shaping the environment — purposefully invading all aspects of the private life of the individual being attacked — and placing the victim in a narrative context. This artificially creates a plot in which an alleged motive behind the torture is repeatedly lauded, while the tortures enjoy every minute of the constant suffering. These are people who take turns, so as not to let the target's mind rest for even a second.

Gang-Stalking or Organized Gang-Stalking are nothing more than the human representation of the Mind Control Technology. They're sociopaths and psychopaths, whose life purpose is unhappiness, evil, negativity and bizarre behavior linked to sex and torture. These are people who revel in the gratuitous suffering inflicted by them. That's why people who torture animals for pleasure should receive special attention. In the future they may become these aberrations called OPS.

First, a story is created. Then they insert the human guinea pig, making them the main character of a hysterical and morbid narrative in which they lead all aspects of their life in a well-constructed plot both during the day and at night. At dusk, a unique event takes places and it repeats itself every night: the operators' special evening performance via SYNTELE (Synthetic Electronic Telepathy) that prevents the target from sleeping. Therefore, sleep deprivation occurs every night. The target will hardly sleep in the silence of the room, but if they succeed, the continuation of this performance migrates into their dreams, now with voices and images. That's where the true face of the gang and its ulterior motives appears. Bear in mind that these contents present bad actors in a bad play, as we saw in the D2K chapter (Volume 1).

Some are professional actors who find it easy to create shallow impersonations and characters. They often emulate the voices of the target's friends and family to confuse them. This is all possible due to the fact that the audios are extracted from the EMRa - Electronic Mind Reading (auditory), intensifying the mental confusion of trying to find out who is behind the voices and for what reason. As soon as the target mentalizes a person who could be this character — that never-ending voice in their mind—, this mental representation is also captured by the OPS equipment along with the vocalization of thought.

The ventriloquist — an OPS actor — starts to imitate vocal mannerisms of this mentalized person, using real properties of that voice that causes immense confusion in the target. They can impersonate friends, enemies, relatives and spouses with extreme fidelity to the originals, using computers and voice-altering programs.

Organized Professional Stalkers (OPS) rely on eternal anonymity and that the victim will die before discovering who they really are and what is truly going on. Each member of this gang has a well-defined role that is adjusted over time according to the theme and objective behind the attacks.

It's possible to find out some details of those responsible for the torture through the characters. After some time analyzing the role that will be played, one can understand to a certain extent their behavior and some nuances related to the person who plays it: age, nationality, gender, intellectual level, purchasing power, accents and the effort in the representation of each character during the torture and harassment day by day in these representatives of evil. The target becomes a prisoner of the perverse wishes and desires of the OPS, imprisoned within their own intimacy.

Using a more archaic language, they'd be known as the defilers of homes. In other words, people who take pleasure in violating the territorial limits of private property and the sacred right to privacy of others at all times in a purposeful and audacious manner. OPS are the base of the pyramid of the MKULTRA scheme. They are the lowest level in the hierarchy. They're responsible for screaming into the microphones and transforming the pre-elaborated script into a rant, putting all the torture content into practice.

They're also responsible for creating the narrative that will be inserted into the target's mind via SYNTELE (Synthetic Electronic Telepathy), and for listening to the return response and reactions to attacks, thus adapting the torture script always based on this feedback and in the adjustment of the target to imposed scenarios. They also bring the evening performance to life and create artificial dreams (films) in their studios. Such dreams will be transmitted when the victim goes into REM sleep or has hypnagogic dreams via D2K (Synthetic Electronic Dream). They also chase the targets everywhere using radio waves, consistently leaving the impression that they're in the adjacent apartments or in the vicinity of the target's everyday locations. They're known as actors who perform street theater. For them, torture excites orgasmic pleasure that feeds back torture in a vicious circle.

It's like they are living with the target. It's a forced coexistence, without limits and with no end in sight. It ruins every moment of the victim's life. Several people dedicate themselves and spend years on end participating in this "mental theater", which includes the victim as the protagonist, but without their consent. Every plot is composed solely and exclusively of materials taken from the target's reality. Their memories are used for the scenario; interactions occur during the re-enactment of events that, in fact, occurred in life. The target experiences, sees and hears these events over and over again. They're narrated and staged by strangers who project them into their mind. These people try to direct the victim's attention to themes that deteriorate their existence, fulfilling the purpose of MKTECH, which is to control thoughts and lead the target to erratic or destructive behaviors.

It's very curious. I never thought that people like that could exist. I was naive, or maybe I just underestimated the power of nature and the diversification of human beings and their natural propensity for evil — within certain occasions — in their actions and interaction with society.

In this dynamic, you have to deal with the invasion of all your cognitive functions that are clogged with several insults and negativity, as operators maximize any natural mishap of life in society — for example, problems at work or at home and financial issues, all arising from social interaction — to increase psychological pain and consequently physical pain. And the significant mishaps themselves become the fuel of the plot along with the natural feelings surrounding their thoughts. All these materials are extracted from monitoring and interacting with the victim's mind, as we already know they're watching 24 hours a day, everywhere the individual is. The victim's physiology and physical and emotional responses to each advance are observed. Operators become omnipresent beings in the target's mind.

Some members of this gang specialize in professional torture, applying classic and new methods used in Prisoners of War (POWs) in dictatorial regimes or in prisons of civilized countries that protect terrorist suspects. Thus, they carry on their sadistic fun, the daily satisfaction dictated by their addictions — the vicious "puff"—, while causing the greatest psychological and physical damage to the victim.

They're very peculiar. They do things that normal people of good repute would never do. First, they're people with shallow intellect and neurotic, cynical, extremely long-winded and boring characteristics. Stalkers are always linked to sexual activity, reflecting on the content of words and actions in dreams. By successfully observing the process return of these data, it is possible to notice that a lot of sexual content is transmitted during dreams, including masturbation, explicit sex, pedophilia, and all the excitement related to the impact of this unholy dream/play on the target's mind that was caused by torture. They're repulsive beings whose main delight is to demolish the lives of others in a kind of primitive ritual in which torturing others becomes the source of insatiable sadistic pleasure.

I will again refer to the ICD-10 Classification of Mental and Behavioral Disorders: Clinical Descriptions and Diagnostic Guidelines to point out, in a more precise way, the nature of Organized Professional Stalkers' behaviors:

* **F52 Sexual dysfunction, not caused by organic disorder or disease** - excessive sex drive. The individual always appears to be immersed in the chaos of torture transmitted via SYNTELE.

* **F65.0 Fetishism** - intensification of arousal, an important or essential source of sexual stimulation, in which case the torture and the suffering of the human subject intensify sexual satisfaction on the part of the torturers.

* **F65.2 Exhibitionism** - tendency to show their genitals to strangers, followed by masturbation, but in an extremely bizarre way. OPS usually show their genitalia and transmit such content via D2K (Synthetic Electronic Dream) straight to the victim's dream. Then they start masturbating or having sex with other OPS members using masks and other paraphernalia to camouflage themselves.

* **F65.3 Voyeurism** - recurrent and persistent tendency to look at people in their privacy, at all times engaged in sexual behavior. This usually leads to sexual arousal and masturbation and is carried out without the knowledge of the person being observed.

* **F65.4 Pedophilia** - a psychiatric disorder in which an adult or young adult experience a primary or exclusive sexual attraction to prepubescent children, usually under the age of 11. This occurs when they use this technology on children, as we're going to see in Chapter 8.

* **F65.5 Sexual sadism** - someone who seeks to pleasure themselves by inflicting physical or psychological pain on another person, generally seeking sexual pleasure in the other's suffering.

* **F60-F69 Disorders of adult personality and behavior** - syndrome of exaggerated behavior due to the psychological state of the individual; syndrome of attention-seeking (histrionic) behavior, despite repeatedly negative findings, there are always obscure motivations for such behavior.

They're also responsible for transferring the atmosphere of the stage into the brain. There is always a heavy aura that shifts the focus and attention, creating a permanent state of negative expectation and causing prolonged stress — only a few can escape unscathed from it, and even fewer manage to deal with these voices that destroy the central nervous system. These people are actors in a new category of 21st century theatrical performance.

They're hysterical and rejoice in making noise. They recklessly conduct experiments on an entire community. Their voices are heard by the target for years on end, wherever they are. These people are also professional stalkers and torturers, a long line of people whose way of life is MKTECH, torture, persecution and death. Some gangs have up to 3-4 generations of members of the same family. Moreover, they're responsible for creating and developing the alleged reason why the victim becomes a Targeted Individual, if any.

They're extremely important for the modern MKULTRA, as the data collected from the experiments conducted by them wouldn't be possible to be acquired otherwise, at least not without the constant and uninterrupted operation of the Organized Professional Stalkers (OPS). They have the alleged reason and motivation for torturing the victim. MKULTRA provides the infrastructure, the means (transmission towers, satellites, programs) and collects all the data relating to the conditions of the victim subjected to intense torture.

Have you, as a decent man or woman, ever imagined being forced to listen to all kinds of moral and sexual harassment? Have you, the head of the family, ever imagined being harassed, on a daily basis, inside your own house? How would you feel? What actions would you take? This is already a reality for many people in the world, and it causes outrage, deep sadness and enormous frustration for not having a way to defend themselves. Victims wonder how operators use this weapon for such purposes and no one — absolutely no government agency — comes to their help.

To make the immersion of the experience as deep as possible, these actors are omnipotent and omnipresent. They use known theatrical performance techniques in the form of a radio soap opera without the possibility of being turned off, which activates the target's imagination. Some examples are shown as follows: simulation of people ready to fight, or actual fights, crowds breaking into nearby buildings and shouting coming from apartments close to the target. As the target never keeps eye contact and OPS remain anonymous, only the content of the voices that stimulate vocalized thoughts and mental images can be extracted. It is the mind creating a context for the voices behind the torture. It slightly resembles an important event in history.

Do you remember the consequences of the radio drama called "The War of the Worlds" in 1938? During the broadcast, the radio simulated an extraterrestrial invasion and caused panic on the East Coast of the United States. The dramatization was broadcast on the eve of Halloween in the format of a news broadcasting and had all the properties of radio journalism of the time, to which listeners were used to: sound effects, ambient sounds, screams, and the commotion of the supposed reporters and commentators. It all gave the impression that it was being broadcast live. It was the seventeenth in its weekly series of dramatic broadcasts featuring Orson Welles and the *Mercury Theatre on the Air.*

The target's imagination is taken to the extreme. OPS explore the same principle used by old radio dramas, which were so efficient that they caused turmoil in an entire city right after the broadcast of the story created by H. G. Wells. Thoughts are driven by sounds, narratives and dialogues. However, this "radio soap opera" is transmitted directly to the individual's brain, transforming them into the main protagonist of the story and with no possibility of being turned off.

As time goes by, OPS begin to use the method of forced intimacy. They start talking to the Targeted Individual as if they were old friends. Then, they try to reverse roles by drawing the target into the problems or scenarios they create based on the narrative. Keep in mind that with each psychotronic attack, the victim creates (in their mind) a likely scenario to contextualize the events staged on the OPS side. In other words, the brain automatically attempts to create the people, the place, the scenery and the unfolding of the "radio soap opera" in a visual manner. This thought is then monitored and recorded, and is later used to embarrass and harass the target in their attempt to understand what is really going on. It becomes one of the most profound experiences in a person's life.

To some extent, they're based on the Aristotelian theory of the four discourses. It consists of four ways in which someone can influence another person's mind through words. The Poetic Discourse is the most used, as its main feature is to address the imagination that captures what it presumes in images. The indecent torturers can thus steal these images and adapt the performance according to the mental premise of the target, at all times focused on a theme that is the main object of the experiments. They play with the targets' mental habits without abandoning the storyline that is composed only of inputs of thoughts that form the individual's mental world.

Targets living in buildings have the feeling that OPS are gaining access to several apartments above or below theirs, or even in neighboring buildings. This makes them feel surrounded, cornered and watched at all times in their own home. However, it's just a feeling in most cases. The Targeted Individual looks around, but ends up finding nothing and no one. Their neighbors won't know what they are talking about, because most of the noise exists only in the victim's head due to the V2K.

One of the most used tactics to make targets apprehensive and disoriented is: as soon as some noise occurs naturally in the apartment above — it can be next door or in the apartment below — where the individual is located, the OPS quickly send a message using V2K and adjusts the auditory perception. This makes the brain believe that the sound preceded the attack, as if the neighbor started to scream at the window towards the target's apartment. A never-ending song or constant noises are also a possibility.

They use normal sounds of an apartment: the opening of a window, an object falling to the ground (and echoing on the target's floor), and so on. By doing so, the victim thinks that the upstairs or downstairs neighbor is part of the torture and surveillance scheme. The idea that neighbors conspire against the target will always be emphasized, creating a potential scenario of self-destruction between them, which can culminate in physical aggression against a neighbor with serious consequences — but for the delight of the Organized Professional Stalkers (OPS).

OPS are extremely patient, as they know tortures have cycles of 5 to 10 years. Long-term intensive recurrent attacks are generally employed. There are other types of short- to long-term attacks, depending on the objective behind the endeavor.

Each voice of a specific channel will play a different role. Some of them pretend to be friends with the target; others play villains. Depending on each case, characters are "recycled". They draw the attention of others with their dramatic and excessive behavior and assertive techniques of emotional expression, making continuous attempts to direct attention. They yearn for usual activities that involve strong feelings, but ignore critical details that will eventually ruin everything.

An atmosphere of self-confidence — the internal sensation of grandiosity — takes place in the environment of intense excitement with the suffering and torture of others, which occurs in an impressive theatrical performance. They wear costumes and clothing similar to children's theater performances, such as animals (birds, bears and dogs) and props. All of this conveys a mess of nonsense to the target's REM dreams, feeding back the excitement charged with loud, noisy emotions. People who specialize in screaming are part of the cast. Yes! There are people who specialize even in that. Punishments and penalties are imposed in certain circumstances. When, for example, the target isn't meeting the expectations of servitude sought by the OPS. They keep their "Tamagotchi" suffering intensely as the fun continues for the operators.

How do you support yourself financially if this role requires full-time dedication? In Brazil, the money to sponsor this endeavor comes in part from the public treasury, the theft of information and the salaries of gang members who defrauded civil service competitive examinations using SYNTELE (Synthetic Electronic Telepathy). The tax paid by the taxpayer finances this type of enterprise. Among other shady initiatives, such as torture for fun, we have illegal surveillance, sexual and electronic harassment, and cyber torture. These are basically the people responsible for the torture content and the enactment of the theatrical presentation and virtual reality created by them to collect data for the modern MKULTRA.

## 5.8 - Winter Soldier, "Programmed" Killer or "The Manchurian Candidate"

The process of turning someone into a "programmed" remote killer is slow, cruel, and it depends on many different factors. At each step, electronic schizophrenia takes over mental biological processes, while the primary processes and faculties no longer belong to the person, but to MKULTRA. After some time, the target no longer remembers what their life was like, and their long memories begin to be interfered with, creating a picture of cognitive chaos.

Brainwashing, moral emptiness, the feeling that you can't fight this attack, the will to free yourself from the technology and regain control of your senses are also a reality. The insertion of false realities/memories, heavy torture, including sleep deprivation, the impossibility of full rest, blocking the flow of thought, and the loss of reality make certain people enter a process of transformation into programmable killers without knowing it. All these factors are capable of making individuals extremely receptive and willing to obey orders that come supposedly out of nowhere in their minds, leading them to believe that this is a God-like power and that it is impossible to fight this misfortune, added to the novelty regarding the interaction with mental processes and the full modification of the victim's personality.

The path to be followed also depends on numerous factors that are practically impossible to discriminate unitarily. However, a preliminary analysis of mental disorders or genetic predisposition to violence, alcohol and drug abuse — or if the individual's profile complies with, or facilitates, their change — will show if the victim is a more vulnerable candidate. This would then require less effort for total transformation.

Even if people are different from one another, it's possible to categorize them into groups by observing certain patterns and characteristics. Experiments precisely determine such characteristics. They also verify the behavioral biophysiological responses according to their conception of the world. Standardized stimuli are presented to the target, so that he or she reacts according to their particularities — their unique personality traits, projecting their inner world to the outside world.

By analyzing the attacks, the way the experiments are camouflaged in the tactics and the protocols followed during the course of time, along with the circumstances in which they're carried out, what caught my attention was how operators try to manipulate the mind using various resources, including temporal amnesia due to acute stress and the insertion of false memories that replace such memory lapses via remote dreaming and prolonged sleep deprivation. I was stunned when I came across something that corresponded exactly to the metamorphosis that all victims went through — each one in a level (some more, some less), but all were drastically affected. People known as Manchurian Candidates or involuntary remote killers of the 60's jumped to my eye when all the atrocity of the experiments weighed on thousands of people. What struck me the most during my researches were the similar tactics used in the past compared to those conducted today with the aim of pushing someone to kill.

I even cited a more recent example of Manchurian Candidate events in Volume 1. Several other similar occurrences also happened in the same way and went unnoticed. Moreover, what was found when the target has easy access to weapons worsens the situation. The lethality of their acts increases substantially. They usually end up expressing all mental violation suffered in these macabre experiments with any available device, such as firearms or knives.

I never imagined that something of this magnitude could exist the first time I heard about it — a subject always very distant from our everyday lives, mixed up in a mass of countless nonsensical conspiracy theories. But my perception changed completely when I witnessed this neuro-weapon being carelessly used. A lot of research confirms that it's in fact possible and feasible to modify behaviors to the point of transforming normal people into evil killers.

Remember: there's a fine line between putting pots and pans over the head — ending up in a mental hospital — and turning yourself into a remote killer. Most end up falling by the wayside. However, those who are driven unconsciously and cross the line of madness without knowing that they are about to become a murderer provide valuable data on the behavior that culminated in this disaster, in addition to causing a lot of damage. Valuable data such as the following:

* Which of the mental attacks caused the instability conducive to the unconscious implementation of the remote killer?
* Which protocols were used successfully?
* At what point in time did the target succumb and commit the act of violence?
* How long did it take from the initial connection to the target's brain to result in the final attack?

* What criteria made up their social bonds, financial health, religious beliefs, educational level, profession, family ties, personality and character?
* Electrical patterns, brain metadata and psychophysiological data collected throughout the whole process.

These data are collected, further perfecting the weapon and its path to the ultimate solution: the full active control over the brain on the first electromagnetic blast.

The transition occurs in a very subtle way. A simple erroneous act will change the target from victim to murderer or aggressor. An error of judgment, one's pent-up anger, the madness of the events and the hatred towards the attackers serve as catalysts for an outburst of uncontrolled rage directed at people close to the victim. A misinterpreted word can mean a march to the end. Knowing that the torture produces this type of behavior, they can lead the target to self-harm or exacerbated violence against someone, even family and friends. They can also cause a slightly more sophisticated hallucinatory behavior in which the target obeys orders to murder someone directly, or has a high capacity to inflict pain and death on many people at the same time.

Many questions are raised especially in relation to the relative ease of indirectly manipulating minds to the point of creating this type of character. The use of sophisticated techniques that "manually" modify fragile memories, inserting a certain theme and maintaining and placing the target in the most adverse situation based always on false memories created by the operators. This produces a false reality that is internalized every day until the target completely loses the notion of reality and becomes a *Winter Soldier*.

Furthermore, crimes are carried out unceremoniously by operators. During investigations into what happened (in all cases without exception), operators are sure that the blame will always be on the alleged mental illness, pre-existing personal issue or something of the sort that is more plausible than a weapon that causes acute, post-traumatic stress, and is able to hear thoughts, to see thoughts from images and to monitor the target's raw waves while inserting maddening and constant voices among other hellish qualities.

## 5.9 - Full picture of the MKTECH universe

# MKULTRA

## MKTECH

- SYNTELE
- D2K
- EMR
- V2K
- RNM

- EMRi
- EMRv
- EMRo
- EMRa

## OPERATORS

- ILLEGAL ESPIONAGE
- ILLEGAL REMOTE SURVEILLANCE
- IMPLEMENTATION OF TORTURE PROTOCOL
- BRIDGE BETWEEN THE TECHNOLOGY AND THE GANG-STALKING
- IMPLEMENTATION OF IMPROVEMENTS IN TECHNOLOGY
- TECHNOLOGY SUPPLY AND ITS MAINTENANCE

## GANG-STALKING - OPT/OPS

- DREAM CONTENT PRODUCTION
- THEATRICAL PERFORMANCE/TORTURE CONTENT
- HARASSMENT, STALKING AND TORTURE UNTIL THE TARGET DIES
- ILLEGAL 24-HOUR SURVEILLANCE
- NOISE IN THE TARGET'S CORTEX 24/7

## TARGETED INDIVIDUAL

Target # Victim/Human guinea pig\Lab rat! Mental prisoner # Electronic Holocaust Winter Soldier/!!* Loneliness ^^Madness $%Hatred(Psychological pain)*\\Fear: Anger -Family- Human Rights Violated$:::Buried democracy! ~~~I'm there, but not really # # # :Did they take my mind??_-Who controls me-?-End of freedom? Struggle*****-Perception #$@!-Wake up-dream [Hacked Dream] fight- sleep- wake up&^^%$]

Figure 5.1 Full picture of the MKTECH universe.

# CHAPTER 5.10

## HOW ELECTRONIC TORTURE WORKS IN PRACTICE, "ELECTRONIC HARASSMENT" OR "CYBER TORTURE" INVOLVING ORGANIZED PROFESSIONAL STALKERS, MKTECH AND MKULTRA 2.0

*"Every breath you take, every move you make, every bond you break, every step you take, every word you say, every single day, i'll be watching you!"*

–The Police.

It is possible to observe the limits of the brain and the human behavior towards it only under intense torture. The stories told here are surrounded by an intricate relationship between the characters of this clash involving humans armed with universal natural forces, and their outcome, cause and effect along the paths to be invariably followed by people affected by this disease-making machine.

All objectives follow the same process, except for the theft of information directly from the mind, which only listens without interacting, or interacts subtly with the target, leaving no trauma or sequels. Below we have the separation of some distinct goals to help the reader visualize the possible paths taken by the targets and all hidden objectives.

* **Objective A**: Revenge, disagreements, attempt to stop the enemy, labor, commercial or land ownership disputes, eviction of a property, political or commercial use, murder, "torture for fun", terrorist attacks, attacks on people from hostile nations, warfare.
* **Objective B**: Theft of information and intellectual property.
* **Objective C**: Creation of Remote Killers (Winter Soldiers), Manchurian Candidates.

* **Objective D:** It encompasses A, B, C and all other primary objectives; the conduction of MKULTRA 2.0 neural experiments that happen naturally in the processes that lead to any objective above.

In this part of the book, I'm going to take you through the processes that the targets are subjected to in their private lives, detailing absolutely all the facts concerning torture since its inception. It'll be based solely on real cases, reflected exactly as they occur with all Targeted Individuals, only subject to slight variations from case to case. This report shows the step by step of a real torture, how each tactic explored in this performance works and results in cognitive chaos, emotional misdirection, personal reengineering and the introduction of a new mechanism capable of creating a parallel reality.

Furthermore, we're going to understand in detail the reason such an event occurred in a certain way and how it affects a person's life. As the situations develop, they become similar — they bring up points in common in the unfolding of the attack dynamics, which is faintly perceptible when analyzing events around the world. Thus, the similarities in the relationship between **MKTECH** (Mind Control Technology) elements, the modern **MKULTRA** experiments, the **OPS** (Organized Professional Stalkers) and the **TARGET** can be observed, both in their conduct and in the type of weapons used.

## 5.10.1.1 – Zero Hour

Before attacks begin, operators surround themselves with as much information as possible about the target and their family. They will get used to the ecosystem that surrounds them during the preliminary information gathering. Then, with the target confirmed and the operation about to take place, the power of reaction or the ability to threaten the integrity of the secret, or the gangs involved in this worldwide enterprise, is verified. Lonely people and couples with young children are conveniently good targets; those with a good family support network aren't. The target profile varies from case to case, but the ideal would be a person who is momentarily separated from their family for any reason, such as moving to study or work, a newly married or separated person, or for some other occasion in modern life.

The next step is to capture as much information as possible and deepen the data collected in all existing aspects. Once the target has been selected, they will be studied in order to find out what traits their personality is made up of: IQ, personal history, family ties, education level, beliefs, background, purchasing power, their entire network of friends — that is, everything that is possible to search in social networks and using other ways (such as data from official government institutions).

With this in hand, a neural signature is created relatively quickly and remotely. This signature is sent to MKTECH computers, where the algorithms responsible for triangulating between the target and terrestrial and space antennas based on this signature are initiated. They maintain frequent contact, a back-and-forth communication with the target's mind, constantly comparing the parameters of the signature on the computers with the electrical signature of the brain.

From then on, the open link and triangulation between these devices will never stop intercommunicating. The process of analysis and interpretation of emotional complexity begins. The personality structure is measured and the cognitive process is observed along with the repercussion of this aspect in the individual's interaction with the environment to slowly steal their "soul" — piece by piece —, until the subject becomes unrecognizable.

Now the individual is officially "linked" to the scheme; their mind is connected to all the systems that make up the MKTECH, making them easily locatable anywhere in the world. Satellites responsible for locking in the target's mind — hacking, detecting their position via GPS and torturing them — reach throughout the entire national territory. You can't hide or run away.

This step takes place, of course, without the person's knowledge. Only when the signal begins to interact with the mind, a slight discomfort can be noticed — a kind of ringing in the ears. **Then, that characteristic noise, similar to a cicada singing, a cricket or the whistle or hiss of a pressure cooker, sets the exact moment when the person becomes a target.** As soon as all devices connect and synchronize with brain waves, one can hear an extremely high-pitched hum that resembles the pressure effect of various atmospheres, under water, at high altitudes or a blow to the ear that affects the eardrum or even that " whistle" similar to those caused in events with exposure to high levels of decibels for an adequate period of time — nightclubs, concerts, electric trios (trios elétricos in Portuguese). This "ringing" can be heard clearly when leaving a noisy area and entering a quiet place. A characteristic, debilitating buzzing occurs for a few seconds followed by a few voices that quickly dissipate — symptoms easily confused with other everyday problems. It is as if some electronic device is tuning or synchronizing with the brain electricity of certain areas responsible for the target's hearing and communication skills. In fact, this is exactly what happens, and this tinnitus is the side effect of this process.

Another strange sensation that can be described when synchronizing with thoughts is a sudden erratic/sporadic visual thought. An image enters the display of mental images, generating an immediate impact on all cognitive systems that reflects throughout the body, as a transient low-intensity tremor and a sudden weird/abnormal feeling of worry and decreased attention, with interaction in the visceral neurons, producing the so-called "butterflies in the stomach". Perhaps it could be the interaction of the wave traversing and modifying the electrical transfer between the specialized cells, but that's just speculation. More likely, the sudden change in mood and interference with brain frequencies will cause this effect.

As soon as the mind connects to MKTECH, the victim feels a strange presence, or a sensation that leads to the synthesis of that thought, in addition to the characteristic high-pitched ringing in the ear. This is a strong indication that something is wrong, even if the operators are completely silent, as the strange presence never dissipates. This phenomenon is similar to acquiring a new sense, a seventh sense — since the sixth, in popular belief, refers to the anticipation of future events — in which we feel as if we have a virtual door wide open in our mind, in which different properties of events that occur thousands of miles away are captured. The victim feels uncomfortable for having a strong feeling of being watched within their privacy. It prevents the observed person from feeling at ease inside their rooms, in their own residence. Perhaps you have already experienced this phenomenon in your life. For instance, when you're walking down the street and something makes you look in a certain direction. Suddenly, you see a random person staring at you. It is similar to that fleeting sensation that leads the person being watched to automatically and instinctively direct their gaze to the observer in the exact position without having first noticed their presence. In the case of the attack, however, this feeling is maintained as long as the link with the technology lasts. This occurs because, from that moment on, there is a communication bridge established in the target's mind. A "door" is open and is about to receive and send data to computers — this channel where the upload and download of data captured by the mind will travel.

We must understand that this is an unprecedented sensation for the brain. The target's reaction leads to this general state, as it is not common for us to receive data through unconventional ways that exclude our external receptors, our sensors that capture the stimuli and compile what we call reality — eyes, skin, ears, nose, mouth.

It is worth remembering that at the time of the classic MK-ULTRA experiments one of its projects was called STARGATE. The name of this enterprise really represents the types of experiments being conducted there. In fact, it is possible to mentally visualize it, in an analogous and metaphorical way, through an imaginary "toroidal vortex" that represents an open door in one end (the mind) and in the other, the remote location of transmissions (computers) through which all kinds of information pass, physically located thousands of miles away. It is a good analogy for what happens in the target's mind precisely because it resembles the opening of a portal in the brain. However, it has nothing to do with an interstellar portal or wormhole as portrayed in movies and TV shows.

The next step is to let the Electronic Mind Reading (EMR) of vocalized thoughts and images record thoughts 24 hours a day for weeks or months, performing naturalistic observation that doesn't interfere directly with the individual's behavior, and controlled observation that allows conclusions with a high degree of generalization and specialization in the course of the process without the human subject being aware that operators are invading their mind, amplifying their thoughts, discovering and searching all the details that are part of their essence.

This phase is extremely important. At this point, they can verify the target's reactions to certain situations at all times. The person is comfortable with their thoughts, not knowing that operators can see and hear everything, so they don't mind thinking normally, as doing this action is the only way to live as sentient beings. We don't know any other way to do this; we just go with the flow — in this case, the flow of thought. It may sound odd, but anyone who has been targeted knows what I mean. The normality of being human is completely lost in this horrible process.

This massive capture of thoughts is similar to intelligence service data capture, where one can later and calmly analyze the raw mass of information using data science or Neuronal Big Data principles. It may not seem like it, but listening to people's thoughts 24 hours a day can reveal all the details necessary for an efficient attack. It is easy to capture (awake) dreams, deepest desires, regrets, yearnings, affective processes that interfere with reasoning, social interactions and internal responses, the most complex human functions.

Observing a scene happening between the target and a friend, for example, checking the target's external reaction in this interaction and simultaneously looking at the truth from the inside (what the target really thinks about the observed situation) is something extremely relevant, as we rarely fully expose our thoughts and emotions in almost every situation. Intentions that aren't observed in behavior, or directed to others, what is faked, how and why is faked or any hidden information is even more relevant than the truth. It often demonstrates what the target's personality is like and the internal reflexes that lead to the responses perceived in their behavior. A real-time analysis of the human is in hand with the debugger[48] mode on, decoding its detailed processing referring to each act, sensation and emotion.

---

[48] A debugger is a computer program used to test other programs and make their debugging, which consists of finding program defects that facilitate access to program

With several months of thoughts and data captured along the lines of Neuronal Big Data, the amount of information obtained is impressive: each month has 43,800 minutes of thought. It is estimated that a normal person has, in a day, more than fifty thousand thoughts of the most varied. If we stop to observe the amount of information that the brain processes in just 10 seconds, and among them those that can be captured by this technology later filtered by several relevant algorithms, an absurd amount of information is generated. In other words, there is enough ammunition to be used during long years of torture. For example, if patentable ideas (images and sounds) with commercial value arise in this sea of words, they will be unceremoniously stolen.

The number of thoughts that are captured every second from a key individual, capable of bringing added value to the data, is a gold mine. Here one can only deviate to plan B. So, the scope of the attack shifts to stealing valuable thought and ideas, which may have a relevant economic bias that exceeds the desire for a 5-year cycle of torture. In objective **B**, operators become a neural parasite, capturing and sucking all relevant data from the target until all that matters is exhausted.

## 5.10.1.2 - Proceeding with plan A...

It is time to start working on the target's memory; to tinker with the "hardware" (brain) and "software" (mind). Therefore, the powerful and surreal Synthetic Electronic Dream (D2K) is activated. The process of transmitting images during the target's sleep begins, making them dream of strange, out of the ordinary contents, and synthetic, disturbing dreams that begin to artificially alter the victim's brain. It's the first contact with the technology that simply takes over the mind and causes an overwhelming and primary impact, as seen in the respective chapters.

Try to visualize everything we've learned in the previous chapters being implemented in a person's brain at the same time, every millisecond of their life, and think about the amount of suffering that is inflicted. Some

---

instructions, the step-by-step execution of a program; suspending the program to examine its current state, at predefined points, called breakpoints; tracking the value of variables that can even be used to generate a suspension or activate a breakpoint.

people start a nightly activity of writing down details of those horrible dreams, imagining it to be something of great value. **However, this action creates long-term memories without them even being aware of it**, which reinforces the synapse of such memories built in a manipulated dream reality and the strong emotions associated with it, ultimately cooperating with MKULTRA's tests.

## 5.10.1.3 – EMR (Electronic Mind Reading), V2K

The next step is to activate the crazy-making machine, a psychotronic weapon with the greatest capacity to cause direct damage, the V2K (**V**oice **to S**kull) — intracranial voice or microwave voice —, in a manner similar to turning up the volume of a television or a sound equipment. They gradually "increase" the volume of the V2K inside the target's mind. The weapon sends to the brain voices of many people in several different channels that creates the same positional effect of common sounds. That's when the target starts to hear strange voices, which at first appear only sporadically, creating the auditory illusion of screams in the distance or bizarre murmuring.

A very common example occurs when the target is located in public places with a lot of people, such as a street market, restaurant, mall, airport, among others. The target hears shouts like *"Heey, there!"* directed at them as their name is mentioned while walking in a supermarket. They look around, but nothing strange is detected; everyone continues to do their shopping normally. As the days go by, the voices emerge more frequently in different locations, such as at work. A conversation in the room is initiated and it sounds like a murmuring in the background. The artificial voices that evoke their name "merge" with that natural "murmur", which is a hallmark of V2K that holds the target's attention, but again, nothing unusual is found. The victim can only turn to a co-worker and ask, *"What did you say?"* or *"Did you call me?"*. But the coworker will answer: no, I didn't!

## 5.10.1.4 - EMR, V2K, D2K (Synthetic Electronic Dream)

Then, the most severe attack begins. Inside the target's house, the victim begins to hear a kind of conversation between several strangers. **It looks like the neighbors are locked in an eternal debate, a somewhat curious and abnormal behavior erupting into arguments, fights or whispers that seem to pierce through the wall of the room where the target is, but in an unnatural way.** The illusion is created: in addition to being able to hear the voices, one can perceive the heavy atmosphere of the discussion together with a characteristic noise in the background that resembles air conditioning or a running engine.

These conversations give the impression that the neighbor is talking loudly, or are similar to the sounds of a party, but they don't behave like normal sound waves. As a result, this conversation and this festive mood, the unpleasant atmosphere, never disappear. The constant noise fills the air day and night. This ends up forcing the target to search for their real neighbors and question if they are really hearing screams and weird fights that never come to an end. After this questioning, they will probably receive a negative answer about the hellish noise that seemed to start inside their rooms.

Operators adapt the auditory illusion according to several variables. Even the characteristics of the physical location where the target is enters this equation. So, this endless debate now seems to come from further afield, or it gives the impression of gradually altering the number of people speaking, the volume of the conversation or the sound quality. When the target changes rooms, or goes to the window, this perception is more latent. This event occurs relatively easily by exploiting the positional sound effects emulated with V2K as we saw in the respective chapters of Volume 1.

As soon as the target enters the house and goes to their room to relax after a hard day's work, the previous noise — the endless conversation with background music, sounds, echo, female voices, a high-or low-pitched voice, sounds that give a context of what is happening in that remote location — returns to the previous high level. And again, it seems to come through the wall and fill the room with sounds that never dissipate.

The target goes to the window, and the acoustic feature of the place doesn't reveal the position from which this evil and incessant discussion is coming. At that point, as they walk through the house to reach the window, the effects change slightly. The screaming continues; however, it gradually changes, differentiating from the main attack that takes place inside the room, which is by far the most disturbing place in the house. In it, intensity and intelligibility of sounds are always at maximum configuration. Sometimes only one V2K channel is used on purpose to generate the brain's interpretation of voices that seem to whisper inside the mind.

Upon returning to the room, identical effects of conversations that leak through the wall come back with the same intensity at high volumes, making it seem that there are real people talking there. This irritating performance projected in the target's mind turns into a routine, accompanied by constant and increasingly graphic and realistic nightmares led by the authors of the voices coming from "somewhere in the neighborhood". This affects the mood of the person upon waking up, which ruins their entire day. From here, this shouting — this party, conversation — will never be turned off. That factor alone might seem downright maddening, but that's just a rehearsal of what's to come. This first offensive lasts around 3 to 6 months and will vary according to the evolution of the target's mental degradation. Some people can't even withstand this initial attack and ends up succumbing in one way or another.

The next step now is to develop V2K content for the target, who starts to feel the true maddening power of the Intracranial Voice. If demons[49]

---

[49] Here I am referring to the demons materialized in mythical figures from the popular imagination and religious cultures. However, when it comes to their philosophical conception of influence on human behavior in the manifestation of evil and its various facets, or its personification in acts included in uncontrollable desires that lead to fights and violence towards others, and mainly focused on entertainment, I believe that these demons or their representatives are very real.

existed and they inhabited our world or our universe, the microwave voice would be their voice.

The continuous discussion of the neighbors in the background includes several voices that seem to come from nowhere and from all directions. Such voices penetrate directly into the mind of the target; others seem to sprout from within the brain itself as if they were part of the target's mind — another "self". The mental representation of these voices trying to constantly create scenarios, such as understanding what is happening and the origin of these artificial sounds, end up occupying a large part of the victim's time along with the natural wear and tear of mental processes that are automatically requested by these weapons.

This type of attack is purposely executed because it is known that there is an imaginative creation process generated by the voice that analyzes the physical characteristics of speech sounds, the physical properties investigated by acoustic phonetics that refer to amplitude, duration, frequency, among other aspects. Our brain is evolutionarily conditioned to create automatic acoustic analyses in fractions of a millisecond, such as: speech synthesis, text-to-speech conversion, voice recognition, speech-to-text conversion, speech comprehension and determination of the meaning of the utterance. We thus do auditory phonetic analysis without thinking about this particular process. Speech is such a complex process that, through the acoustic, articulatory and perceptual analysis of its components, it is possible to identify not only the content of the utterance, but also traces of the speaker's identity — sex, physique, age, background, ethnicity, "race" —, easily capturing their current emotional state.

Based on the automatic evolutionary process of message decoding, the brain will naturally and unconsciously analyze the characteristics of the voice such as tone, timbre, cadence, modulation of pitch and volume, which has the power to influence the emotional state of the target, to persuade them and to create mental projections, interpretations and the most diverse scenarios in a matter of seconds, resulting in thoughts that interpret such voices. Thus, all qualitative data are generated by conceiving an imaginary structure assisted by recognition images that try to synthesize everything that is happening at the location of the transmissions.

These processes require a lot of mental effort to occur. Designing such internal scenarios that are purposefully assimilated causes great fatigue. Operators and OPS are professionals and know in depth the torture protocol as well as the natural consequences of certain attacks and how they will reflect internally on the target. Given the infinite variety of individual characteristics, this exteriority of factors creates different representations in each person, but the central premise remains the same and can be used in all victims. This, dear reader, occurs in fractions of seconds in the human brain, with total control of the information that is sent. Without barriers, it becomes trivial to perpetrate constant attacks, overloading the entire system. **In this way, the V2K will inevitably drive the target crazy in their own home.**

Well, we now have the endless conversation as background sound and its atmosphere purposely created to generate expected negative effects on the target, a gathering of elements captured, processed and interpreted by the brain. The voices that act clearly and directly in the mind pursue the target at all times, kicking off the process of diverting the train of thought and attention.

Difficulty falling asleep due to voices that stir the thought, and the adaptation and intended change in the way of sleeping with the TV or the radio on all night in an attempt to drown out the voices and the dream content that is being systematically modified — creating vivid memories upon waking up — start the process of brainwashing. The artificiality of D2K is notoriously reflected in the quality of life, which already shows a visible and significant degradation. This will affect the target's existence, especially their work and interpersonal relationships.

At this point, some more time has passed, adding up to around 7 to 8 months of ongoing torture. Several changes in the target can already be noticed. One that stands out is the occurrence of pathological characteristics known as schizophrenic symptoms. The target, however, is unaware of such changes. Medicines, alcohol, drugs, psychologists, religious places and charms are also used to try to get rid of this "whammy". Advice from friends during an open conversation in which these voices are cited always ends up the same way. The answers may be as follows: "Spirits! You need the healing touch", "Gods or Demons, cross yourself", "It's someone's negative energy, rock salt bath!" or "The evil eye!". Even something related to an alien attempt to make contact as a response to the phenomenon is put on the agenda. If the target follows such advice or seek a definitive solution solely in these metaphysical, religious, mythical and mystical matters, the results will be absolutely null.

The small improvement in the natural state of the mind when resorting to this type of faith or cult of any kind can cause a momentary well-being and help gather strength to go on. However, the beneficial effect of this type of activity alone is not able to cope with the influence of electromagnetism on the brain; only the complete knowledge of these degrading weapons can fight themselves.

Deterioration in the target's behavior is already discernible. Catatonic schizophrenia is also noticed, due to the intense clash between the target's voices and mind, whose interaction occurs constantly through vocalized thought. Furthermore, loneliness in the face of the situation leads the mind to increasingly immerse itself in constant discussions and dialogues with the voices, and distances it from the common reality bit by bit.

## 5.10.1.5 – SYNTELE (Synthetic Electronic Telepathy)

From now on, the complete torture begins. The reaper of human lives enters the picture: the weapon that makes the victim repeatedly wonder if what is happening can actually exist in our reality; if there is any logical explanation for all the content that is unfolding inside their mind. At this moment, several visits to the depths of the most intimate and private thoughts take place, an observation by processes of analysis of all this disgrace followed by a slow, complete degeneration of the human being.

A "cognitive surgery" is performed, in which all aspects of existence are exposed from their private life, to emotions, perceptions and motivations. The focus on personal maladjustment is visible in the attacks. Now, in addition to the voices that hammer the brain 24 hours a day, every moment the "big news" emerges through the stunned paralysis; a mixture of irritation, fear, stress and bewilderment. **The voices now interact with the silent thoughts within the mind**. This interaction causes a rupture between the before and after of this event.

First impressions of this interaction are truly upsetting. The target doesn't know how to deal with the voices that seek information inside their memories, "encouraging" the brain to create constant flows in the concatenation of unconscious and involuntary silent thoughts, in which the main idea grows as it passes through cortical areas that instill on the information parts of life experiences that, in short, form the "self" and adds a range of information in a single automatic response.

Cognitive processes and the way they interact with the world, their relationship with the environment in all aspects of their existence, are radically modified. SYNTELE is ultimately activated in the target's life. The volume of sounds and voices that reach the brain is intensified and remains at a level that will not decrease under any circumstances. The actors change, the number of channels and the 3D effects of V2K are improved.

From that moment on, the attempted murder by torture or suicide due to acute stress comes into the equation. It's latent and intentional. Now the survivors have few options and one of them is to embark on a story created in their mind by the operators, in which its fragments are being gradually instilled day after day. Memories altered at night inside the dream via D2K strengthen the daytime narrative. Everything is carefully elaborated, based on known techniques from the Kubark[50] manual that have been adapted for this modern attack. At this point, the path to become a Winter Soldier can be glimpsed and followed if you focus on Objective C. It all depends on the target's mental and physical strength and support network — financial, family/friends —, their cognitive stability, their sagacity in understanding this complex phenomenon, and

---

[50] KUBARK was a codename used by the CIA to refer to itself. The codename KUBARK appears in the heading of 1963 CIA documents that describe interrogation techniques, including what the CIA called "coercive techniques". This manual is the result of years of torture that we've seen in the previous chapter.

their willingness to live or succumb to total mental breakdown due to constant stress.

At this very moment the mind is lost. Most programmable killers commit atrocities; their brains in the hands of hateful people are now completely dominated by this technology. It's possible to verify that the target is becoming a real-life biological experiment. This perception can occur due to countless external factors, but it usually starts by perceiving the nuances of MKULTRA 2.0 experiments and their impression on the power of attacks and in the way in which everything is meticulously conducted in order to hide the experiments within the experiment itself.

There are some mandatory actions that may be taken: fighting for control of the mind with all your remaining strength and resources; dying from acute stress; becoming seriously debilitated by strokes or problems resulting from the attacks; going completely mad and checking yourself into a mental hospital; or becoming a Manchurian Candidate. There is actually an allegory of technologies being used in the target's mind.

## 5.10.1.6 - EMR, V2K, D2K AND SYNTELE

Stepping up one notch in the attacks, the Synthetic Electronic Telepathy (SYNTELE) provides a complete turnaround in the target's private and psychic life. Some of the victims convince themselves that they are completely schizophrenic, because now the voices that inhabited their mind start to interact with their thoughts. The Targeted Individual is aware that their mind is indeed being heard, read and manipulated. Torture and experiments go to a whole new level. The target knows there are people trying to kill them, inflicting as much suffering as possible and performing a range of thought experiments with a variety of electronic devices. The originality, both in the technology and in physical and psychological changes that cause emotions added to all this, creates a unique condition that is cowardly taken advantage of by the system operators. At this point, emotional maturity and the power to overcome primary stimuli are put to the test.

Nonetheless, whoever reaches this stage (there aren't many) is considered a winner. Most will have already given up: they will be on their knees in some sacred place, banging their heads against the wall in mental hospitals; others will be scratching the ground with their fingernails until this digging process tears away the flesh and exposes the bones; or maybe they will simply end up brain-damaged, completely doped with strong medicines of all kinds wrongly prescribed by professionals in the field who still don't know this type of device and its power in altering human cognition.

Assimilating to SYNTELE is not the simplest job. The insanity and hesitation in revealing to someone what is happening is a constant part of life. A recurring question arises: how to share this event with people close to the target or if the target should even think about it. *"If I [the victim] don't even know, will my brother, mother and friends know how to handle these events if I tell them what is happening?"*. The answer is no, unfortunately. Don't involve your family members; this will only make it worse. Seek help from a group or someone who knows what's really going on, and then get family and friends involved.

In simple terms, it'd be like comparing the target's brain, the equivalent of a processor on your PC, with a central core that processes conscious thoughts with five subdivisions or parallel processes, each responsible for their respective distinct stimuli that come from the external (vision, hearing, smell, touch and taste) and internal (cognition) common reality that organizes all this, supplying the mind with the flow of information that makes up our understanding of reality. Now imagine that psychotronic weapons, when used today, require a generous portion of such processes and occupy a large part of the subprocesses responsible for vision and hearing, in addition to overloading the central nucleus. They force the brain (processor) to create more internal processes or parallel nuclei to deal with this volume of information that attacks it. The effect resembles a kind of parallel reality which the brain is forced to compute and try to put in order. In the individual's mind, it ends up producing the feeling that there are two processes: one that will manage the common reality and the other working in parallel within this artificial psychotronic reality.

Between one reality and another, they end up colliding, generating erratic behaviors, outbreaks of anger, attention deficits and exaggerated movement of the mouth. The silent thought slips away, and is visible on the targets' faces as if they are talking to themselves, an act that immediately makes people around them perceive the electronic schizophrenic behavior. Most of our behavior is unconscious because our awareness, or perception, is limited in the sense that we can only focus our attention on one or a couple of things at a time. It's impossible to be aware of everything that is happening around us. In most cases, this compulsory processing recruits too much mental resources and irreversibly affects normal processes as a whole.

As for the attacks, the voices slowly interact with more intensity and depth within the target's thoughts. At this point, many questions start to arise, such as *"How is this happening?"*. Some targets — when having contact with SYNTELE — thought that maybe they could be thinking too loud, perhaps out loud, inside their own minds. This kind of absurd questioning is natural when faced with the Synthetic Electronic Telepathy, that acts in an open channel of the thoughts and reception of voices. *"Am I thinking out loud? Is that why they can hear me?"*

A process of self-knowledge also begins. It's when the target becomes aware of their vocalized thoughts, or internal thinking, and its importance in cognition. Besides, the hearing goes through a process, in which much of the attentional effort is directed at the sounds that reach the ears along with the microwave voice that appears to be somehow sharper. Changes in the way we focus on certain processes also take place. Our mental effort of attention starts to analyze and focus on brain processes that we previously didn't give due value, for example, how the thought is generated, the memories accessed and the result of the dream manipulated in the mind.

For the target who is aware of the events, more profound questions arise, especially about how to deal with cognitive violation and hacked mental processes, which are part of the cortical mechanics of the natural creation of certain thoughts and memories. Changes in thinking and difficulty in memorization become more and more exhausting and jeopardize life. Anxiety with decreased quality of selective attention and impairment in executive functions also become frequent symptoms due to constant psychotronic torture.

This emotional charge, the extraordinary nature of the interaction of these weapons with the brain, acute stress, sleep deprivation and incessant noise in the cortex, generate the deterioration of reactional mechanisms, the end of volition, and openness to suggestibility, steerability, passivity and submission, in addition to the feeling of inferiority and insufficiency.

The primary objectives are quickly fulfilled by the torturers. At this point, the absorption of the attacks by the brain already affects the target's daily life, leading them to no longer see themselves as the same person as before. The difficulty in organizing and managing time and meeting or fulfilling commitments becomes a constant. In the next chapter, we're going to delve into advanced torture techniques. We're going to see the reason for each phase of the attack as reported here.

At this moment we realize how vulnerable and exposed we are by this hacking of the mind. Some more naive targets fall into a deep sadness for society and for the future of their children and grandchildren. However, we must remember that humans have adapted to everything until the present day, so we can also survive this episode of social transformation.

Another point that becomes perceptually relevant is the attention that the target gives to the formation of their thoughts, that is, to their flow of thought, and to what extent they let their thoughts immerse into the memories, exposing their cognitive privacy. They become keener and more aware of the inner processes of mental mechanics.

Needless to say, the feeling of knowing that a stranger hears your thoughts 24 hours a day and gives you feedback at every moment changes your life. Within this perception, a doubt is created and persuade you to overcome this misfortune. You keep creating a world that is easily modified by unexpected external and/or internal stimuli. An instinctive process of self-preservation is initiated, and targets adapt their thinking as to hide information and not access it so easily when forcefully requested through V2K interaction. The beginning of a particular mental "firewall" starts to be developed.

## 5.10.2 - My house is the worst place ever

Positive feelings coming from a general atmosphere of optimism leave the way open for the individual to carry out their activities in the best possible way. Without stress, the environment conducive to what we call happiness is created. This is how we think about living and working. Negative feelings associated with a toxic environment that are imposed by external agents, such as technology operators, within any environment that the target goes to — especially at home and at work — cause severe, harmful effects. This is due to the use of techniques based on the classical conditioning (Pavlovian or respondent conditioning). Your home becomes the worst place to be. Whenever you head home, you will instinctively remember the discomfort, stress and all the sleepless nights. This will motivate you to stay as long as possible away from this place artificially generated by OPS and psychotronic weapons.

The indescribable feeling of losing your own home enters your thoughts amidst the chaos. This is a known strategy that causes anxiety and anticipation of conditioning actions, a stimulus association capable of provoking deep emotional reactions. This tactic is widely used by Professional Organized Stalkers to traumatize and to create negative artificial associations to certain everyday actions, such as the simple act of going home, using the bathroom and sleeping. The target is always distanced from what they belong, where everything is conditioned to harmful emotions, psychotronic torture and memories of events that impresses the mind.

Technology operators make the target's home an uninhabitable place. They expel the victim under the intensity of incessant attacks, generating an unpleasant, repulsive feeling or the desire to flee the house; a feeling of escape. There is severe emotional suppression. This is also one of the hidden goals of the experiments. Operators have access to inner emotion before it even manifests itself externally. Remote EEG waveforms (Electroencephalography, Volume 1) help to physically verify the electrical configuration, causing illness, impairing performance, and interfering with the normal functioning of intellectual and motor faculties.

Faced with all these strange facts, some more resilient and curious souls understand that the time has come to try to unravel the secrets as an act of survival. The initial reaction, however, is always based on topics we are familiar with. So, the first thing that comes to mind — in an attempt to assimilate the feeling of being watched, those loud and annoying voices that somehow interact with your thoughts — is the possibility that there is surveillance equipment hidden in the house or in the car, such as cameras, microphones and speakers.

Encouraged by the voices that take advantage of such mental attempts, operators heat up these thoughts, driving them according to the target's mentally created story that is increasingly dark, depressive and deep. So, after this psychic driving, they say: "*We placed cameras, microphones and sound equipment in your home while you slept/were out*". This type of driven thinking keeps echoing throughout the day, mingling with the paranoia of being constantly observed that escalates to levels never experienced by most people, and leading to a permanent state akin to persecutory delusion and delusion of reference. The person feels threatened by supposed enemies who are watching, spying or trying to annihilate them by various means; they feel watched by passers-by on the street and think that co-workers are always conspiring against them, directly and indirectly. The most incredible thing is that these symptoms are legitimate as they were created electronically during the course of torture. MKTECH has the power to simulate, emulate and properly create various disorders of the human mind known to science.

Some targets even go to police stations, but don't know how to explain what is happening to the authorities and are unable to file a police report or a lawsuit against the people spying on them. OPS know that they can carry out their experiments and remain hidden from everyone, including using the element of surprise as a primary resource that prevent any reaction due to a total lack of knowledge of this technology and disbelief in the existence of something capable of having complete access to thoughts and dreams. After all, it is impossible to imagine that an explanation as absurd as the one given is responsible for all this terror.

Having voices in your mind everywhere, every single day of your life, is a fact so extraordinary that makes the most skeptical target at some point wonder if it all comes from some alien technology; atheists may believe that such voices are related to deities — it's even considered that the target died and is trapped in some timeless zone. Hopefully, this kind of fantastical daydream will probably fade as this weapon is disclosed to the public.

Targeted Individuals report that they got rid of several personal objects while trying to locate the source of the voices that appeared to come from microphones built into such items. They are constantly encouraged by OPS that incite absurd thoughts and attitudes (like looking for built-in microphones inside their belongings: cell phones, glasses, handbags, clothes, backpacks, etc.). The voices claim that they placed a chip in their belongings while the victim slept, ultimately leading them to destroy their objects in the search for such a device responsible for capturing thoughts, in addition to filming and listening to the target. The victim then gets rid of their belongings to stop the madness, including throwing their cell phone into a lake, for example, but unfortunately to no avail.

The voices, that are amused by the confusion and torture, lead the individuals to erroneous conclusions. Confused, an individual reported that they thought that small cameras and microphones were installed in the screws of his sunglasses. Even after the victim eventually threw away all their belongings, the voices didn't disappear — they continued to observe in real time the target's every move and to comment on them. *"Hey, we've placed cameras and microphones in your bags"*. At this point, everything is in the garbage bin. As they return home, however, the dense, negative atmosphere with the same intruders is back in full force. SYNTELE (Synthetic Electronic Telepathy) in its "input" or "downlink" into the mind is capable of generating sound interpretations that defy reality.

## 5.10.3 - Car

Voices are now clear and come from inside the car as the victim drives to any location. Keep in mind that, as operators can hear, "see" and even know the target's body position in space and their current thoughts, the target is encouraged to believe that there are cameras, microphones and mini speakers in the vehicle. The acoustic signature of the V2K, which manages to blend in with the sonic properties of the environment, begins to confuse them. The noise of the engine, the wind blowing on the window, the air conditioning and the radio sound mix with the voices and create a unique auditory feature.

Cars have a very peculiar acoustic environment, which can be experienced in any vehicle. Whether in an automatic sedan with the windows closed or in a Volkswagen Beetle with the windows open, targets convince themselves that someone installed surveillance equipment there. This line of reasoning is actually easier to accept than the fact that there are weapons that interact with the brain as if they were applications or programs on your cell phone or computer. Some individuals go so far as to ask their acquaintances to get in the car, put their ears close to the air vents and then ask them if they are hearing voices coming from that specific spot. *"Can you hear them?"*. In this particular situation, the acquaintance becomes really concerned about the sanity of the target in question.

Some individuals report that they spent hours and hours a day driving around in their cars in order to follow the instructions of the voices, such as *"Meet me at such a place"* or *"Can we meet somewhere to talk?"*. Some say they drove in circles all night long, day after day, aimlessly. There was always the promise of a meeting to negotiate an end to the torment, but the victim could never physically find anyone connected to the voice. This is yet another macabre "joke" from the Organized Professional Torturers that operate the technology.

At this advanced point of modern MKULTRA experiments we have the constant background noise generated from the remote location of the operators that creates a permanent negative atmosphere, fights, insults and a characteristic and cyclic noise that serves as a carrier for this background atmosphere. The main voices penetrate the mind with unique characteristics through V2K and SYNTELE, coming from several different channels, always amplifying thoughts and adjusting the verbal attack in order to impede the target's cognitive freedom and to divert attention to the reality they want to instill in the mind. This makes the target afraid to think, initiating flow control and access to memories.

Microwave voices will definitely become part of the target's life. Dreams are manipulated to provide the visual component of the story they are creating, to torture and insert false memories via D2K, instilling in the target's mind some painstakingly created reality for some nefarious purpose. All external receptors — with the exception of smell and taste — are under the control of MKTECH (Mind Control Technology), including internal processes such as vocalized thoughts and visual images.

At this point, the battle for life and death, and the mastery of the mind officially begins.

Negative and recurrent emotional states start to affect cognition in decision-making. This is also a process capable of adapting to environmental stimuli (negative or positive), and the ability to face adversity using functions such as attention, memory, judgment and thought. Some targets report that they have abandoned their cars in airports, malls, street markets and stores and returned home by taxi in order to get rid of the "electromagnetic bubble" that they thought formed around their mind, however, without success. After all, the noise is at the same frequency when they go back to their residence. The purposeful disillusionment and the succession of disappointments are emphasized with the aim of leading the target to an ordinary state: they're one step away from becoming a programmable killer. The target begins to understand what the loss of intellectual and physical privacy is like — in other words, the feeling of sharing their intimacy with others —, as operators keep on implementing the tactic of total loss of acquired rights (similar to concentration camps or totalitarian regimes), automatically generating a deadly hatred of the situation, with no possibility of being expressed, since the enemy is completely virtual and materializes in the form of voices, images and dreams.

## 5.10.4 - Bathroom

Remarks about their intimate areas, particularly concerning what happens inside the bathroom of a house, are made by OPS whenever the target goes there. I repeat, every time they sit on the toilet or take a shower, they will be harassed, even by people (voices) of the same sex who constantly use profane and offensive language. It's a clear degrading sexual harassment used to provoke and generate natural reactions that the lack of privacy inside your own house, of your own bathroom, would cause in any human being, and to simultaneously damage one's honor or reputation.

One of the individuals reported that he showered with underwear at the beginning of the attacks. According to him, this cycle lasted 4 months due to the size of the breach of his privacy. **Keep in mind that OPS can see what the target sees with a degree of detail that is still uncertain as it depends on many factors, mainly on the technology version.**

A recurring event that takes place daily in our lives without us being aware of its complex dynamics is the audible variations in air pressure generated by water falling on the bathroom floor while being inserted in a resonance box; the noise coming from the shower. The sound of water hitting any surface generates high levels of decibels — just remember that we can hear the waves crashing or the sound of a waterfall from far away. This happens in our daily lives too — when we are taking a shower, for example, and by any chance someone tries to talk to us through the door that separates the bathroom from the rest of the house. Normally, the person taking a shower will start screaming and won't understand what the person on the other side of the door is trying to say. The noise of water in the bathroom is too loud. And we don't usually notice it. For the dialogue to continue, the shower will probably be turned off and the person will ask *"What did you say?"*, as they've returned to normal decibel levels, which now makes it possible to understand the message being conveyed.

Generally, bathroom walls have a flat, regular and rigid surface, which absorbs less and reflects more sound. Thus, when emitted, the sound will form a perfect return image that was originally generated. This structural configuration also causes the person to hear more intense sounds reflected from the walls (resonance) and the lasting, lowest tones in their ears.

The noise generated by the water hitting the floor in a cyclic way and the acoustic characteristic of the environment (a bathroom) are perfect carriers for the V2K. The microwave voice blends in perfectly in these conditions. As soon as the voices reach the victim's skull, they will hear extremely loud and shrill screams. This greatly confuses the individual and easily invites them to get out from that place as quickly as possible. The end result of this bathroom-shower-water interaction mixed with V2K is maddening, similar to reports of targets inside an airplane (Volume 1).

The most amazing thing is that, when the shower is turned off, and the noise of the water is extinguished, the screams that echoed in their head cease altogether. The "noise" decreases as if by magic. V2K is one of the most complex phenomena to understand within the MKTECH universe. In chapter 2.2.3 of volume 1, there is an illustration with all the components that produce amplification within the mind. And this example leads to the chapter in which I explain this extremely disturbing phenomenon.

This event (in the bathroom) also exponentially increases the sensation of being watched. So, some people decided to relieve themselves with the lights off or/and using towels on their laps to cover their private parts. Introverted people who care about their privacy feel extremely violated by this process. Extroverts won't act very differently either.

The bathroom becomes the worst place in the house, the most "watched" one. As the operators know what the individual is doing, even when targets are fulfilling their physiological needs, they will talk about particularities of the body and very intimate things about the person with the help of any thought captured regarding such biological functions. By now the victim's privacy has been completely violated and outrageously destroyed. They've become a prisoner inside their house — inside their own head and life.

## 5.10.5 - Personal Data

Just as biological needs are shared unintentionally and constantly, so are computer passwords and financial banking activities. Any online privacy, classified or high-value information is also easily stolen. Let's look at the confusion this causes in behavior.

Some victims report that they thought there were cameras inside their house. So, they began to use blankets in an attempt to cover their hands, preventing the (supposed) installed cameras from visualizing their hand movements. Unfortunately, we now know that covering your hands and keyboard in order to protect your passwords is a vain attitude. As the password is mentalized, a microwave voice captures it and sends it back to the target's mind, showing that operators can see and copy everything, which purposely worsens the paranoia.

In another case, the target desperately called a friend and asked him to log into his personal accounts — emails, social and work networks and websites — and simply delete all the important information contained therein. He was worried that the people behind the voices would steal sensitive information from his personal and professional projects. In the end, a huge amount of damage is done.

## 5.10.6 - Trips

Some targets with more capital may choose to travel to unwind — to have a little bit more of inner peace, even for a minute, since they haven't heard the sound of silence for a long time. Nevertheless, that's not what happens. They'll step into the airport, flight-call-like background music will start playing and will amazingly sync with every step taken by the target. With each step taken, a tenuous music, activated in a timed manner, will start playing. It is as if the airport floor is made up of pressure sensors similar to the Walking Piano that is activated with the weight of a person and immediately triggers a maddening sound that is very different from the sounds coming from the loudspeakers that announce flights and passenger names. The most curious thing is that, the acoustic characteristics of the background music remain intact in volume, as well as in spatial and directional interpretation, sound quality and the illogical priority of the mind to the detriment of other sounds captured in the environment at any part of the airport that the target goes, contrary to all laws of physics that governs mechanical waves.

The target heads towards the plane thinking they will be free from the attacks; that the airport is a kind of safe haven due to the radio equipment, radar and reinforced security. However, as the plane takes off, here comes the second surprise of the day: the voices will not only remain, but will be extremely loud, given the characteristics of the Synthetic Electronic Telepathy (SYNTELE) with the Intracranial Voice (V2K). Now the chance of becoming a Winter Soldier or completely freaking out in mid-flight is very high.

At this moment, voices and sounds can no longer be blocked with earplugs, cell phones, TVs or radios. Nothing overrides microwave voices, not even music or headphones. Nothing works! In some cases, such devices only intensify the torture. The goal now is that, every millisecond, the noise in the auditory cortex added to the sound of the engine will create a powerful torture and mental dominance, with no chance of being turned off. The psychic driving created by operators in this favorable environment may drive the already exhausted target to despair, culminating in an attack on the crew, passengers or on the aircraft itself. In this way, the victim becomes an unconscious instrument of MKTECH's ulterior motives.

Upon disembarking at the destination and heading to the place where the target will be staying, the third unpleasant surprise arises: the victim notices that the same screaming that originally seemed to come from their neighbors returned with full force. The attack keeps at the same intensity and the auditory illusions remain intact, such as the screaming in the background and noises and voices coming from the upstairs or downstairs rooms. Some Targeted Individuals report that they even went up to the upper floors of the hotel they were staying in the hopes of solving the mystery. But when they reached said floors, the voices didn't seem to come from the direction of the rooms as expected. They only saw several closed doors, apparently without anything abnormal or strange happening, which prevented them from acting against the people who lived in those places.

This psychotronic dynamics takes place all over Brazil, regardless of where you stay. It may be the 4th floor of an apartment in Ipanema, a penthouse on the 15th floor in São Paulo, a hotel in downtown Brasília, transit hotels next to Alberto Santos Dumont airport, a secluded inn by the sea in Bahia, the Iguazú Falls, some houseboats in the Amazon river, or even during visits to the top of Christ the Redeemer itself. In any of these locations, regardless of the spatial arrangement of the scenario, the audible effect of the neural weapon will always have the same acoustic characteristics generated in the brain.

Also, it's worth mentioning that depending on a series of elements within the operators' objectives, such as the reaction they want to see on the target or what type of immediate changes in their behavior pattern they want to observe, they can opt for some different attack strategies. First, operators may give the false impression that they have turned off the equipment. In this case, they reduce or suppress the volume of the voices inside the mind when the individual travels to another state. As soon as the target feels they're finally free from this dreadful experience, operators reconnect all devices as if they had never left their house. The emotional impact is massive, and it can immediately cause a fatal nervous breakdown.

The second option is to never stop the attacks. The voices will accompany the target from their departure from point A to their arrival at point B at maximum intensity without any intermittence, loss or decrease in the quality of the signal. This clouds the victim's interpretation of reality and increases their susceptibility to performing violent acts or suffering severe neurological damage such as strokes.

The third option is even crueler. Once the target lands at the destination, the voices and sounds drop to minimal levels or disappear, waiting for the target to settle into their temporary place thousands of miles from their home. The victim is visibly shaken and terrified by the events that made them travel, so they hope to relax in a distant environment, to finally have their peace of mind back. So, operators wait for the target to sleep and, within their dreams, the attack with macabre films is restarted. Upon waking up, the operators' advances take place in a harsher way, and with higher intensity, in a totally different scenario and environment. Now it looks like the culprits are located in nearby buildings, screaming, threatening, laughing and rioting once more. Given the circumstances, the target gives up and fell in a deep depression. Here, the complete mastery of the attacks takes place, as the target's negative inner state has reached the desired setting: torturers can conduct any kind of mind control experiments now.

The individual even wonders if the attackers knew about the trip; if they rented rooms in the vicinity for that purpose. *"Who's following me? How did they arrive before me?"*. This also occurs when the target returns home. As soon as they do, the attack continues and causes a huge disgust with life in general, a decline in all cognitive areas, as well as feelings such as tension, fear, panic and apprehension.

Now the pressure reaches a level where mental flexibility and resistance to outside interference — the focus on reaction within the drastic reduction of volition — reflect in basic planning, that is, in effective performance. An adaptive behavior to extreme situations will be required. The target is aware that they will never be free of the attack, no matter where they are located! This is a sad realization. At this point, a few years of attack have elapsed and those who have managed to survive are no longer the same. Despite this, they resist to maintain what is left of their sanity and essence.

Far from their residence, which was abandoned due to the attack, the target finally realizes that all of this was in fact one of the operators' main goals.

## 5.10.7 - Paranoia

The truth is that no daily activity will be free from MKTECH and OPS (Organized Professional Stalkers) influence. The sadists and murderers who command this technology start a morbid activity of reward and punishment. As soon as the right state of mind for the transformation into Winter Soldier is captured, some requests to release the brain become even more problematic, such as: *"Kill that dog and we'll stop bothering you"*. As the requests get more and more serious, the torture completely transforms the target's personality, deconstructing the perception of themselves.

In this process, combined with the paranoia and the consequences of the attacks, the procedure of putting targets against neighbors, employees, friends and family members automatically gets to the point of committing acts of extreme violence or promiscuity, since requests of a sexual nature and incentives to murder are now part of the torture routine with the clear aim of verifying how far the target is able to go.

If the victim lives in a house, their paranoia becomes increasingly severe. They start to feel watched 24 hours a day everywhere they go. They will be suspicious of everything and everyone, and will begin to analyze the neighbors' activities, as the V2K has the ability to create perfect positional sounds in the brain as explained in previous chapters. This makes the voices and murmurs seem to always come from certain directions — that is, directed at the neighbors who have nothing to do with the situation. Even if the target knows their longtime neighbors, the mental illusion that they are spying on the victim inside their own house will be maintained by the encouragement of the voices that come from the weapon. The operators will say that the target is being watched through the cracks in the walls, through the window or through the smallest holes in the tiles, in short, in every crevice of the house.

Some targets end up covering all windows with smoked glass during a psychotic break. They carefully look for a crack through which they can be observed, or they even remove light sources, such as transparent tiles. Other targets report doing everything in complete darkness at home. They only performed any kind of private activity at night, in total darkness, but also to no avail. After all, it's possible to guide yourself through several other advanced monitoring systems within the MKTECH, keeping the target's thoughts and body image captured by the return of EEG waves.

Some victims report covering their windows with newspapers in utter desperation for the return of privacy, obviously to no avail. They searched the entire house for surveillance devices, which is the most natural thing to do, and found nothing. The intention now is to make the target highly uncomfortable, and it is very effective. The target starts to avoid everything and everyone, especially their house and car, as the volume (intensity) of the attack is always greater in the residence, especially in the bedroom and bathroom. The target also feels compelled to stay in cheap hotels to get rid of the chaos of the house.

In a recent case, a victim ended up mutilating themselves with an ice pick. They pierced their eardrum with it because they could no longer stand the torture. In fact, that's what they said to the doctors who took care of them: *"The devil's voices won't leave my head".* Other subjects begin to go deaf on purpose, listening to music on headphones at maximum volume to reduce the influence of the brain's decoding process in interpreting the verbal communication forcefully received by V2K every millisecond. In these cases, the targets were attacked for a maximum of 3 weeks.

Operators and torturers have as their main goal to disrupt the target's life, not just in major events (crucial school exams, college presentations, civil service competitive examinations — and the study that precedes such events —, in the exercise of their profession and in daily commitments), but also in other spheres of life as the continuous interference undermines resistance and the will to live.

Remember when you woke up on the wrong side of bed and didn't want to talk to anyone? This possibility doesn't exist for a Targeted Individual. They will be forced to wake up hearing the voices of several unknown people — to be systematically harassed. If they wake up worried about something, this concern will be the agenda for the day.

Another technique widely used to further degrade the health of the target takes place during meals, especially with those who have the habit of eating at the table with family members, in a more traditional way without the TV or radio on. Within this scenario, targets suffer at every breakfast, lunch and dinner. As soon as they start eating, the most demeaning attacks begin with the intention of diverting attention between the transmissions and family members at the table who interact with each other and with the food, letting through any hesitation, vociferation or tics regarding electronic schizophrenia in responses to the psychotronic attack. This draws the attention of the family who perceives the abnormal behavior and later raises doubts about the target's mental health. This attack leads the victim to swallow the food in a hurry so they can go to a place where their attention can be scattered, especially with the use of a device that produces sound. In other words, they get extremely stressed when eating, which can cause indigestion, weight gain, a swollen belly and various other diseases.

The behavior — added to other problems caused by torture — worsens the condition of the victim who already shows symptoms of psychopathy, constant irritability and impatience, which can trigger an uncontrollable outbreak of anger for no apparent reason. Keep in mind that this is just the impact of verbal abuse and violations of all kinds, which amplify the outrage. In some cases, the target may tend to dialogue with the voices in order to understand what operators really want. However, the hidden truth of the events will never be said or expressed; only more acts of violence under the cloak of the unknown — of the sense of the surreal that these modern neural weapons can provide — will be carried out.

Targets end up developing unhealthy habits and manias. As real human subjects, one or more areas will be seriously affected, creating a psychopathology with its behavioral dynamics usually found in the abrupt change from normal to excessive. The target will notice this transformation if they do a self-analysis. For a certain period, they will see that the change was abrupt if compared to their daily behavior in the past, more specifically before the beginning of the psychotronic attacks inserted in this scheme of modern MKULTRA experiments in an artificial and involuntary way, which was created by external agents.

It's a terrible struggle to maintain normality and naturalness in the face of social occasions, a difficulty in maintaining one's identity and at the same time trying to find out what is happening. Many things are meaningless. A person's sense of hopelessness only increases. Operators make the target share such events with as many people as possible, forcing them to ask friends and family members for help. Unfortunately, these moments are actually celebrated by the perpetrators.

The atmosphere of the physical location where the attacks are produced is easily perceived by the target's mind. The characters, faces, attitudes and other human traits that have been previously associated with the voices, along with the flow of automatic, unleashed imagination that has also elaborated the most diverse theories using visual memories and the voice of the mind, are captured and used as recorded scenes to create "Déjà vu" and "Déjà Rêvé" phenomena (Volume 1) with the Synthetic Electronic Dream (D2K). At this point, the victim tries to guess who the attackers are. The most diverse speculations (mentally) created serve as ammunition for attacks, because in this process new information — that were probably not originally captured by the attackers — emerges, such as childhood memories, some recent squabble or disagreement, or a person with whom they had a relationship who may be behind this feat.

Episodes of adversity will be exhaustively repeated, preventing the individual from overcoming momentary mishaps of life. They're experts at locking the target in a past memory and keeping them stuck in this event that should be forgotten. After all, adversity is part of every human being's life. The way and speed with which these adversities are overcome will decide how the person will continue to move forward with their scars incorporated into their self and the memories resulting from such event. Simply put, this is how human beings manage to go on, otherwise we would spend our present and future only regretting the past, and we would automatically get depressed.

So, this tactic profoundly affects the individual, as they now have two very strong sources of stress: the problem itself that has arisen in the collective reality, and MKTECH, which uses this situation to aggravate verbal attacks, recalling the event and creating projections of the future in a recurrently pessimistic way. The person is completely undermined with all their cognitive aspects deformed together with their personal, deep, private, intimate space immersed in the experience. Overall, there are places that welcome us, others that repel us. MKTECH turns all physical and abstract locations into a kind of repellent.

Changing volition via thought modification and the inclusion of false memories, as well as instilling a reality that only exists in the hands of operators, is a constant in electronic psychological warfare. The target starts to live only the reality imposed by MKTECH. Torture and psychotronic attacks manage to overcome the target's ability to react, incapacitating them. Ultimately, they are at the mercy of whatever is inserted into their mind via V2K/SYNTELE — one of the main responsible for this type of intensive torture.

## VI - Tips. Rule #1 for psychotronic warfare targets

**Never ever believe anything that the voices are trying to tell you**, either in the made-up scenario or within the reality they're trying to instill in your mind. Always ignore the content of the message they're trying to get to you through modified words and dreams: both are part of something bigger — the complete cognitive degradation. Don't advocate for the operators and don't materialize the recurring themes in your mind; otherwise, you will bring to life something that only existed on computers, in the minds of the content creators and in your own head. By exposing the topic to people around you, you're doing exactly what they want, and you are once again manipulated without realizing it. Every narrative developed in the attacks is designed in such a way so that the target, by accepting its premise, loses the battle. As soon as the content is vociferated or assimilated as truth and exposed to others, your disadvantage in neural warfare is increased and the chance of continuing to live with the initial invading idea incorporated as truth in your life is high.

From the moment that your thoughts absorb the attacks and work with future projections based on propositions artificially instilled by the set of content used in the most varied attacks, a unique "road" is created, leading to paths projected by the operators and which have already been plotted before the target even received the first radio burst in their mind. Your free will based on rationality and conscious human choice will no longer be so conscious. This is also one of the many areas of neural reprogramming and complete personality modification.

It seems that the victim's energy is completed drained during the process. The "people" with whom the individual spends most of their time influence their decisions and absolutely affect their life in a negative and destructive way. Some targets reported that they started screaming, begging to be free: *"What do you want from me??"*, culminating in a never-before-seen wailing followed by a nervous breakdown. At this point, TI use hypnotic and anxiolytic drugs to sleep and reduce anxiety. Unfortunately, they end up choosing this path to minimize the torment that puts them on their knees or makes them cry in the fetal position as they wait for death to come. The harmful effects of the MKTECH invasion and torture are potentiated. Some psychotic breaks cause the target to destroy the entire house, leading to self-harm. It's in this kind of self-destructive behavior orchestrated by the operators that the real fun of the sadistic OPS (Professional Organized Stalkers) comes to a head. Some targets reported that they kicked their furniture and broke objects, but that didn't lessen the intensity of the voices or the attacks with neural weapons. They spent months kind of exploding internally until they had a nervous breakdown. Don't forget that operators use the human evils created by themselves to hide from everyone.

The experiment of creating and maintaining another reality was a priority since 1950, and its consequences can be felt in the modern tests carried out today. Needless to say, this goal has been already achieved. Now the focus is on creating more and more concrete realities directly in the waking brain. Mental abstraction, semiotic thinking consolidated from the representative thinking of the symbolic nature of the word are some of the artifices used to create this artificial "reality". In addition to conditioning learning — one of the primary components of this new type of attack —, physical and mental isolation and its structural chain of family/friends and financial support is included, destroying what we know as positive psychology and optimism.

The scheme follows a strict protocol with steps and techniques already defined and new ones being implemented for effectiveness tests in order to be part of the myriad of new torture procedures adapted to the reality of a complete cognitive violation of the individual. Targets find themselves involved in a battle that has been fought cowardly across the world for a while — that is, a prelude to a greater war to come.

## 5.10.8 - Techniques employed by operators around the world

At the lowest layer are some tactics that follow international guidelines and that are used in every country where these attacks thrive. OPS torture protocol comes into play: a mixture of pseudo-stalking with techniques of persuasion, intimidation and mental manipulation, combined with the natural power of the technology, the effects of psychotronic torture on the target's brain, as well as the misinterpretation of the stimuli resulting from these torture techniques such as sensory and sleep deprivation.

Stalkers who keep targets under surveillance set up a system that impresses in many ways. Every time the victim flushes the toilet, for example, a car will honk in sync, or they will possibly hear a noise or a knock on the building structure. Other possibilities are that someone will run through the street shouting topics related to the privacy of the targets, or some people on upper floors of the building will deliberately use their flushing devices and showers simultaneously.

Acts like this can occur at the beginning of the attacks as a physical perception of the aggressors in the format of street theater — under the aforementioned conditions — to generate a greater degree of realism. Targets now sense that they will be hurt, or even tortured at will, if they don't obey the orders. Among the reports researched and those that I could verify empirically, some were conflicting; others somewhat ambiguous.

It is possible that a few years ago the use of individuals to surround the target and make them feel trapped and observed in their own residence should have been more common and intense, as technology operators would be able to rent houses around the target's locations and to use common areas of the building to make noise, such as elevator shafts, stairwells or water and sewage pipes running through all apartments. Thus, they could make direct noises that reached the Targeted Individual, but without being noticed by the neighbors. Remember: noise is one of the means of torture most used by stalkers. So, the access to nearby properties allowed them to make noise during the targets' daily activities as if they were leaving the house and slamming the door quite hard, generating a characteristic bang that immediately captures the victim's attention. For some periods, the target will experience a number of "coincidences". For example, every time they go to the bathroom in their bedroom, the water in the upstairs or downstairs neighbor will start flowing at the exact moment they begin to urinate and will eventually stop when they finish. This mainly occurs at dawn. Targets are suddenly aware of the high number of people entering and exiting the apartments next to them, accompanied by enough noise to grab their attention, but not enough for a complaint to the landlord or to the police.

This event based on this type of attack occurs all over the world with grotesque similarities between them, leading to a very serious reflection on the behavior of the aggressors, who are methodical and use consistent protocols and tactics common in terrorist cells spread across the corners of the entire planet.

Attacks related to street theater produced by the operators physically ("Gang Stalking"), that is, where the target resides, should have been used a lot in the past. Today, however, this is very unlikely to happen, as it is expensive to transport equipment and it requires transportation logistics and human labor. It would certainly draw some attention at some point, especially during the years in which this torture is perpetrated. Don't forget that human beings are spacious, have quirks and consume an enormous number of resources, so taking several people to different locations isn't practical. There may be one or another member of the gang that sometimes does this with the aim to scare the targets even more, but it isn't very common. The likelihood of a cataclysmic encounter with the subject in these physical attacks is low and this type of event is by no means a priority for stalkers these days.

Experiments are personalized and depend on the profile of the victim in question. This personalization takes place in the context of the target's private life activities, the particularity that surrounds them, their routine, taste, preferences, etc. However, there are certain types of torture that are common to all human beings, and which will generate a strong emotional reaction. The very surreal nature of the weapons and the conduct of attacks as reported creates the desired effect on most people. So, it is possible to predict all initial processes of the target's behavioral reaction when following such protocols.

This situation happens due to the attention that is directed to common audible events in which the target is able to perceive details that previously went unnoticed, automatically associating them with the operators. The great villain here is the V2K (Microwave Voice). The device is so advanced today that it is able to blend in with all the ordinary noises that are part of the natural cacophony of a big city, for example, and that is only noticed after the hearing becomes "sharper". Along with that, we have the persecutory paranoia that the victim feels due to intense torture, cognitive and sleep deprivation, and ultimately, the deprivation of their own life.

As the attacks intensify, the symptoms get worse. Since the attackers are able to capture all the audio "heard" by the TI, it is quite simple to synchronize noise events that occur in the neighborhood. A crack, an object that falls on the floor above is immediately followed by screams and insults that come exactly from the location of the initial noise. With V2K one can simulate directional sounds very effectively. Therefore, it is impossible for the target to distinguish between a legitimate sound and a microwave sound and its emitting source.

In apartment buildings, targets will be subject to the noise from "construction work", such as hammering or banging on walls. This type of activity is a recurring complaint, but what I could observe is that the city is always growing and changing. Construction works are being carried out all the time and these annoying sounds — hammering, for instance — may seem like another act perpetrated by the operators, but they aren't. Since technology operators are parasites of the worst kind, they even "appropriate" sounds that aren't generated by them to improve their torture (their voice), taking advantage of the sound waves propagated by other people.

Stalkers will sometimes make noises that indicate they are moving in sync with the target's movements, in the apartment above or below theirs. This creates a lot of mistrust regarding the participation of neighbors in the performance — the alleged noisy activities —, with no one around complaining or talking about it, creating the false sensation that OPS are everywhere at once.

Secrets surfacing all the time, guilt, isolation, breakouts, and extremely negative and unpleasant inner states, give rise to a state of complete degradation, both in the target's life and in the lives of everyone around them. With the target coming apart at the seams, the way is open to perpetuate the basic ritual of attack.

* Stalkers strive to destroy the target's social and family ties and friendships. If the target inevitably describes the harassment, they will likely be labeled as "crazy";

* Then, they work to disrupt personal and business relationships by destroying the victim's reputation;

* **They can eventually do a genuine street act, but 99% of the attacks are totally focused on electromagnetic/psychotronic, increasingly advanced weapons.** Although it gives the impression that OPS are acting, that's only the atmosphere coming from the remote location as it reaches the victim's head;

* An intensely repeated performance;

* Transforming previously pleasant locations into unbearable and uncomfortable places to live, forcing the target to move;

* They use the natural sounds of the neighborhood to blend in via V2K and simulate that they are part of that noise and that such noises are part of the scheme. In other words, there will always be noisy neighbors, parties, construction noise, cars braking, motorcycles and loud music coming from different angles;

* They always give the impression that they are at close quarters to the victim due to the effects of V2k;

* They pretend to be the upstairs or downstairs neighbor via V2K;

* Traumatic experience associated with violence and torture, personality organization. This cycle will occur during the entire time the Targeted Individual is brain-hijacked by OPS;

* Sexual and moral harassment, in addition to OPS taking morbid pleasure in what they do;

* Everything that normally affects a person in terms of thoughts, inner self motivation, personal fulfilment— points attacked by operators — are used to discourage targets, as well as to destroy any kind of motivation;

* They recruit all sensory channels capable of producing a huge diversity of sensations, as OPS objective is to make the target frustrated until the very end;

* Simulation of "voices" coming from everywhere: from the upstairs or downstairs neighbor, the building next door and the middle of the street. However, they are, in fact, mostly fake. No one but the target hears them. In rare exceptions, some gang members appear periodically in the area and start shouting in the street, taking advantage of other natural sounds, such as the noise coming from crowded establishments or tumultuous football fields to camouflage themselves as they know that the target's attention is fully focused on the event. For everyone else, though, this is just another noise coming from the street;

* They use tactics similar to those performed in the program called COINTELPRO[51];

* Then there's the attempt to deconstruct and reconstruct the person's personality based on the ideas behind MKULTRA goals — the desire to transform the victim into something they are not, and never have been, using torture, sense deprivation, brainwashing, stimulus blocking, modification of dreams,

---

[51] COINTELPRO (abbreviation derived from Counter Intelligence Program) was a secret program created by J. Edgar Hoover, consisting of a series of illegal and clandestine operations conducted by the FBI. The guidelines were: exposing, deceiving, provoking disagreements, destroying credibility, as well as neutralizing the activities and leaders of any movements listed as threats to the national security of the United States.

alteration and creation of memories and microwave physical torture;

* As the years go by, some thoughts that were previously recorded at the time of data collection may become the subject of invasion and attack. Some embarrassing or disconcerting thought that has been captured will be used later to cause emotional discomfort on the target, which can cause momentary physiological changes.

Dear reader, keep in mind that the aforementioned strategies — with all the techniques studied in previous chapters — occur simultaneously.

## VII - Tips for the targets

**DON'T ISOLATE YOURSELF.** The tendency is to start isolating yourself from all social life — to become extremely introspective, right? Wrong. This will actually make life easier for the aggressors, since they will maintain a channel of communication for long periods, and will never be interrupted. I understand that talking about the subject is still not ideal nowadays. However, try to pretend to have a social life so you don't end up swallowed by the alternative reality constantly implanted by them.

## VIII - Tips for the targets

Remember: a priority of the scheme is to harm you with words. So, they usually focus on intellectual weaknesses that are constantly elevated while strengths are demoted. Don't fret over this tactic or try to convince the mind invaders otherwise, for that is their intent. Above all, never change your default behavior to satisfy alleged deficiencies in your intellect. If this happens, you will be masterfully led through the perverse paths of mind and behavioral control tests. So, watch out!

Start thinking in other languages if you speak one or more of them, especially when reasoning in terms of silent thought, or voice of the mind. In other words, think only in that language: this confuses the system maintenance bots that work with localization, as their neural network is configured to learn and work with meanings in the target's native language (and its nuances). In this way, thinking in another language will disrupt OPS plans, who work mainly on the extracted thoughts and the interaction with their vocalization. If you are Brazilian, you may think subvocally only in English or Spanish, for example. Until the program and OPS adapt, you will buy yourself quite a bit of time and the human component behind the technology will probably not be able to interact with your vocalized thoughts. If you think in two or more languages interchangeably, in addition to improving your cognitive skills, it will make it extremely difficult for them to steal your vocalized thoughts.

There is a promiscuous, symbiotic relationship between MKULTRA and OPS. They always try to convey an air of credibility or legitimacy in the supposed reason that caused the beginning of the attacks; that everything that is happening is normal and that it is the target's responsibility (role reversal), a divine call or any other reason that will be systematically repeated to be internalized. MKULTRA profits by completely decentralizing torture worldwide, constantly renewing creativity to keep test subjects busy, while OPS enjoy the performance, revenge, theft, rape and anything else that fills and satisfies their morbid fun.

## IX - Tips for the targets

The fact is that everything is done so that the targets take desperate measures without proper thought, jeopardizing their own lives. No one will physically attack you — at least not like they say. The weapon itself is capable of destroying your brain while you worry about the actors and the advanced chatbots[52] responsible for the voices.

The hysteria is thoroughly designed so that the target lives immersed in it. You can never get rid of it. Over time, the hysteria naturally leads an

---

[52] Chatbot is a computer program that tries to simulate human conversation. The goal is to answer the questions in such a way that people have the impression that they are talking to a human, not a computer program. After submitting questions in natural language, the program queries a database and then provides an answer that attempts to mimic human behavior.

individual to make irrational and emotional decisions, living only for the maintenance of their basic needs.

The core location of the stalkers is based on a remote facility, such as farms not far from the city, but enough to not draw the neighbors' attention, as they spend years, 24 hours a day, screaming hysterically into the victim's cortex. That's why there is an entire infrastructure set up to carry out the torture — a video recording and image editing studio with computers, Chroma key technology, virtual reality devices for interaction with D2K dreams, an antenna for direct contact with satellite or the internet that connects to satellites to transmit these sounds and images to wherever the victim is. The place can be considered a confinement of kidnapped minds; a new category of kidnapping that is based on the use of this technology is emerging on the horizon. It is not for nothing, as was said at the beginning of the book, that MKTECH is also known as a "The brain hijacker".

There are countless cases and details that can be discussed over hundreds of pages. But hopefully I think there is enough here for people to become aware of how experiments disguised as psychotronic attacks are conducted along with the techniques used in Electronic Psy Ops (electronic psychological warfare operations) and Electronic Psychological Warfare, which work the mind and the "spirit" in subtle analyzes together with heavy attacks.

If in the future the target has this level of awareness — of what is happening to them based on the knowledge acquired in this book —, the attacks, as they occur today, which are very much based on the complete lack of knowledge of this entire parallel universe of MKTECH, will become unfeasible. Operators will be forced to change their strategies to continue ongoing experiments.

The truth is that we fear what we don't know, and fear is one of the main materials for conducting experiments.

## 5.10.9 - Side effects of the attacks

*"The psychic prison that hinders cognitive flexibility when you try to find a new frontier for whatever direction thought strives to project is stopped by an*

*invisible wall – a block. There are 'eyes' in your thoughts, and the human filters that judge their discernment and perspicacity every millisecond stifles any attempt to emerge new ones".*

-**FELIPE SSCA**

After years of intense torture, some successful targets achieve emotional development and maturity with the learning. Therefore, some effects of torture are attenuated, mainly those within the first year of the attack that are recorded in the memory and will never be forgotten — the reaction to these emotions too. Some victims fail and succumb quickly though. After all, this is a cowardly process, in which there are generally several individuals against one. The people who take turns in the attacks are part of a gang with international ramifications loaded with taxpayer money embezzled from civil service competitive examinations, or money from remote killings, theft of commercial information, industrial espionage, sale of confidential government information, paid stalking and illegal surveillance. It is by far **the world's largest human torture scheme of all time.**

In this way, it's simply impossible to escape unscathed. Even those who managed to stay alive — one way or another — end up living with severe sequelae. They get similar symptoms during and after the attacks, as we're going to see later on. For instance, many nervous tics can be spotted in victims; some of them experience body tremors, such as in the head and in the mouth, and their behavior is always distant. Furthermore, they often talk to themselves. Over time, all this becomes increasingly apparent.

I will again refer to the ICD-10 Classification of Mental and Behavioral Disorders: Clinical Descriptions and Diagnostic Guidelines to, in a more precise way, list the disorders caused by this experience of war both during and after the struggles.

### F43.0 - Acute stress reaction

* Traumatic experience involving a serious threat to the individual's physical integrity or safety (struggle, assault, rape);

* Narrowing of attention and constriction of the field of consciousness;

* Decreased attention, inability to understand stimuli and disorientation (flight reaction or fugue), autonomic signs of anxiety and panic;

* Anxiety, anger, despair.

### F95 - Tic disorders

* Involuntary, rapid, recurrent and non-rhythmic motor movement, meaningless vocal production, including neck-jerking and shoulder-shrugging, with no apparent cause, use of repeated words.

### F43.1 - Post-traumatic stress disorder

* Delayed or protracted response of an exceptionally stressful, threatening or catastrophic nature that causes pervasive distress (victims of torture and terrorism);

* Lived experiences in the form of intruding memories (flashes) or dreams, occurring against the persistent feeling of numbness and emotional blunting, recollections of the trauma, fear and avoidance of situations that generate memories, automatic hyperarousal;

* Traumatic neurosis.

### F43.2 - Adjustment disorder

* A state of subjective distress and emotional disturbance usually interfering with social functioning.

## 5.10.10 - Peace and Quiet

Intrapsychic tensions, the more accentuated and more difficult to bear, the more concealed this suffering will be, thus creating interpersonal barriers, denying others their internal sufferings to avoid even more wear and tear, as the daily stress of the mental struggle against the invaders of the mind that try to access all the target's functions and to manipulate them as well as to destroy them demands all the victim's strength.

If the atmosphere is extremely troubled at home and people such as spouses fight, the target will hardly be able to withstand it. This opens up two battlefronts: at home and against the perpetrators. Therefore, peace and quiet must prevail as much as possible between family members who do not know what is happening and the target who suffers crippling attacks on a daily basis. Even if we're living with others, we're lonely creatures, isolated in the intimacy of our own silence. Solitude, although it doesn't seem like it, is important for us. We need it to reach concentration. Only those who manage to remain immersed in the constant flow of thoughts can go a long way. Remember the library posters that asked for silence in order to keep the place calm and quiet, extremely conducive to savoring the knowledge that is more easily absorbed that way? Well, they were there for a reason. Silence and positive, inner solitude are luxury items for test subjects who are deprived of these privileges even in places like libraries. This is just one of the numerous negative effects of the constant noise in the cortex.

A restful night that you didn't value before, or wanted to spend in clubs or weekend entertainment, becomes another luxury item you'll never get back. Today we miss those nights of sleep we despised, especially as youngsters. The link between the present and the past that is established with the help of memory and intellectual processes are affected, modifying the way the brain evaluates and measures these phenomena. The representation of both no longer has the same power in the present.

I close the chapter with the wise words of Arthur Schopenhauer in his book Aphorisms on the Wisdom of Life: *"A man can be himself only so long as he is alone; and if he does not love solitude, he will not love freedom; for it is only when he is alone that he is really free."*

## X - Tips for the targets

Doppler effect: the greater the impression of hearing sounds within other sounds (V2K) and the more intense the feeling of the Doppler effect in vehicles passing on the street near you as you walk means that the technology is operating at a high level, and the likelihood of voices and shouting getting louder, thus preventing sleep, is high. The ability to perceive the level of attack within the sound waves that move away and become increasingly low-pitched serves as an indicator of attack intensity. The greater the perceived effect on vehicle noises on the street, the greater the intensity of all MKTECH equipment — the more powerful and clear will be the artificial remote dreams, the noises and the shouting that insistently come from SYNTELE.

# CHAPTER 5.11
## ADVANCED TORTURE TECHNIQUES USING MKTECH

*"If you know the enemy and know yourself, you need not fear the result of a hundred battles. If you know yourself but not the enemy, for every victory gained you will also suffer a defeat. If you know neither the enemy nor yourself, you will succumb in every battle."*

— Sun Tzu.

Some advanced techniques described in this chapter come from well-tested and widespread procedures used to extract the greatest possible amount of information from a given target. This set of tactics is known as coercive counterintelligence interrogation for resistant sources, and causes physical and psychological pain to those who are submitted to intense moral suffering and distress. These interrogations are often conducted on uncooperative sources, whether it's a prisoner of war, an intelligence agency spy, or a terrorist. It results in an external pressure of sufficient intensity to penetrate the mental defenses of any human, highly educated or otherwise.

Every coercive interrogation technique is built to perform mental regression, create painful situations and make the tortured continuously face a new insurmountable challenge, causing complex situations that **always lead to repeated responses such as frustration, fatigue, emotional disarray, pain, total deprivation of sleep and anxiety.**

After being subjected to this type of torture, people may end up revealing information that they wouldn't have disclosed otherwise. These techniques are put into practice when the person being interrogated is placed in a kind of interrogation room together with their tormentors. Despite these techniques having proved to be extremely efficient, it was never possible to confirm the veracity of the reports with 100% certainty. With the advent of MKTECH (Mind Control Technology), the same techniques used in person are now conducted remotely with the use of advanced neural weapons, enabling the use of every element we've seen so far. Consequently, now it is possible to confirm whether the answers given by a target are true or false. OPS (Organized Professional Stalkers) use these same techniques combined with the latest methodologies that are managed through this technology, so the same foundations remain intact, but the target no longer needs to express their thoughts in words; the words can be captured in their vocalized thought, which is extremely tiring to control and requires superior cognitive self-control and combat experience in psychotronic warfare.

The ease of not having to deal face to face with the victim while torturing them to death, without having to bear the consequences of physical interaction (such as judicial consequences if this aggression were conducted in person) attracts a lot of cowards who spend 24 hours a day using a screen and a virtual representation of the victim's experiences. They treat the target as if they were a pet, a modern slave or a macabre human Tamagotchi fueled by hatred and destruction, and use their brains to run experiments of all kinds, modify REM dreams and end the target's life little by little — remotely, of course — and without suffering any damage whatsoever. After all, the victim is just a virtual representation on a screen and their brain, the computer, is responsible for processing the torturers' day-to-day fun.

To make matters worse, the weapon leaves no visible traces like the traditional ones that use propellants that cause perforations in the body. The only scars are present in the victim's soul, in addition to the neurological damage of an apparently circumstantial nature that creates a unique sense of security for torturers. They establish a regime similar to that of prisoners in the worst prisons in the world. They have their own rules, completely outside the law. In other words, there's a clear path to direct and sadistic torture carried out in its entirety and until the target collapses. This all takes place within the individual's sacred right to privacy, freedom and choice. The weapon acts in the opposite way to conventional torture: instead of inflicting physical damage to the body, as usual, it directly attacks the brain — and, consequently, the body — in a long-lasting systematic process, capable of inflicting more suffering than classic torture.

In spite of that, torture techniques remain similar to those applied in the past in prisons, wars and dictatorships. The foundations (concept and base) remain the same. However, it is now possible to delve deeply into thoughts, into the human psyche — the heart of cortical processes generated by the brain —, and verify the entire reaction to a certain torture not only expressed in physical suffering and external exposure (as a kind of regulator); it is now possible to verify how the current state of the brain affects all intellectual phenomena and what state and immediate factor of some of these phenomena evokes sensations in the human being. To put it another way, it is possible to measure — to capture reactions and processes like never before seen in history. One can internally see the mental processes that triggered classic torture reactions, opening the possibility of new observation and interaction of information decomposed and scattered throughout the brain in the form of thoughts.

All this, of course, under an immeasurable systematic affront to the early days of freedom and total invasion of privacy, in which everything happens inside your own house (in your intimacy), and in which it is possible to evoke intense physical suffering as well as psychological suffering, keeping the foundations of torture intact and surpassing the intellectual limit of the individual — no date, no location, no time.

I present to you the advanced torture techniques performed on humans and used by the MKULTRA 2.0 experiments.

In the previous chapter, we've seen the causes and consequences of this attack on victims just as they occurred. Now let's delve into why a particular technique is used in a specific situation and what is behind its use. I'm going to use plain language, combining the explanation of the tactics employed by the enemies with some tips for the targets, but overall, we can safely say that it affects each of us — after all, you, the reader, may be learning about the subject for the first time through this book, right? This chapter is then an overview that covers just about everything we've seen so far, however, in a more in-depth and lucid way, based on the tactics implemented in the psychotronic warfare that takes place in the mind of each Targeted Individual.

An infinity of old, already established torture techniques merges with new tactics in a constantly changing process, creating several new possibilities within this modern attack — cognitive processes still in the exploration phase with direct observations on them. For this reason, the most effective attacks for a given situation are always pointed out by the AI on MKULTRA 2.0 servers. The AI filters what is valid from what is not, including the most modern tactics that are being explored at that very moment, the **Electronic Psychological Warfare** or **Psy Warfare**, that are nothing more than electronic psychophysiological unconventional warfare tactics using psychotronic electromagnetic weapons.

In fact, there are special units increasingly specialized in this type of warfare heading towards a new future. In this future, such tactics and weapons will be the cornerstone of a new category of warfare and the epicenter of new human paradigms. The relevant data will be explored in detail, mainly to be used in the battlefield or on the implementation of new procedures in the field of counterintelligence, or will even to be included in the attack itself. Such data will be accessible and available to government defense and espionage agencies, which will be part of protocols to be drafted, adapted and implemented.

Another term that we must get used to is **Electronic Psy Ops**: electronic psychological warfare operations, in which localized and surgical attacks depend on all coding/compilation of experiments conducted to assist these weapons and tactics directly in the field.

Keep in mind that, in addition to the process being carried out, the result obtained from the torture will shape a new range of successful procedures performed in the field. You, human subject who is dreaming electronic dreams, are part of a worldwide experiment, whether you like it or not — and whether you're aware of it or not —; and you're contributing to the elaboration of such procedures.

Experimental attacks — deep and advanced techniques from real situations in which these experiments take place, thus improving tactics to be used in battle — are also part of this "game". For example, working on the collection of information directly from the mind of a relevant enemy that is of importance to a successful attack or manipulating an internal opponent using neural reprogramming tactics, consequently turning them into a remote killer who commits acts within enemy territory while making them believe that they're doing something impressive. Little does the victim know that they're being manipulated and that they're ready to attack their own allies or directly alter cognitive processes, which artificially modifies their way of thinking through changes in the perception of external events, along with relevant sensations that follow rational and unconscious outcomes.

## 5.11.1 - Advanced tactics used by OPS using MKTECH

Some tactics may have the same causes and consequences in the target's behavior and nervous system, whether applied together or separately. So, I suggest you use this chapter as a reference. You can also read the tactics in chronological order or at random. If you prefer, you can also search for specific techniques. I decided to specify each of them, even when they appear more than once when combined with each other.

### 5.11.1.1 - Macabre and resilient companion

Forced coexistence with unknown people in your mind. First tactic: the arrival.

To begin with, we are never alone. There is no absolute isolation; we're all sharing our lives with one another. The people you live with on a daily basis — friends, co-workers, family members, boyfriends and girlfriends — certainly influence your life in a positive or negative way, affecting all your daily activities, either hindering you or helping you. As time goes by, the consequences of such interactions are grouped into blocks that make us who we are, and end up becoming more apparent. If someone is hurting you, negatively affecting your life, you have the option to remove said person from your life, or to walk away so that this negative interaction disappears, thus improving your well-being.

It is important to highlight that this action usually takes place in an established democracy. When an individual becomes a target, however, they start to live with people that will accompany them at every stage of their life, from the intimacy of the bathroom to bedtime, including inhabiting the content of their dreams. So, the option to chase away the voices doesn't exist. The target is therefore affected by this unknown, macabre and resilient companion. From the moment the voice appears, it will only leave when the victim is completely adrift or dead. Its only purpose is to influence the target in a very negative way, destroying the quality of their life, sucking their energy, joy and all positive feelings in order to torture them to death, collecting as much data as possible in the process, profiting from the experiments and, of course, having fun with it all.

Several people who take up residence in the victim's mind (not metaphorically, but concretely) open a permanent communication channel via SYNTELE (Synthetic Electronic Telepathy) and D2K (Synthetic Electronic Dream), transmitting everything that happens in the physical facility where they're located to the target's mind in the form of sound, images and signals that generate physiological changes in the brain. These signals stimulate a configuration in the mind identical to the one pre-filed in the system, and insistently try to make the target converge to that state using all resources available. The target is also tracked by them, wherever they go.

OPS have a negative influence on all aspects of the targets' lives, especially their professional and personal life, preventing them from sleeping, thinking and talking normally, which includes the targets' internal monologue. Their freedom and their sacred right to privacy are completely and purposefully destroyed. Gathering data on human behavior in the face of the complete lack of privacy within their own residence, the worsening of the victim's condition, is also part of the ongoing experiments. This strategy is painstakingly designed to subject the victim to one of the principles of thorough questioning and efficient torture to control the environment.

## 5.11.1.2 - Controlling the environment

In order to have the ability to alter or break a behavior pattern using physical or psychological means, that is, to ensure cooperation according to the interrogator's wishes, it is necessary, among other things, to have complete control of the environment.

The total control of the conduct of torture and the convergence to desired paths voluntarily or involuntarily depends entirely on the control of the environment where the target is located. For this reason, operators remain in the target's everyday environments at all times. Their home, work, sport practice locations, places for leisure, relatives' houses — in short, the presence of OPS can be felt everywhere. The impact of the ubiquity of operators implemented by psychotronic weapons causes a false sensation on the victim: a feeling that they may be being followed, as perpetrators have an organizational and logistical ability to deduce which locations the target will visit to attack them there. This gives the impression that this group is very well organized and has complete control over the environment, including during visits to public and military facilities, for victims are still able to hear the voices coming through SYNTELE in these places. This strategy provokes a feeling of degradation — of being inferior economically, physically and mentally —, paving the way for the success of all MKULTRA 2.0 experiments.

## 5.11.1.3 - End of privacy

Freedom is closely linked to our faculty to make decisions — to deliberate the goals that we seek to achieve in the real world limited by moral, social and governmental rules. The novelty of this type of restriction of freedom, which takes place directly within mental functions before they're effectively turned into behavior, sets a precedent for profound reflections. That is, operators completely remove the possibility of acting or carrying out acts. After all, even before such acts take place, the thought has already been captured and its probable actions detected and rendered useless. At this point, it is very frustrating and disturbing to know that everything that you've planned in your mind to counter MKTECH enemies are useless.

They remove your freedom to choose, the way you make decisions or experience your own existence; the freedom of the imagination and all other external factors that interact with you. Operators take away our most precious power to transcend the laws of physics within our minds — to make dreams come true — as waking dreams are the first natural event that doesn't happen anymore for a Targeted Individual, along with dreams of future plans and night dreams. Then comes the rest: it jeopardizes the possibility of interacting with people, for example, the freedom taken away from the target's own privacy and home. What would humanity be without dreams? Without privacy? Only a tormented, empty soul at the mercy of the operators' whims remains. And I'm not talking about the restriction of the body's freedom in space, but the mechanism that controls its movements, which is the mind. The target now only dreams of regaining their former cognitive freedom, of getting rid of this despicable weapon that finds no physical barriers in the common infrastructure of a city.

Someone asks them, "What is your dream?".

And they answer, "Having the right to think privately again!".

During this connection that cannot be broken — with the channel open 24 hours, data entering and leaving the network indiscriminately—, privacy is a word that has long ceased to exist for targets. The bewilderment of experiencing this type of aggression in a democratic state of law and on their own property leaves them exposed to the conduct of experiments.

## 5.11.1.4 – Bathroom: taking a bath without privacy

The bathroom is a favorite place for Organized Professional Stalkers (OPS), as they use the invasion of privacy as an instrument of torture. They know that their actions will create constant stress, as it violates at once the target's space, their privacy within their own home and, even worse: the privacy of the room they use to satisfy their physiological needs; a place they can freely examine their genitals, for example. When doing this, the voices narrate every particular act and observe the victim's physiology second by second. With their private moments being sent to who knows where, it is impossible to remain impassive in the face of such an affront to individual freedom. It is in fact a deplorable and overexploited act. Every trip to the bathroom, without exception, will be accompanied by voices talking and narrating everything targets do there. Operators are strategically prepared to wreak havoc and violate any shred of intimacy.

The bathroom is then one of the places where OPS can be heard more clearly. They insist on keeping the transmission power at its highest level. That is, any biological need will be accompanied by various screams and grotesque sounds coming from V2K when it comes to body interactions. When taking a shower, for instance, the phenomenon of resonance or reverberation is very accentuated. The victim can clearly hear the operators' voices echoing in the water (and in the echo of the bathroom), which becomes deafening. It is so loud that the target cannot hear their inner voice, only the voices of the V2K torturers who try to control them as if they had any power over people. A kind of modern slavery, with the end of privacy in what should be the most private place in the house. And all of this is done deliberately to cause discomfort and to force victims to share their physical intimacy with strangers.

## 5.11.1.5 - Your home becomes the most unbearable place in the world

Constant harassment 24-7-365 makes your home the worst place in the world. It is so bad that operators are able to drive the victim into bad thoughts: they consider moving, or staying as far away from the house as possible so they don't have to constantly deal with unwanted intruders. Everything is carefully planned. **So, never be afraid to return to your home despite the feeling of revulsion that was artificially created by evil tactics.** Furthermore, don't be tempted to walk away and wait for peace to return: it won't happen. It is exactly this feeling and erratic attitude that operators and OPS expect. Therefore, there is no use in moving or traveling; operators don't need to be exactly where the target is. So, one must find one's way back home — just like it was before.

MKTECH is scattered throughout the national territory with advanced weapons designed to attack the human mind at any moment. Just one electromagnetic wave is enough to do harm. When the victim doesn't really know what's going on, they usually want to move or sleep away from home, thinking they'll relax and get rid of the microwave and radio frequency (RF) voices that devastate their mind, along with the insults and moral and sexual harassment that they're forced to endure day by day.

In this way, operators successfully drive people out of their homes. The individual develops such a negative feeling about their own house that they end up not wanting to go back there. They see their residence as a bad place, and totally lose their reference for comfort, rest and privacy. I consider this premeditated attack to be the shadiest of all, based on the total disrespect for any residence in society, which attacks private property and family privacy and, most of all, our democracy.

Some targets reported that they felt real hatred for their house, especially at the beginning of the attacks when they didn't have the faintest idea of what was going on. That's why this technique is used to forcibly remove people from inside embassies and from properties at the behest of others.

## 5.11.1.6 - Surveillance 24 hours a day, 7 days a week, 365 days a year

The desire for privacy is an inherent characteristic of human beings since ancient times, especially at the individual level, within their own space. But for someone who has contact with this modern weapon, the word privacy is just a vague memory as the surveillance and invasion has made them distance themselves quite a lot from its original feeling. The importance of its meaning was even lost due to the conviviality of strangers inside the house.

The truth is that, the simple presence of an observer — consciously or unconsciously — already changes the posture of the observed person from the most subtle details for the adjustment of conduct to a completely different behavior that usually occurs when the individual is alone. In other words, the perceived presence of an intruder watching you changes the way you act. It even happens to the most extroverted people, especially in the initial period of invasion of privacy. This surveillance, over time, together with the torture, ends up driving the target's behavior towards the interests of the operators.

When combined with this new tactic of cognitive surveillance, the feeling of being watched by strangers affects, disturbs and modifies the behavior, in addition to causing extreme psychological damage, demoralizing, tiring, degrading, and even leading to death or suicide. Showing that the details of the victim's thinking is observed at all times — namely, everything that goes on in the individual's mind, all their acts inside and outside their home — results in abnormal, repeated behavior often observed with difficulty adapting to behavioral patterns imposed by the invaders.

So, the control of private life advances, paving the way for the end of it by focusing directly on the source — our brain. But I hope society remembers the value of privacy before it's too late for that.

## 5.11.1.7 - Cognitive deprivation

Playing with human memories: the newest game of our time! Operators have far too much time to scrutinize the targets' thoughts and memories, as they use a large team for this purpose. With the help of selection algorithms that determine relevance of facts, they spend the whole day commenting on memories and thoughts that the target had on a certain subject. They analyze their automatic physiological reaction to certain topics, making the victim a prisoner of their own mind. So, the TI avoid thinking deeply about any subject, especially regarding their childhood, spouses, fights or a topic that exposes their privacy, negative memories from the past and what they represent in the composition of their being, their self-identity, their personality.

Like it or not, knowing that someone is listening to your thoughts causes a range of behavioral reactions in the system that controls the thought. For example, creative thoughts are one of the most affected. Creativity is not the gift of geniuses, as many may think, but of everyone who is willing to practice some specific type of conduct. In fact, we're creative all the time — not only do we create extraordinary things, but we also use creativity to solve problems. That's what operators do: they restrict all creative and inventive minds, including the ability to solve casual problems. As there is a crowd watching over your life, some may feel deeply intimidated by the harassment of every daily act.

Directing your mental efforts to avoid distressing memories, thoughts or feelings that can be picked up by operators and later used also contributes to mental deprivation and inhibition of the flow of thought. This act is extremely tiring — a sense of guilt and unease washes over the target whenever they let a memory slip. Thus, in the next cadence of thought fragments, an interruption of reasoning and memory access will recruit a large amount of brain processing to prevent the continuous flow of memory access and its cascade of sentimental display.

As if that wasn't enough suffering, the target is still forced to deal with constant changes in behavior and diversion of attention to solely focus on the subjects sought by MKTECH. With each thought considered inappropriate by the subjective filter created by the operators, that is, with each execution of thought inside the mind that goes against the behavior projected for the target, there is a tormenting and disabling increase in the volume of the attack, both in frequency and in the sensation of sound intensity. These actions intrinsic to systematic attacks are known as neurofeedback.

If the target naively tries to talk them out of it, or if they try to negotiate a release from the electronic captivity, less amplified/painful responses will be momentarily received. This will generate false positive feedback to the brain, creating submissive behaviors — the victim becomes an electronic minion. This is a technique known as contradictory stimulation and is capable of creating this type of behavior, leaving the target at the mercy of the attackers. Regardless of how the feedback feeds the brain's sense of reward, always try to ignore both.

Our brain is evolutionarily prepared to unconsciously and constantly seek rewards — to try to protect ourselves or avoid threats to our social, physical or psychological integrity. Acute and prolonged stress episodes that produce a feeling of insecurity cause cortisol secretion, modifying cortical dynamics and altering attention and consequently generating a decrease in the performance of everyday activities, as it keeps the focus on the threat.

Well-being — that feeling of peace and joy — releases dopamine, which improves memory, focus and attention in normal activities, such as studying and retaining content. Armed with this basic knowledge, the attack directly addresses the emotions and the prolongation of stress that destroys the channels of communication with the outside world, affecting and modifying the target's behavior permanently.

## 5.11.1.8 - Decimate the will/Self-esteem/Motivation

A main feeling that must be curbed at any cost: ending the volition in any activity that is not linked to the reality they're trying to convey through MKTECH's torture content. If you eliminate the will, the game is over. Motivation depends on how a person perceives the state of affairs that influence their behavior.

Regardless of the phase they're going through, people with high self-esteem are confident with themselves. In other words, they're more assertive, ambitious and successful in life. On the other hand, people with low self-esteem tend to be socially anxious and ineffective. They feel threatened with interpersonal relationships due to a lack of trust in their own judgment and opinion, and are also less successful, both academically and popularly.

Self-esteem and willpower are part of our lives as human beings— we wouldn't get out of bed without them — and these are exactly the feelings that will be buried as deeply as possible inside the target in the very first moments of the attack. Any remaining motivation to even fight back or to investigate what is really behind those voices that hinder, control and reverse the normal flow of thought is consequently eliminated. Positively nurtured self-esteem generates greater confidence in one's own personal resources and in solving problems, producing satisfaction. This type of person is more likely to be more competent and self-determined. The aim is then to create the antithesis of completeness: discouragement through intense pressure and tension arising from psychotronic torture.

## 5.11.1.9 - Moral and sexual harassment

Anyone experiencing harassment of any kind is likely to react in a similar way. Generally, the human being doesn't remain impassive in the face of an event of this nature. We have personal resources to interfere with the environment and react to an attack (physically or psychologically). However, the natural reaction that occurs sporadically in most people's lives causes harmful, negative consequences, such as elevated stress and extreme tiredness after the maximum peak in reaction magnitude and the emotional state related to the act during this zenith. Overall, the event that caused this incident is left behind and life goes on. When this moral or sexual harassment is taken to the extreme, that is, taking place inside your home, in your thoughts and dreams, it causes severe damage to the person beaten by anonymous criminals and advanced chatbots.

We always have to keep in mind what goes on behind the scenes of these attacks. The main objective is the full knowledge of the human psyche, the total control of the mind and the very interaction with this weapon in order to make it more efficient. This attack proficiency for a complete picture of psychophysiological chaos invariably goes through the use of words in a native language, recognizable codes for the mind that demodulates the information that comes with the weapon.

This is a more efficient type of attack than just random noises. Words, as we know, have a profound impact on the listener. They can remain echoing and directing thoughts for a long time. Therefore, the attack using strong, demeaning and purposefully disrespectful words and phrases causes three direct disruptions: a) first, they follow the protocol of making constant noise of any kind in the auditory cortex; b) then, there is the content that will make sense of the noise, the interpretation of words, which triggers cognitive processes, decodes communicative signals and reacts to the received message, including triggering the images of the imagination to make predictions and hypothetical scenarios based on the content of the attacks that activate automatic processes in the human brain; c) finally, while interacting with thoughts, they use their advantage to harass and have fun in the whole process of torture and verbal attacks, causing more indignation, and directly interfering with the will.

In all areas of human communication, it becomes critical to consciously choose words that create the effect we are looking for in a given situation. This is done systematically with repetitive and methodically designed words that do as much damage as possible while within a context that guides the target through a backstory. In the meantime, operators have fun with the way the target processes attacks (how they deal with the information and memories stolen from their mind), and are amused by the victim's confusion and increasingly deteriorated behavior due to moral harassment via SYNTELE.

The perpetrators criticize the victim's weaknesses at every second, and diminish their strengths or render them non-existent, often treating them as an ordinary skill or a skill inferior to theirs, as they convey a powerful feeling that they have control of everything in every way. During this constant harassment, there are ways of — calmly and slowly — testing and implementing different theories linked to human cognition and its predictability in the face of certain events. One of them consists of inserting systematized contradictory stimulation, alternating between humor and threat, or between threat and the insertion of "gentle" characters who supposedly defend the target during staged attacks that are manipulated by the very ones who work for them in order to create confusion and disorientation as they limit the target's response, and force them to unconsciously give in and obey the attackers' decisions.

Constant sexual harassment also follows a disturbing script of attack due to naturally demeaning factors about the subject that are impossible to be rejected and are continually reiterated by operators. It violates the principle of free disposition of one's own body, causing deep embarrassment, since targets don't even have the chance to defend themselves physically (or virtually) against this type of attack. It is also used as a means of intimidation and to cause systematic repulsion, affecting life as a whole, especially regarding sex, which, among other things, may even result in lack of sexual activity during the course of the attacks.

Once again, operators surround themselves with all relevant information regarding the target's sex life — with every intimate detail of their relationships, but now with something that only this weapon can do: detect the details within the thoughts that activate or repress sexual desire, verifying subvocalized mental images and thoughts during past sexual acts, thus gathering a range of highly private information.

First, it is necessary to determine the target's sexual orientation and use a narrator, an actor for the V2K/SYNTELE/D2K voices. This actor has to be appealing to the target — the opposite sex, if applicable —, so they can harass the victim sexually. This person will then speak words against the victim's honor inside their house, using terms that the target isn't very familiar with — highly disturbing words. Those who don't tolerate (or cannot coexist with) this type of conduct that causes high stress levels may experience very intense emotional reactions. If the same situation occurred, for example, in a public location with the harasser and the harassed person in the same physical space, it would probably end up in fights, or worse.

The consequences are even more devastating for women. For them, harassment will come from all sides, systematically hitting their character and morals with grotesque attacks using words loaded with sexual innuendo at completely inconvenient and unpleasant times, or using memories of their loved ones to accentuate verbal abuse with the use of abhorrent voices that provokes immediate repulsion and disgust.

Pornographic comments involving close people are also included, which are professionally placed to generate thoughts about the topic. Automatically, a virtual image of the narrated scene is created in the target's mind and causes a lot of emotional pain and embarrassment for imagining this sexual absurdity. These are techniques that use vulgar details, sexist comments about their physical appearance, mainly regarding some detail that bothers the target, about their private parts, among thousands of other possibilities. And all of this occurs with the use of psychotronic weapons. Their objective is to drive the target crazy with the constant noises in their auditory cortex. Thus, moral and sexual harassment takes place 24 hours a day for years on end, and the negative psychic reaction, no matter how much it diminishes, never ceases to exist.

Operators resort to the person's own memory to harass them, such as something that involved a problem from their past that causes pain — a fight, a breakup, an event that is relived over and over again both in words and in mental images. Thus, it interferes with the person's willpower. As much as a target doesn't demonstrate a predisposition to violent reactivity or even if they present positive characteristics of self-control and sobriety and the person isn't easily carried away by verbal provocations, the psychotronic weapons that emanate voices from all directions captured by the target's mind, like a radio picks up a signal without the possibility of being turned off at any time, will cause accentuated cognitive problems very quickly, in addition to the abrupt change in their peaceful behavior. Torture in its purest form!

Fortunately, humans are equipped with some natural defenses in the quest to adjust to this "new normal" — an attempt to neutralize external threats. It all depends on your stress reaction and physical and mental strength, which will be put to the test every second. Defense mechanisms will be activated, the sum of everything ends up undermining the change in perception about the low points of life, degrading into personal wear and tear. Going through stress and overcoming it is a sign of being alive, but experiencing an organic unbalance with exaggerated emotional discomfort in one's own personal life deteriorates health and quality of life.

Keep in mind that the harassment is verbal and visual (during dreams). They are deliberate acts that together harm and hurt as moral and sexual violence alone disrupts the human psyche. When conducted with the use of this weapon as the content of the incessant noise in the auditory and visual cortex its efficiency scales to higher levels.

## 5.11.1.10 - Fear

Feeling is what inspires emotion; emotion is a response to the way that we deal with feeling. Fear is a response to the feeling of threat. In this scenario, it is necessary to revive the fear moment by moment to sustain it for long periods. Operators may threaten relatives or keep the target's mind on constant alert, most of the time inducing the feeling of loss. When they threaten to kill your child or a sibling, for example, the fear followed by a future projection of the event in mental images — that is, the continuation of life without the person who was cowardly murdered — stimulate reflections, regression and automatic intimidation, which is the booster of a false superiority of MKTECH's operators over the person being tortured. Therefore, keeping the fear alive is the key to successful torture in all aspects, provoking anxiety and driving the target's behavior in the face of the reality instilled in their mind on a constant basis. After all, fear is the most primal of our emotions.

On top of that, attention can amplify the capture of information by making it more discernible. Emotion can diminish it or even block it, preventing it from being perceived. Fear is one of those feelings, capable of maintaining complete focus on the triggering factor. So, operators have efficient methods of detecting the target's fear and its intensity, mainly based on data from mental projections about the event generating the reflections triggered by the absorption of the threatening story used to instill fear in the target. In that sense, each aspect can be used separately to amplify the fear, always attacking something important to them.

As the fear naturally fades away, the operators' primary attack factor disappears. The effectiveness of attacks and in obtaining information using the remote polygraph become innocuous, as we saw in chapter 3 of Volume 1. Without the paralyzing fear, the effects of torture mitigate and the effectiveness of the experiment decreases.

## 5.11.1.11 - Tension

The act of threatening to kill — opening up the possibility of inflicting pain at any time — is usually more effective in the long run than the sensation of pain itself as this type of threat can be continuously executed. The same principle applies to other negative feelings like fear. The threat is only taken seriously and has the expected result in the behavior of the target if they believe that OPS are capable of carrying out threats such as executing family members, among other verbal threats. So, the stress from threats and psychotronic torture alone can cause a lot of tension. After a certain period, death threats even become a form of relief, if possible, given the dimension of the suffering. It destroys trust in the future and in the present, preventing balanced conduct that affects work, personal life in all aspects and alters our life's sense of direction, ultimately eroding its wider meaning.

These tactics are aimed at preventing personal growth at any cost, destroying the target's productive life, immediately causing social withdrawal, especially among familiar peers. They dismantle their relationships, preventing positive external influences that could help them to block attacks and understand what is behind this phenomenon. The surreal power of this event can now and then lead the target to accept their fate based on the voices that never cease, on the extraordinarily disturbing dreams and on the voices that interact with their thoughts every millisecond, leading to the total loss of mind control. They cause self-destructive behaviors, especially related to addictions, such as alcohol, tobacco, licit and illicit drugs. They keep the victim in a kind of virtual solitary confinement: the Targeted Individual is deprived of everything and everyone, leading to isolation. At this point, the victim tends to turn to superstition or the existence of a being of higher consciousness. Then, the subconscious begins to face the new reality based only on corrupted memories, fears and concerns. The distorted reality due to the lack of positive external stimuli imposed by MKTECH on the target's mind reflects the way they deduce the common reality, projecting their internal reality externally, thus creating a distorted fear based only on the real threat that lies ahead at any time, which improves their instinctive behavior while letting it flow naturally in the face of the situation.

## 5.11.1.12 - Maintaining emotional superiority over the victim

Hardly anyone goes through their short period of existence on this planet without going through mishaps — some mild, others more serious —, including small internal struggles in maintaining cohesiveness (their spirit), in case a day-to-day event makes us sad, anxious, or thoughtful. We don't always live under blue skies, right? So, that's what operators do.

Any mundane mishaps of life are used by them to try to gain more power (mentally).

As operators invade your mind, privacy, sleep and thoughts, ruling almost every aspect of your life, they'll use these mishaps as ammunition whenever they wish. They'll dwell on the issue for days until its effect is exhausted, as certain setbacks and memories are cataloged and ranked by the level of destruction that the event is capable of causing internally, along with its capability to generate the previous emotion if uttered long after the event occurred or even the speed with which the effect is lost if used repeatedly.

This sense of superiority is present in every second of the target's life and can cause anxiety.

## 5.11.1.13 - Anxiety

Anxiety is a vague, unpleasant feeling of attenuated fear and apprehension characterized by tension or discomfort derived from the anticipation of a danger from something strange or unknown, a future event with unspeakable consequences that bring negative thoughts and is capable of causing pathological anxiety — the long-awaited emotional loss and decay in all aspects of the target's life. It may cause chronic stress.

## 5.11.1.14 - Chronic stress

The ulterior motive of psychotronic torture experiments is to develop chronic acute stress along with psychic exhaustion — to generate stress with maximum intensity for long periods until the total loss of personal resources. When accumulated, stress can cause depression and serious physical problems. So, operators focus all attention on negative aspects.

Chronic stress may lead to a state of heightened electrical excitability and damage the hippocampus, causing memory loss, mental breakdown, severe neurological problems and disorders of all kinds for prolonged periods that provoke indescribable frustration and terrible sensations. There is a loss of interest in life, a fatigue that generally prostrates — a motivational state of the organism associated with a need for rest. It may even lead to death in some cases.

In this case, chronic stress is caused, in part, by not being able to fight back the attacks and by the purposeful violation that is painstakingly designed for the person to collapse due to the rage that is naturally generated from a hysterical scenario, leading to inevitable illness, including psychosomatic disorders. It causes a generalized discomfort and terrible consequences, such as ulcers and strokes. Imbalance, hatred, accumulated resentment: all of this works like a biological time bomb that sooner or later will impair the target's physical health.

Remember, this state occurs due to feelings that are generated basically using psy warfare.

## 5.11.2 - Electronic Psychophysiological Warfare Protocols (Psy Warfare)

### 5.11.2.1 - SYNTELE + D2K = a havoc combo

Complete deprivation of stimuli induces stress, and stress distorts the judgment and logical reasoning, causes anxiety, neurosis, and psychotic breaks. Sometimes the false impression that operators will decrease the intensity of the attacks is created, thus reducing the level of anxiety and bringing back the feeling of well-being and relief. However, peace never lasts. In fact, the attacks are intensified by the minute. The conditioning is also there: any sound, shout in the street or noise from a neighbor, is already associated with operators and MKTECH attacks even if it is unrelated.

In this combo, which is a part of the target's daily routine, the victim feels as if they are in a Nazi concentration camp, or in a communist regime, perhaps in the worst prisons in the world or in places similar to Guantánamo — all of this inside their own house, which clearly resembles the experiments carried out in the 50's, 60's and 70's.

Can you imagine being dragged out of bed in the middle of the night? Imagine waking up to shrill screams inside your head, noises so loud they can burst your eardrums, speed up your heart rate and stun your mind. And all this without being able to regain consciousness from sleep. At this point, you're still being subjected to increasingly constant screams and noises that will never stop, depriving you of sleep and forcing you to live in isolation. You fear being killed — in fact, you start to die slowly — and feel unable to carry out important work and social engagements as you should due to these attacks.

This is the routine, not of regular prisoners or prisoners of war: this is the daily life of a Targeted Individual. In addition to being on their own property, not having been officially arrested and not having any conviction against them that would supposedly serve as a pretext to demote them and accept being transformed into a human guinea pig, the victim is subjected to this type of degrading and absurd regime.

First, the target will be violently awakened. But before that happens, their dreams will be utterly filled with content that purposely causes maximum stress and involves veiled threats to their family. It gradually destroys their honor and works on matters that the target doesn't like or abhor. Usually after dreaming these exhausting nightmares that physically affect the body in a post-processing by the brain, an abrupt awakening will follow. At this point we can already verify the use of advanced techniques of psychophysiological electronic warfare implemented in a sophisticated way. They're hidden in a deeper layer of experiments and torture, overshadowed by the explicit violence of the acts that are routinely followed, in which the perception of events is directed only to this outermost layer. The most valuable one goes unnoticed by most targets.

After that, the target has to deal with three parallel realities: the reality that we all live in, the reality created in dreams and the reality of the voices and thought images. The latter is broadcasted like a radio soap opera, deceiving and daily inserting a new step that permeates the entire reality of the target unconsciously and that only exists thanks to the persistent insertion of fragments of information created by MKTECH. After each attack with this combo, which is the most important in the course of MKULTRA 2.0 experiments, the target will be taken over by a self-destructive feeling imposed in a premeditated way. After all, they were woken up by unknown voices inside their own house and this will probably ruin their whole day. Faced with this cowardly act, the target will ignore all other relevant aspects of life. They won't be able to focus their frustration, or to go after the real culprits who are thousands of miles away, thus preventing justice from being served. This situation may lead to a massive heart attack or stroke given the constant physical stress caused by neural weapons.

Faced with extremely traumatic experiences, Targeted Individuals tend to suffer from chronic tension and eventually a breakdown. The fight is for survival in a mental and physical battle waged against a weapon of war designed to destroy the human brain. Post-traumatic stress, dismay, distraction, oversensitivity, sleep disturbance, excessive fatigue, and permanent threat of death guide the target's actions.

Then, confronted by the first very deep-rooted negative experience, which causes traumatic memories of great power, the days that follow the negative expectation of the possibility of the same torture events happening again cause a huge emotional impact, fully modifying the daily behavior of the target. To make matters worse, the attacks, in addition to repeating themselves, gradually increase in intensity, startling the target who thought they'd already experienced the maximum power of the weapon.

So, the attacks rise to the highest level, mainly at night, when they reach their peak and can last for months. It'd be the equivalent of trying to sleep with multiple neighbors screaming into loud speakers aimed at your window all night. This attack at maximum level (night and day) is known as Swarm Attack (chapter 5.11.2.11). The act of sleeping, as it normally happened in other times, will only be possible if you enter a shielded place, or by ripping your own eardrums.

As soon as this daily journey declines towards the end of the night, that is, early in the morning, it leaves the target exhausted and in a deplorable state. At dawn, the microwave voice continues its work with the AI using even more disturbing sounds or with the second round of gang members taking their places at their remote base in an ongoing process, leaving no room for even a second of rest. The act is repeated every night for as long as the target lives. It may take years for the target to collapse, die, or commit suicide. It becomes a slow, painful death, but one that generates a lot of data for the worldwide network of experiments carried out across the globe on many different targets.

The MKTECH scheme was designed to immediately overwhelm targets and convey a feeling, a clear message, that those people and their technology possess an enormous amount of power that should under no circumstances be confronted by anyone. In the end, they can perfectly convey this meaning to the target.

Those involved in this scheme have well-established guidelines on how to alter the "software" installed in our brain, modifying, inducing organic reactions at their lowest level in order to manipulate automatic and semi-autonomous features that are innate, even at a level below consciousness, in which the essence of perceptions is still in their purest state. It'd be the equivalent of working with machine language (in computing terms). To this end, it is enough to directly and indirectly manipulate human sensations, emotions and memories.

Remember that sensation is awareness of the sensory components and dimensions of reality; perception anticipates the sensation accompanied by meanings attributed by us as a result of previous experiences. Both work as a result of a process based on external and internal stimuli.

We receive information from the body in the form of sensations. We're able to interpret such analyses through sensations — such as our mood and well-being — which is the set of several connected factors that cause this abstract effect and the perception of energy level and disposition. The modification in these factors is something that completely alters our process of receiving and interpreting information, leading to the wrong reading of the data that our mind analyzes, thus performing system analyses that will not be interpreted or read correctly.

## 5.11.2.2 - Reliving the past

The human mind is made up of experiences and memories associated with good or bad feelings, which creates the concept of memories. The ability to employ electronic resources to manipulate such memories, wrongfully accessing them and working with the originality and unthinkable nature of acts that violate the mind with the unknown and the perplexity of having private processes exposed, easily achieves the desired goal of altering core mental variables that were previously inaccessible if accessed directly. Operators explore, for example, personal experiences from childhood and adolescence that are constantly deliberated on during the adulthood years. This sends the target into a tedious regression even in situations that demand full attention in a given activity and lead to cognitive inhibition at all levels.

In order to undermine internal forces at any cost, attackers work exclusively with memories that have caused distress. Thus, they force targets to constantly relive them, overcoming a brain defense that works in reverse to eliminate harmful memories with the aim of keeping them in oblivion — or at least not so accessible. When using this tactic, the past is kept to the detriment of the present. The re-elaboration of sad memories of place or times is prevented, suppressing potential new happy memories from the present day from consolidating.

Keep in mind that loss — pain — is inherent to humans. Everyone without exception has gone through or will go through difficult times involving sad feelings. Perhaps these feelings will never be overcome with this defense mechanism of emotional blunting linked to memory. Despite sadness, one lives and goes on with life, connecting this pain in the composition of one's "soul" and one's being, serving as an experience to guide oneself in a more improved way in this complex and dynamic world that surrounds us.

Emotion, on the other hand, is the most defining characteristic of humans. So, constantly activating it — or enhancing an emotion that is already there — is the main purpose of the experiments conducted by the operators behind the technology. As we're governed, in part, by our emotions, we use different strategies to prevent a certain underlying feeling from setting in, executing cognitive changes that are not appropriate for the moment. So, this resource exists to prevent the emotional impact from spreading and altering the individual's entire physiological condition. Once placed, an emotion continuously affects the individual. Depending on the intensity of it, we may not be able to hide external reactive behavior, both conscious and unconscious, from others and thereby we put all of our attention into the emotion created by the event. Autonomous processes inhibit or regulate emotions. The prefrontal cortex is responsible for inhibiting the activity of the amygdala that deals with emotions.

With full control of our memory, auditory and visual circuits, operators are able to cause and maintain negative emotions using various artificial means. They don't allow inhibition systems to anticipate emotion before it sets in. Once there, the emotion will be systematically fed by electronic equipment — images, sounds, dreams and by reading physical and mental information —, accurately assessing whether the negative subject is really being effective.

To make matters worse, as soon as the negative emotion is inserted and modifies the target's behavior — that is, recruiting all their mental defenses —, immediately a new attack evokes other memories along with different negative scenarios. This causes the target's attention to jump from one subject to another, processing and reflecting on them, splitting their mental defense system and dissolving them into multiple fronts for full control of the human's mental mechanics. At this point, several emotions and feelings are expected, such as anxiety, anguish, personality disorder and unhappiness, disruption of balance with the environment, alteration of the perception of conscious awareness that can be accessed via introspection and attention to events, even leading to irrational decision-making based on emotional beliefs. It finally reaches the sweet spot: to transform the target into a derailed train of emotions about to collide with any physical object of their everyday life.

This delicate and complex balance composed of countless abstract and concrete parameters can be quickly destabilized during the direct access process It is thus able to explicitly manipulate mental data with advanced psychotronic attack techniques brought in directly from the **Neuroelectronic Warfare divisions**. Uninterrupted attacks that violate the inside of the mind by this havoc combo (SYNTELE and D2K) first reveal their hidden interests inserted in tactics concealed in deeper layers of the attacks.

### 5.11.2.3 - Night attacks, nightmares and screams

A technique widely explored by OPS is to remotely alter dreams. They exploit the most sensitive nightmare that will modify mental and bodily functions and follow a narrative in a reality and context that makes sense to the target. As the artificial nightmare causes the brain to recognize its content while being projected as a real threat to the dreamer's ego, there is a violent rush of adrenaline, tachycardia, increased body temperature, sweating and mental confusion. As soon as the victim is awakened at the height of the synthetic nightmare via Synthetic Electronic Dream (D2K), as we saw in the chapter dedicated to this technology in the first volume of the book, OPS cowardly take advantage of the temporary mental confusion to start threatening them through acting (street theater) using V2K. They attempt to extract information using the reaction of vocalized thoughts with certain themes from the victim's personal life.

At this point, the target feels trapped inside their house, as attacks with these weapons confuse depth perception and sound direction. The sensation that all the neighbors are conspiring against the victim is intense, especially from upstairs neighbors. If the target lives in a house, the noises seem to come from neighboring houses, as the screams can be clearly heard anywhere — inside and outside the residence. This completely stuns the victim, making them take actions they'd never take on another occasion, such as attacking someone — a neighbor, for example —, self-mutilation, hospitalization (psychiatric hospitals), or taking a lot of drugs. At worst, they could end up killing someone or themselves. MKULTRA in its purest form.

### 5.11.2.4 - Sleep deprivation, the mainstay of modern electronic torture

As it is known, sleep is as vital as breathing, eating and hydrating. Our body needs sleep to recover as a whole. We know we become sluggish, indisposed, and error-prone when we haven't slept for 24 hours. After this period, the system begins to go into rapid decline in terms of cognitive functions. A state of sleep deprivation that lasts for long periods can lead to death, but not before going through a process of complete loss of mental faculties.

Sleep deprivation is a technique that has been embodied in torture since time immemorial. It causes serious damage to the body and affects the ability to stay active in everyday situations, such as dealing with the intense and natural stress of modern life. It is then considered one of the central pillars of torture and is widely used in current interrogation techniques.

You can leave a person without food for weeks and not achieve the efficiency that arise due to sleep deprivation when it comes to disorienting someone and driving them crazy, making the victim susceptible to cooperating with the torturer by telling everything they know. In the case of MKULTRA 2.0 — conducted with neuroelectronic weapons —, sleep deprivation is also the crown jewel. Remote deprivation occurs through deafening noise and a neurotic environment capable of keeping sleep at bay while maintaining the process of mental decline.

The strategy of manipulating dreams, of distorting sleep, is fundamental to the process of behavior modification, neural reengineering, psychic driving, instilling false memories and creating multiple personalities. Without complete mastery of the sleep state, the results of the experiments would be ordinary, and some types of mental manipulation attacks would be null and void.

With this premise already prepared in advance (every single night without exception), the victim will only be able to sleep with the TV or radio on or with a sound-emitting device in order to divert some of the attention from auditory functions. The attack with Synthetic Electronic Telepathy (SYNTELE) using the Intracranial Voice (V2K) at maximum power makes it impossible for anyone to sleep. The noise of people screaming in the target's head, among other sounds, have different intensities that trick the brain, keeping it active and motivating feelings and emotions that accentuate the alert state that is not compatible with the sleep state.

Sleep-inducing centers (hypnogenic centers) in the brainstem, forebrain, and other regions where neurochemical stimulation leads to sleep are affected by the whole process of electrical/psychological modification. Simply put, the brain cannot rest. Now imagine this going on every night for years. In addition to natural noises, there are words that accentuate the whole process, making all the victim's problems travel through their mind, raising unnecessary feelings and perceptions of the issues generated strategically to worry, cause anxiety and make it impossible to rest.

Even if the victim manages to overcome this initial barrier and cross from the waking state to the sleep state when unconsciousness begins to take on and assumes the characteristics of the sleep state, a peculiar configuration is generated in the torturers' monitoring systems that indicates that the victim is sleeping. At this exact moment, they wake up the target, reversing the sleep process, synchronizing and changing the frequency of the neuronal movement of the brainstem, more specifically the thalamus, causing the individual to immediately return to the waking state. It's a similar process to someone screaming very close to your ear. This scream seems to be amplified in the mind and causes countless sensations: despair, anger and helplessness, for example, due to the cowardice of attacking unconscious targets who are unable to defend themselves.

After overcoming this obstacle — that is, if the victim miraculously manages to sleep and enters a state of relaxation and REM sleep —, the operators will once again alter the natural dream with D2K and transform every dream into a nightmare that will violently affect the individual physically and psychologically, as we discussed in the chapter on D2K in Volume 1.

There is, therefore, the technique of waking the victim with terrible nightmares every night. Upon waking, the brain is normally still in the process of regaining consciousness and is more sensitive to word content and language processing. So, as soon as the target wakes up, drowsy and with their heart pounding, the piercing voices that play along with the target's thoughts via SYNTELE return and seem to come from inside the house. They try to keep the target scared at all times and constantly remind the victim they will never be alone anywhere.

When we fight with all our strength to sleep and end up having all our mental processes manipulated and electronically altered by strangers, we wake up with the same effects, but even more intensified than before. We take a lot of time to orient ourselves in the world. So, if the target doesn't have outstanding mind control, they will easily fall into the hysterical scenario generated in the course of the torture.

With this in mind, OPS immediately send death threats, negative thoughts or anything that verbally attacks the victim, usually commenting on the dream content sent via D2K the night before to show that they can play with all their cognitive systems, maintaining constant amplification of attacks, without ever dissipating or extinguishing the discomfort. Operators also convey a feeling that they're unreachable and will continue the attack and torture until the individual dies as they vigorously use the elaborate narrative that gradually achieves MKTECH's goals. Ultimately, they prevent the victim from taking any coherent action and from having any hope of returning to a normal life.

In the field of new psychotronic tests — direct and physical manipulation through electromagnetic waves that interfere with bioelectrical mechanics of the brain —, there are several topics to be tested and verified with results still unclear. Since this is a recent problem, there are still too many unknowns involved in the MKTECH universe. In any case, I hope this book will help clarify some points and make more people start researching from now on with these 10 years of knowledge on the subject.

Sleep deprivation causes several cognitive, behavioral, hormonal and neurochemical changes. It greatly reduces the person's longevity and causes serious disorders. Performance in their everyday tasks is completely degraded and responses are slow. There is also an increase in reaction time and errors, memory problems, irritability, symptoms that lead the target to hopelessness, to give in to torture and, of course, to decline, to succumb to hysteria and become another degraded mind.

This is in fact a practical way to inflict immense damage to the target. All over the world, reports of psychotronic attacks always follow the same pattern. The attacks become intensely unbearable at night to purposefully achieve complete sleep deprivation. The voices and noises are accentuated; the volume and frequencies are increased mainly at dawn. During this period, the operators' activity becomes more intense. Every night there is a very elaborate performance with rancid actors behaving like nocturnal creatures who hunger for positivity, thus leaving a trail of decay capable of invading the dreams of targets.

Sleep deprivation is essential to cause mental stress, raising the possibility of nervous breakdown. At any time of any day, regardless of how tired you are, the torturers will use the dream manipulator. As soon as you fall asleep, your thoughts and dreams will begin to be replaced by those transmitted by them, causing the victim to be woken up at most every 2 hours.

Anger and hatred begin to overload cognitive processes. But that's exactly the kind of reaction operators expect. They want the victim to be consumed by unbridled hatred caused by the deliberate violation of all aspects of their life (private/cognitive life, the time for rest, work, among others) and by sleep deprivation. In other words, they turn targets into zombies. The victim is always awakened by artificial/simulated nightmares in the middle of REM sleep, in which the memory of the dream becomes much more vivid. Over time, this makes the victim no longer know if some details really happened in real life or if they are part of an artificial dream.

This leads to a disorder linked to dissociated states — the target can't discern between sleep, dreams and reality. The disorder occurs due to the intrusion of three main states of consciousness: wakefulness, REM sleep and non-REM sleep. This is yet another bizarre mind manipulation experiment.

## XI - Tips for the Targeted Individual — Psychotronic Warfare Tactics

Try to maintain a healthy diet, social relationships and especially a physical exercise routine, one of the most important factors of all, to balance the continuous electronically debilitating induced stress and its harmful consequences to your body. In addition to de-stressing, burning calories, improving and strengthening mental and body structure, exercising will help beat your attackers during sleep.

Don't give up, enjoy the rare moments of positive thoughts and fight with effort and dedication. Keeping this routine will draw the combat out with great chances of victory against an enemy much more powerful than you. Always remember that return waves generate negative structural changes in your brain. Meditation, for example, is able to do the same, but naturally and not in a negative way, sharpening focus and reducing anxiety caused by stress. So, try to practice it to condition your mind to progress within the chaos. After all, MKTECH is capable of measuring the capacity and the moment in which the target is immersed in an activity that requires full concentration, a kind of internal flow. So, fight. Don't let them take you away.

## 5.11.2.5 - Psychic driving

Psychic driving: an attempted murder by acute stress over long periods of time, based on heavy psychotronic torture, which makes targets start banging their heads against a wall within one to two weeks, mutilating themselves as a way out to get rid of this unbearable stress. Some people scream, *"Get out of my head! What do you want?!"*. They punch objects, try to rip the steering wheel out of the car while driving, screaming, startled by the constant presence of voices that echo inside their brain. Targets do all this in an attempt to raise the awareness of those responsible for the technology — to show the intentional harm they're inflicting on ordinary people who don't understand the diabolical level of these practitioners of modern torture.

Self-mutilation, breaking objects and moving away from home are consequences expected by torturers. Modern psychic driving is a torture technique that consists of the continuous repetition of a set of stimuli for long periods. It completely changes the person's personality, creating memory and emotional triggers. This systematic repetition of stimuli can, over time, lead to brainwashing, personality modification and various mental pathological conditions in which it is possible to "install and uninstall" human parameters, creating another "self" quite different from the original. This happens gradually; every day a little of the essence that made up the original personality is lost. This includes making the target believe in the reality created by the operators in order to go with the narrative in which the victim is included as a main character.

For the techniques from the past to be implemented, it was necessary to have the individual conducting the experiment and the human subject in the same physical space. The target would then be doped with drugs, electroshock and other procedures would be performed repeatedly for months to achieve the expected result. Today, remotely, that is, in different and distant physical locations, thousands of miles away, it is possible in a much more efficient way: a) to conduct the target's behavior inside and outside their privacy; b) change some features of their character; and c) cause unimaginable trauma and mental confusion with the direct manipulation of visual memories, only using the maddening stimulus of the V2K intracranial voice.

This driving causes temporal amnesia. Targets create a parallel reality due to the repetition and introduction of false memories, loss of awareness of oneself, as well as frustrations in solving problems inside and outside the experiments, which results in the deterioration of their inner self and leads to general degradation and total decline. This opens the door to the complete restructuring of the character or parts of it.

Thus, targets are able to keep these intrusive distressing memories and relive a traumatic experience at all times. This generates systematic, negative and exaggerated emotional states, in addition to the inability to feel positive emotions. Nothing seems to cause personal satisfaction anymore. Events that once attracted attention, symbolized happiness and well-being, are no longer the same. This discovery creates a void, a lack of interest in participating in relevant events and an alienation from others. The body's continuous natural attempt to recover the well-being and internal emotional balance essential for a full life is completely hampered. That is, personal strengths and limitations, questions about the individual's own value before society, family members and themself. For its full functioning and accomplishment, modern psychic driving depends on the combination of all the techniques studied here being successfully implemented by the torturers.

Destructive tactics, even if they aren't clear when in contact with the target, are implied in the most visible dark intentions behind the strategies. So, the most attentive people are able to see what is not there. For example, the way one directly manipulates the target's mind and body leading to the reengineering of the brain (within our biological structure) reveals hidden agendas in the most straightforward attacks.

This systematic repetition of processed facts within the narrative encompasses all the points created to be internalized, instilled and automatically assimilated by the human guinea pig. At one of these points, the victim will have their mind working unconsciously from the narrative infused in the memory that was created via systematic dialectics and insistently reinforced by visual images transmitted in REM dreams. This type of brainwashing and personality modification occurs naturally in the brains of those affected.

Even the most resilient, skilled and mentally and physically strong person ends up losing a good part of their identity along the way. Those most aware of the artificial reality created around them can be tricked into performing highly disturbing and erratic behaviors in the blink of an eye. This technique was widely used when torturing prisoners of war with repetitive sounds. Yet, it's nothing compared to what can be accomplished today with the advent of psychotronic weapons.

Nowadays, the result of this methodology has taken on such proportions that these acts should be considered crimes against humanity due to their lethal ability to affect several people at the same time, thus becoming a weapon of mass destruction.

## 5.11.2.6 - Thought filter

Target-private, strictly unique, internal and highly personal thoughts that should never be violated pass through subjective filters used by operators. At each moment, constant feedback hampers the rhythm of reasoning. Over time, a certain type of thought is buried deeper so that it doesn't surface under any circumstances, altering the cognitive brain mechanics and obtaining important results found in the MKULTRA 2.0 experiments — for example, how the human subject manages not to internally change the flow of thought under any circumstances by microwave voice stimulation. Sharp eyes and ears of observers attentive to the response and processed stimuli verify if mental defenses have been broken or if the response generated by the stimulus is some kind of secret or something the target deems important.

Exposure and lack of cognitive privacy occur all the time. It limits thoughts, restricts their reach and, just like a football crowd, operators cheer with every thought to give the impression of being in control of cognitive processes and their abundance. They can, for instance, discourage certain paths and encourage others. They usually let silly and useless thoughts (related to the torture) naturally flow while creative thoughts that contain information that might confront the gang are quickly rebuffed. Any thoughts that may contain information that could be used for financial gain will be stolen.

In this war, as time goes on, the subjective filters of the operators manage to completely alter the normal functioning of the target's brain, preventing the search for knowledge in all areas of knowledge and making their intellect shallow, futile, silly, and harmless.

## 5.11.2.7 - Keeping the mind busy, "clogging" the primary reception systems with useless content

Perpetrators fill the victim's mind with useless things, trivial matters and nonsense, and with highly disturbing details involving intimacy. They repeat the same aggressive and disrespectful words and phrases routinely, using every possible external stimulus to do it, thus creating a kind of blockage of the primary functions of capturing external stimuli, which prevents the functioning of attention in relevant everyday matters, stresses the brain, hinders learning, reduces emotions and exacerbates negativity.

Successful communication of a more complex set of ideas and thoughts is also prevented. The immersion of ideas and themes is hindered. The way technology gets the brain's precious full attention ends up drastically reducing social interaction and conditions the target to go deeper into solitude and complete immersion within the torture content and interaction with operators and OPS (Organized Professional Stalkers).

## XII - Tips for the Targeted Individual — Psychotronic Warfare Tactics

**Emotional control: keep this term in mind. Emotional control is a superior quality of the human being.** It consists in controlling your emotions; in creating an affective regulation to regain control of the situation that was altered by immediate impulses caused by external stimuli, despite the power of the attacks and the set of factors compulsorily converging to the evocation of self-destructive feelings such as hate, resentment, hatred and a desire for revenge against the attackers.

Every time one of these feelings becomes screams of revolt, physical violence to people who have nothing to do with the attack and destruction of property (such as furniture) in order to vent the stress and emotional charge, OPS scores an extra point: 1x0 for dirty games.

Therefore, control yourself, focus your anger on physical exercise in proper locations and don't submit to the attackers' will. As difficult as it sounds, this kind of explosive emotion is to be expected. Another tip for absorbing the torture is to try not to think about the harm they're causing you at the moment. Wait for the right time. Whenever you can, find out who is behind the attacks and remember everything they did to you. You can use this as your advantage and start fighting back, triggering the events that will contribute to the end of this model of torture around the world.

You have the right to defend yourself, to fight for your life! Sometimes the best defense is a good offense. So, do what must be done! This war can only be won if everyone does their part.

## 5.11.2.8 - Using hatred to destroy the target, to make them "implode" in their own rage

Making the target "explode" from the inside out is one of the most dangerous and damaging possibilities of this weapon. Having someone at all times preventing you from sleeping, thinking, dating, working and studying, that is, from living your own life, will quickly drive you crazy.

Deprivation and excess are essential elements in the torture process. Every now and then targets will find themselves screaming in rage in front of the mirror in the hope of addressing the torturers behind this evil weapon, but with no ability to vent their anger.

As feelings and emotions play a central role in intelligence, some say that we're made of pure emotion — in fact, some of us really are. So, SYNTELE (Synthetic Electronic Telepathy) is able to access memories and evoke different emotional states. It is a weapon that manipulates the codification of reality, interposing itself between the capture and interpretation of stimuli, using all known strategies that in fact instinctively modify a person's behavior, violating their privacy everywhere. In short, all basic, primitive feelings are affected! There is no way to reverse millions of years of evolution in such a short time. Targets will always feel heightened emotions like anger and fear. In addition, the main function of this weapon is related to the attention deficit.

## 5.11.2.9 - Ability to cease attention and divert the focus

One of the most perverse characteristics of this weapon lies in the fact that it hides the real reason for the format in which the attacks are conducted. During the course of this book, we've seen the connection between psychotronic weapons and targets and the slow revelation of an ulterior motive camouflaged in imperceptible layers. One of the major attack points of this weapon is related to perception and focus, which are the logical entry point to comprehend information coming from the environment and later its organization in our minds.

Contrary to popular belief, the brain works to switch the focus on the activity that is currently being performed. Thus, it is able to switch it from 2 to 8 times per second, precisely so that we can, for example, deviate a moment from the current activity on which we focus and "defocus" ourselves, allowing us to immediately stop the ongoing activity and become aware of the environment around us. In this way, we're able to react at once to an unexpected situation, should one arise. On average, the brain fluctuates between focus and distraction every 250 milliseconds. Therefore, in order to stay focused on our activities, we must constantly readjust it so we can ignore other distractions that constantly reach our brain from everywhere.

MKTECH devices are able to capture this event and translate it into a sine wave format. So, each time the brain stops focusing on the momentary activity, the AI launches SYNTELE attacks synchronized with the wave that represents the end of the focus. As the brain readjusts for the next activity, it will invariably pick up on the psychotronic weapon's stimuli, removing concentration from its important task to embark on the sordid narrative of voices and noises that permeate the mind. Consequently, victims need to work on their concentration skills on a daily basis. Yet another advanced psychotronic warfare tactic.

The voices that interact with the target's thoughts explicitly intend to hold all of their attention and consume their mental resources, luring them into the content of such conversations and the reality that is predicted within the subject. They rely on the unconscious range of effects resulting from this interaction that will immediately hamper focus and attention of ordinary physical reality, causing enormous damage to the target's analysis and judgment. Keep in mind that attention plays a fundamental role in the execution of daily activities. After all, the environments that surround us are full of constant stimuli and various distractions.

The ability to learn — a vital resource for human existence within society — is also hampered. Learning promotes a systematic growth of the individual, and concentration is a must for this act to be carried out. As this technology is designed to mess with the target's concentration and focus, it prevents the individual from filtering countless information according to their needs and manipulates it through mental operations. If this, in a regular situation such as studying a certain subject for a test, is already difficult to be achieved, imagine when your brain is connected to a network that sends audible and visual data non-stop.

Even though we're totally focused on important events, some intrusive thoughts naturally enter our "mental cinema": a bill we have to pay, plans for the weekend, the memory of a loved one, among others. We all eventually experience this in our lives: that inner distraction where you find yourself immersed in random thoughts and realize you forgot about the world around you. This weapon does exactly that, but infinitely more powerfully than the brain's standard distraction process. It forces the shift of focus, insisting on visual thoughts by stimuli via D2K which, even if not demodulated correctly in the waking state, are still able to influence the order of mental images that pass through our mental "projector".

Let's not forget the unparalleled power of SYNTELE, which inserts voices and sounds of all nature, activating antagonistic mental processes and capturing any external sensory data — for example, the content of the test that the target was studying for. Attention modulates the quality of the stimulus learned by mental processes. It is the basis of all our actions. Through it, we select which information will be used, disregarding secondary stimuli.

Attention is, therefore, a complex act and requires several factors to happen. Psychotronic attacks act directly on such processes. In general, all types of attention are affected, especially the selective attention, which is the ability to focus on just one important stimulus and suppress the rest, which induces concentration. Focusing on the teacher and not being distracted by secondary stimuli is a good example of it.

**Sustained attention** is the ability to focus on a given activity for a long period of time, and is greatly impaired by interruptions due to secondary stimuli. In this case, the shift of focus is abrupt due to the magnitude of the stimuli coming from psychotronic weapons. This leads to the induced adaptation of very clear brain dynamics and forces a mental habit (in daily processes), which maintains our ability to respond to more than one task at a time and uses **divided attention** — divided into MKTEC attacks and in the reality around us. It can also recruit **alternating attention**, which is the ability to switch between tasks.

Thereby, with this high-level ability to manipulate one of the most important processes in humans, a constant surveillance is exercised and it seeks to detect the appearance of some valid stimulus for the target. Orientation and concentration directed towards the attack itself are maintained, altering alert levels, diverting attention and spending energy, also compromising several functions directly linked to attention, such as motivation, fatigue, mood and anxiety.

## 5.11.2.10 - Classic brainwashing techniques

The same techniques employed in the original MK-ULTRA, aimed at brainwashing and behavioral change, are also applied to targets through psychotronic weapons, mainly based on the theories of Russian neurophysiologist Ivan Pavlov (1849-1936), the so-called classical conditioning (Pavlovian or respondent conditioning). This technique basically consists of producing stimulus and responses, shaping human character through punishment and reward. However, the technique combined with the contradictory stimulation has a devastating effect on human behavior. It does just the opposite: it breaks the chain of conditioned reflex, either punishing behavior that was previously associated with reward, or rewarding behavior that was associated with punishment.

Operators work with sensory data constantly processed in the target's mind along with continuous pain and stress. From time to time, they create situations that seem to help the target, making them relax and enjoy a false sense of freedom. At certain times, they cause the same situation, however, causing subsequent stress that can last for days, and exacerbate intolerable anxieties until personality change occurs. They even lead the target — who know perceives MKULTRA operators as allies — to a kind of inversion of values and encourage them to attack friends and family members.

They exploit the brain's vulnerabilities and its organizational mechanics. This attack is constantly used within one of the many layers that make up the unit. In addition, there is the sending of continuous auditory stimuli via V2K, and traumatic events created via D2K are included. Therefore, this great dissonance is combined and weak, positive stimuli become negative due to increased sensitivity to stimuli. It uses the most subtle brainwashing techniques that exploit the ultra-paradoxical effect discovered by English psychiatrist William Sargant who studied prisoners coming from Chinese concentration camps after the Korean

War. The target becomes receptive to stimuli to which they were previously immune to, whether the artificial psychotronic stimuli themselves, or stimuli from the surrounding environment. For example, a noise upstairs can cause extreme paranoia.

Most individuals were taken by surprise and ended up falling into the traps of these stealthily managed classic techniques without the unsuspecting victim realizing their devastating consequences. The "transparent" insertion of this methodology within the attacks leads to cognitive dissonance, paranoia and psychosis at the end of a carefully planned path, thus establishing control of the flow of stimuli and information that quickly breaks down the target's mental defenses.

## 5.11.2.11 - Intensity of attacks during long periods of exposure to microwave voice (V2K), Synthetic Electronic Telepathy (SYNTELE) and Synthetic Electronic Dream (D2K)

As the intensity of the SYNTELE attack varies over the years of MKULTRA torture-based experiments, adapting to the situation and the particular condition of the target, the intensity can either increase or decrease gradually or abruptly. It always depends on the strategy currently employed by the operators and the hour-by-hour, day-to-day conditions that the human guinea pig is exposed to. In this way, it prevents organic adaptation and keeps the target under a rigid kind of electronic halter.

One way of indicating how this continuous variation should be used is through the cyclic conditioning effects of punishment and reward. If the target isn't fulfilling the objectives of OPS — if they aren't following the plan —, the intensity of the attack will be increased in such a way that it will make the victim submit to any (in)human whim. Once coercive cooperation is reestablished, attacks may decrease in potency. Unfortunately, the technology will never be turned off completely. After logging into the system, operators are expected to keep torture at unbearable levels (above 5), just enough to drive a person mad in a short period of time. This leads the victim to spend the entire day at war with the invaders of the mind, which culminates in the act of "talking to themselves" to exorcise such evil beings that have settled and now inhabit all mental domains that once belonged only to one owner.

## 5.11.2.12 - Activity levels

- 0-No activity (rare).

- 1 to 3-Reserved for stealing information and intellectual property straight from the target's mind. Small sporadic V2K attacks may occur to stimulate the subvocalized thoughts and mental images concerning the subjects that are the search object. Small clicks in the head and high-pitched sounds in the hearing circuits may be felt and easily mistaken for natural events. Sometimes, subtle background music can also be demodulated, or low voices causing the target to reason, even if unconscious, about the topic that must be snatched from their mind.

- 4-Mild activity: beginning of the interaction with the intracranial voice. Only V2K is used here.

- 5-Moderate activity: SYNTELE, light interaction of voices with thoughts.

- 6-D2K - synthetic dream activation.

- 7-SYNTELE + D2K.

- 8-Severe activity: SYNTELE+D2K+V2K.

- 9-Extreme activity: SYNTELE+D2K+V2K+TELEMETRIC EEG+REMOTE POLYGRAPH - operating at almost full capacity 24 hours a day, 7 days a week.

- 10-Maximum activity: all components operating at full capacity.

- 11-Physical microwave attack capable of altering the electrical flow of voltage-dependent ion channels.

Prior to level 10, just before this attack turns physical, the weapon starts to be called "Directed Energy Weapon". It causes skin damage, burns and permanent physical and neurological damage. At the threshold of level 10 lies the terrifying Computerized Swarm Attack.

## 5.11.2.13 - Computerized Swarm Attack

Computerized swarm attack is the most devastating type of attack within the entire MKTECH dynamics. All the characters behind the torture unite in a collective action, similar to an attack by an African honeybee colony on an unsuspecting invader. These bees attack simultaneously with unimaginable ferocity and chase their victim for almost 50 yards, guided by the gases generated by the rapid metabolism of oxygen, the result of rapid breathing. In the field of psychotronic weapons, this attack recruits all humans, operators and OPS with the help of an evil AI that runs the show.

The AI manages to extract maximum performance from other processes running in parallel to deliver the worst cognitive experience ever — worse than anything a human is capable of seeing, hearing and feeling. The Artificial Intelligence can, for example, extract the maximum power from another algorithm responsible for telemetry and triangulation of the waves that will reach the target. It recruits the maximum power at all frequencies used. In addition, it demands the highest performance from the equipment and exclusivity in its use, in its bands and spare equipment in the event of any failure. It also uses all available V2K channels and performs all sound directional perceptions that the brain is able to interpret simultaneously at its maximum capacity. That is, the ability to gather more resources to achieve the desired goals of neural devastation and chaos.

The technology is also capable of providing the target's level of adaptation in the face of the attack and the progress in achieving certain goals by monitoring mental and physical responses at each moment, analyzing behavioral signals that break the harmony of the integrated functioning of their thinking and acting, previously established by statistical parameters in charts and in the database. The perpetrators thus extract maximum performance from all MKTECH modules in a bizarre relationship with the victim who desperately screams as they enjoy the result of each electromagnetic blast processed by the target's brain.

The attack can last for hours, days, weeks or months — in some cases even up to 1 year depending on the target's resistance to absorbing mental hits without being knocked out. Consequently, when the level of the voices in the target's mind works at maximum power, the Synthetic Electronic Telepathy (SYNTELE) merges with horrible night dreams (D2K), which are the most intense dreams possible to experience as a human being.

In short, AI and humans work until the internal and external clutter is so great that the target is no longer able to put the world back in order on their own. If the attack doesn't kill them, it will undoubtedly change their life forever. In the meantime, this whole process will be of great value to the MKULTRA 2.0 experiments as it captures unprecedented data for the scheme, thus integrating details into their databases regarding the limits of a human being under extreme physical and psychological stress. The brain embodies this new flow of artificial information as part of a model of induced reality, a world created by OPS.

As the individual reacts to the multiple, more intense external stimuli that one can experience in terms of electronic torture, the attackers decide whether to go back to "normal" levels — back to 8.5, maybe 6. Such an act depends a lot on the objectives to be achieved at the time. Remember, there is still the action of obtaining information without revealing the real intent. As the seemingly baseless attacks occur, they occasionally insert something pertinent to the experiments into this mass of content — a film in specific dreams, sensory data sent to the subject's mind with nuances of scientific driving of some kind of experiment that goes unnoticed at first for most targets.

Maintaining the swarm attack for long periods is costly for operators in terms of human labor and maximum use of infrastructure, even allowing occasional tracking by some equipment that is scanning the electromagnetic spectrum at the moment.

Inserted in MKTECH technology, we come across the most perverse side of AI used as a weapon to kill and torture humans. For those who thought that this type of technology only existed in movies, think again.

Among other things, the swarm attack is capable of bringing the target into a state known as electronic narcosis.

## 5.11.2.14 - Electronic narcosis

Electronic narcosis is a condition that shows symptoms generated by long-term psychotronic weapons attacks. This includes the result of the combined attack of all MKTECH tools with the aim of killing and inflicting as much pain as possible, especially during the night when sleep deprivation produces unique effects on the mind.

After a prolonged period of swarm attack, it is possible to reach the state of narcosis early in the morning if the individual survives without going mad or having a stroke in the process. By dawn, in the first rays of the sun, disbelief is exacerbated. The target feels completely anesthetized with levels of stress never experienced before. Time seems to move at high speed. During the early hours of the morning—the time when attacks increase—continuous high-intensity events transform the dynamics of mental processes so unrecognizable that it is possible to enter a state of acute, disabling and anesthetic stress. Usually, when this state is successfully reached, operators surreptitiously and simultaneously turn on the microwave weapon used in the embassy attacks (chapter 8). Keep in mind that these are weapons of direct attack: its transmission alone is capable of altering voltage-gated ion channels, causing neurological damage and unique sensations, such as the sensation that the place is "shaking" and your brain jiggling. It is also possible to feel the electromagnetic waves passing through your body and interacting with your organism, causing effects such as the inability to remember events and electronically generated dissociative amnesia, in addition to the sensation of having an electric current passing through your body, nausea, headaches, etc.

These harmful effects can be achieved with or without direct microwave attack, but their use increases the damage and accelerates the degradation process of the organism. Both technologies can be used simultaneously. If operators don't want the target to know that they have been attacked by a direct attack weapon, they use the effects of narcosis that occur in swarm attack cycles, as such effects get mixed up at this point. Some targets are attacked only by direct assault weapons; others are attacked by weapons that interact deeply with the human mind. The rest of them are attacked by both. In any case, the result is catastrophic.

By the time electronic narcosis sets in, SYNTELE keeps massacring the target with V2K to the point of being able to physically damage the speech and hearing systems in some cases, which causes the victim to develop the symptom known as phantom concussion. Don't worry, I'm going to provide more information on the subject in Chapter 8. At the same time, D2K works at its maximum capacity with antennas and satellites producing radio waves at high power levels. It takes advantage of a unique and extreme configuration — the target sleeps and wakes up repeatedly —, and there is even an intercalation between the two states without conscious perception, making the brain demodulate in a tenuous way the images that are destined for the visual cortex when in a state of unconsciousness. It is at this maximum stress, as the target is in a deplorable state, that the intrusive mental images of the transmissions overlap with those of the current thought. These perceptual images and representations provide content to the thinking process and the flow that leads to judgment that is already completely compromised, erratic and with little sense regarding the content and fluidity of its course.

Due to intense tiredness and after sleepless nights and days, your body and brain begs for sleep. The mere closing of the eyes takes the mind automatically to the next state between wakefulness and sleep. In the event reported below, we can see how the state of narcosis is special as the brain picks up transmissions destined for waking dreams:

After nights of swarm attacks, the electronic narcosis was already "placed" accordingly, the target decides to sit in front of a computer. Suddenly, a visual thought of a bulky person appears. This person is running towards the ego in the dream (first-person point of view) — a widely used technique, as we saw in the chapter on D2K of Volume 1. So, the image of this physical attack triggers neural mechanisms similar to a jerk reflex in real life. Even the sensation of being moved at the exact moment of the virtual shock can be felt without the force of the real impact. The reaction is an instinctive behavior — a leap, an awakening, a spasm — back to being fully alert. This type of event occurs in the midst of SYNTELE attack in which the brain is about to collapse from the excess of electronic stimuli, and in the very genesis of thoughts that now captures D2K images during the waking state.

Yes, it is possible to insert visual thoughts into this dubious state using the weapon designed to manipulate dreams, thus causing the disruption of the flow of images. In fact, now the assimilation with the brain, with the conscious mind, is very feasible, altering the architecture of coherent and understandable thought, modifying the associations of ideas and modifying the internal content with images sent by psychotronic weapons, distorting the stimuli captured from external reality and putting them in the background to provoke a reaction similar to what one would have experienced in real life. At this point, the target is extremely vulnerable with regard to brain reconfiguration and is a serious candidate for committing violent acts. They may also even be recruited to be a Winter Soldier.

Electronic narcosis always occurs after a long period of attack at intensities ranging from level 8 to 10. The amount of time it takes for this state to be reached depends on numerous factors relating to the target and the environment in which they are inserted.

## 5.11.3 - Advanced and complementary techniques employed in the use of SYNTELE

### 5.11.3.1 - Mind bridge, electronic transfer of vocalized thought

Mind bridge means amplifying a thought of a close person and sending it via V2K to another person's thought, creating the illusion of hearing from time to time a voice that mixes with the sounds that come from the physical environment in which the target is located. In other words: you sit down to lunch with someone and suddenly this person brings up a topic you were also thinking about at the moment. The content of the vocalized thought can be transferred from one person to another. Just amplify it via EMRA - Electronic Mind Reading (auditory) from an individual A and send it to individual B. This is a technique used on several occasions to confuse, drive mad, create a fictitious bond between two people to meet a certain criterion or to create a false impression of someone with telepathic powers.

## 5.11.3.2 - Hidden interlocutor

This is a vile tactic based on compromising the integrity of a person while making them disseminate the ideas of OPS so that they can become the agent of the enemies. This artifice is extremely dangerous, heightens paranoia and is capable of provoking potentially violent situations that can even end up in murder.

Generally, they use someone close to the target, a real person, to serve as a parasite. This individual in turn will spread gruesome ideas that operators want to bring to the surface in order to reinforce a narrative or argument. So, they hack the brain of this person and quickly drive them crazy. This leads the person to erratic behavior and, most importantly, to a behavior observed in all victims without exception: the act of talking to oneself and screaming at "nothing". In this case, they're reacting to technology operators via SYNTELE, thus reinforcing this behavioral dysfunction to the point where victims need to pretend to talk to someone on the phone to justify their actions.

Subsequently, the target begins to scream and becomes the personification of the aggressors. Everything that is sent via SYNTELE will be uttered in words by the interlocutor. This is a devious tactic, but extremely relevant as the main target is under electronic surveillance and now someone close to them will also be watched. Thoughts and attitudes that lead to verbal aggression, invasion of privacy, including depraved thoughts or thoughts that provoke this type of violent behavior, are instilled. This can also be a potential catalyst for a likely outbreak of violence, as it will then direct the anger that the target feels towards the attackers into something palpable, physical. The interlocutor will be an intimate enemy. The target and the parasite may eventually end up killing each other, and then MKULTRA's mission will be accomplished without raising suspicion.

## 5.11.3.3 - Directed murmur

Ambient sound — very common in many places of everyday life that house a cluster of people— naturally make noise. The most common are conversations that leave a trail of cacophony that usually reaches the ears as a normal background sound. You don't pay much attention to it; the focus remains on the conversation with the people you're interacting with at the moment.

A good place to notice this noise is a crowded restaurant at lunchtime. As we saw in the V2K chapter, V2K has the ability to generate a sound interpretation of the accompanying voices identical to the acoustic signature of the environment where the target is located, and is indistinguishable from the surrounding murmurs reaching the ear. So, in the midst of several voices, conversations, laughter, fork and knife scratching a plate, or plates being piled up, that is, every dynamic that involves a restaurant generates characteristic noises of the place that are easily associated by the brain. Suddenly, a clear voice is heard from within the murmur and says something related to the intimacy of the listener or those who are directly close to them. The auditory sensation makes the target believe that such voices come from people at nearby tables. The action is then repeated wherever the target goes, causing massive discomfort and the feeling of going crazy. Paranoia and attention diversion are also associated with it.

## 5.11.3.4 - V2K audio for everyone

There is a frequency common to most people, which is not supported by the set of frequencies based on the Neural Biometrics of the target. In this case, the microwave voice can be detected by any mammal or animal with a developed brain — like domesticated dogs and cats, for example — in the radius (amplitude) or range of the signal.

This frequency is sporadically used to hit people close to the target, a few-second display of the V2K's power. The target who manages to reach a certain level of knowledge regarding the technology has the opportunity to witness firsthand the impact of the microwave voice on an individual who has never had such an experience before. A simple burst of a powerful voice echoing in the environment for everyone to hear —for fractions of seconds— is able to change a person's emotional state in a significant way. The affected person will probably ask, "*What was that?*" or "*Did you hear that?*". The uniqueness of this type of attack shows to the target the grandeur and intensity of the microwave voice in the minds of people who had no contact or aren't aware of the existence of this device. It automatically generates conjectures of how was the first contact with this weapon and how this person is still alive and overcomes obstacles that seemed insurmountable in the initial moments of the attacks.

When V2K is demodulated by several people in this tactic of intimidation and demonstration of force, we are thus able to see the fear in their eyes, the immediate change of mood and the disorienting feeling about the transient and sudden event. This can be strategically repeated using other MKTECH modules, such as the Synthetic Electronic Dream (D2K). This module can reach people directly linked to the TI and preferably who are literally in front of them so that targets can observe with their own eyes a relative desperately waking up and complaining about a horrible nightmare or that same person jerking in their sleep in an attempt to try and silence the supposed neighbors who scream and interfere with their dreams.

## 5.11.3.5 - Crossed thoughts

This is a bizarre tactic that consists of transmitting in real time what the person interacting with the target is currently thinking — from one mind to another — during a certain period, that is, while the interaction takes place. Many standard channels of microwave voice torture work incessantly with the operators' voices. One of these channels is used to send fragments of the vocalized thoughts of the person who is close to the target, forcing the victim to listen to some sporadic thoughts of relevance coming from other people.

For example, when the target greets an acquaintance, relative or friend on the street, SYNTELE immediately makes available one of its input and output channels dedicated to the thinking of the person with whom the victim is talking at the moment. It then "streams" the vocalized thought of that person with whom the target interacts with directly into their mind via V2K.

Imagine what it's like to have a conversation knowing that they hear your words and vocalized thoughts; that they're processing the entire moment in which numerous daydreams are captured, including the person's ability to distract from the conversation.

Automatic association processes work freely within our minds. No one is ashamed to think. Although we don't express every thought in the form of words, they're there — "traveling through the brain" — for the most diverse purposes such as maintaining coherence or the association of ideas.

What before was an internal process inaccessible to external agents, today becomes a public process with the help of the mind-invasion technology. It works like this: a V2K channel transmits the thought of the person with whom the target is physically interacting at the moment. And let's be honest: one of the most disturbing and strange sensations is hearing someone say one thing and thinking the opposite.

In short, unimaginable scenarios like these take place inside the minds of those involved in the scheme. A real emotional discomfort and a series of disorders are caused by operators. Having your processes exposed like this is an unprecedented invasion of privacy — like nothing experienced before in human history. Therefore, seeing this happening in front of you, or inside your own mind, during social interactions is unforgettable, to say the least.

## 5.11.3.6 - Echo in thought

A technique widely used to break the train of thought of the target in question is to echo every subvocalized thought that occurs. As soon as the operator receives the thought via EMR (Electronic Mind Reading), the exact content that the victim thought is sent back using V2K, but with a delay of 1 second. This confuses the brain and causes the victim to lose any line of reasoning, whether solving a problem or writing a text, for example.

Reading is one of the most affected cognitive activities. As soon as the target starts reading a text, an intruding voice will replicate the content that was read moments earlier by the target's internal voice. It's a kind of audio feedback that live music bands use: the vocalist wears an earpiece on stage so that they can hear themselves and the rest of the band. In the target's case, however, this feedback is sent via microwave hearing and its content is composed of intruding voices that are processed inside the mind. Or to put it another way, the last words read silently are replicated, completely destroying the comprehensive understanding of the information and ending the reading process along with the absorption of content. It's an extremely maddening and unnerving act.

## 5.11.3.7 - Electronic gaslighting[53]

This is a well-known technique of psychological manipulation that is systematically employed and included in every attack. The tactic consists of convincing the target that their sanity is compromised, creating distortions of reality through psychological abuse and memory modifications via SYNTELE and D2K. Soon, the victim will be disoriented and unable to determine if events that occurred in concrete reality really happened or if they were manipulated with systematic attacks.

---

[53] The 1938 play Gas Light and its film adaptations, released in 1940 and 1944, motivated the origin of the term. The plot concerns a husband trying to convince his wife and others that she is crazy by manipulating small elements of her environment and later insisting that she is wrong or that she remembers things incorrectly when she points out such changes. The title stems from the dimming of the gas-powered lights in the couple's home, which happens when the husband uses the lights in the attic while searching for hidden treasure. The wife notices the dimming of the lights and discusses the phenomenon, but her husband insists she is just imagining it. The term gaslighting has been used since the 1960s to describe the manipulation of one's sense of reality.

## 5.11.3.8 - Artificially-induced suicidal ideation

Psychotronic weapons and their unique ability to induce the most diverse states of mind are able to assess the target's susceptibility to suicide, detecting behavior patterns to identify a suicidal person. They make use of the precise pattern recognition of artificial intelligence to accurately conduct and diagnose whether such intellectual parameters are fulfilling the requirement of being a potential suicide.

SYNTELE and D2K are able to easily remove the target's reasons for living even if they've never shown any tendency to do so. The factors that lead a person to attempt against their own life are not fully understood, but it is known that it may be the result of a mental condition. Thus, by emulating such diseases in the mind of a healthy person, the attack with extremely negative thoughts is initiated. It simulates the result of social stress and leads to a guilt feeling that will never dissipate, followed by excruciating mental pain that would only be relieved if the brain was turned off and stopped processing external and internal signals.

## 5.11.4 - Torture content

Generally, OPS and MKTECH operators will send content with words that make sense to the victim. They may use repetitive music or painful noises. But as we've already seen, words when interpreted are able to access multiple areas of the cortex at once, so they are more effective at stressing and distracting the target. Consequently, operators start talking non-stop in a kind of babble about the observed and private life of the target, and highlight the key word or words that catch the victim's attention to take them away from the daily activity in which they're currently engaged in.

They fill the victim's day with this type of content and usually use stimuli that triggers anxiety, disgust and frustration. That is, they use negative stimuli that cause the elimination of the perception of external signals, altering internal factors of cognition and inducing their perverse presence to perceptual constancy, to the perception of the object and its properties as something constant, as one becomes familiar with it.

The desire to be overcome by ordinary feelings, guilt and negative self-criticism makes the target fall into a decadent condition in which they create the illusion that no one is capable of overcoming such challenges. Victims accept that they are at the mercy of events, and sustain an empty perception, a lack of expectations for the future, which decreases their self-confidence and leads to extreme subjectivity. In other words, these attacks create a range of mental obstacles, preventing clear reasoning.

There is a lack of interest in all everyday activities as Targeted Individuals no longer feel worthy of finishing tasks or of being capable of making something happen. Willpower is completely undermined due to the false sense of "superpower" coming from the operators if compared to the impossibility of the target to modify the scenario created and imposed via MKTECH and psychotronic torture. The outlook is overly pessimistic. There is difficulty in concentrating and, in the end, the targets can do nothing to change the current situation. Their completely limited resources eventually lead them into a spiral of total failure, starvation, suicide or defeat as they wait for death. The attacks are designed in a way so that the target projects a degraded self-image in their mind. They feel they are of no use in this world. Constitutional rights do not apply to them and, as a disgusting being, they must be subjugated by this weapon and accept their fate.

Each element of the mind control technology converges towards the goal of generating heightened cognitive dissonance, the loss of the targets' confidence in themselves, artificial modification of self-image and transformation into a neurotic person, suffering from depression and who was electronically induced by modern mind-control weapons. Operators tend to create distressing situations during the events in a deliberately hysterical atmosphere as targets face horrible death threats against themselves and their family.

The use of visual aids — such as the inclusion of grotesque montages (scenes) of their loved ones being killed during D2K dreams that convince the brain that those things actually happened — usually makes the target take the threats very seriously. Themes that are repeated over and over again create emotional triggers or repressed memory triggering, increasing the possibility of the victim becoming a remote killer, as has happened with many others in the past.

Torturers create a suffocating, corrosive atmosphere in which all the surrounding air appears to be toxic and contaminated by the negative waves created by the system (telemetric EEG). In addition, the atmosphere of fights and profanity modifies the real world, inducing the target to be careless about their own safety. The victim becomes extremely hostile. Negative emotional outbursts and their behavioral reactions harm everyone around them, including themselves. In this way, the individual's motivational cycle is completely broken and offensive aggression is triggered. As this aggression can never be directed at the remote torturers, the target's health gradually declines. Those who are unaware of the struggle going on inside the victim's mind may notice this change in behavior.

Making the situation more intensely negative than it actually appears to be is another recurring strategy built into attacks. It leads to a phase of dissatisfaction, raising a set of negative (low self-esteem) feelings and erupting into episodes of aggression, and later recruiting automatic processes called adjustment mechanisms, which may worsen or improve the target's overall condition. It's a kind of revelry capable of conjuring up everyone's worst inner demons. It sends the target to an immense darkness composed of the satisfaction from the accomplishment of evil deeds, leading to the sacrifices of the flesh and the spirit and with severe consequences for all involved.

## XIII - Tips for the Targeted Individual — Psychotronic Warfare Tactics

If you're constantly under attack from psychotronic weapons in the same way as reported in this book, you've undoubtedly been chosen to be a target of modern experiments. You're likely to experience the situations described here, some more pronounced than others; in any case, buy and have a safety net installed on the windows of your home immediately. Nets that prevent children from leaning on parapet are very useful, as they also keep you from getting hurt when looking for aggressors on apartments above or below yours with the true certainty that you'll be able to find them this time and put an end to this endless nightmare.

The chances of an accidental fall or an episode leading to suicide are high, as both behaviors are constantly encouraged by operators and by reconfigured mental feedback signals that lead to extremely negative brain electrical and physiological configuration. It's similar to the effects of a "bad trip": you can run blinded by hate and end up falling. So, think about it, as the urge to look everywhere for those annoying voices that seem to come from apartments above or below yours is very strong.

## 5.11.5 - Altering raw brain waves to modify the victim's behavior

MKTECH has the ability to create and register moments in a person's life by recording the target's neural parameters when they go through difficult times and replicating the exact same configuration, thus stimulating the brain to return to the configurations with those same parameters during the negative event. These parameters are captured in the electrical patterns of the brain waves by EEG re-radiation. It's staggering: the bad feeling comes back suddenly without the previous causing factor being present. Having to deal with the monitoring of their brain configuration that represents the feelings already mapped every second of their life is a routine for every Targeted Individual. Even the program's own AI does this constantly. It detects a positive configuration in the victim's general state — known as a positive state of mind —, and automatically attempts to revert to a negative neurophysiological state.

## 5.11.6 - Consequences of persecution, torture and long-term exposure to the technology

Disorders according to the ICD-10 classification of mental and behavioral disorders are as follows:

* **F60.0 - Paranoid personality disorder** — Mistrust and a pervasive tendency to distort experiences by misinterpreting them as hostile and dismissive.

* **F60.1 - Schizoid personality disorder** — Preference for fantasy and introspection. The target starts a process of introspection. They must be aware of responses or attacks coming from MKTECH operators.

* **F60.5 - Anankastic personality disorder** — Insistent and unwelcome thoughts or impulses.

* **F60.6 - Anxious [avoidant] personality disorder** - Persistent and pervasive feelings of tension and apprehension.

### F22.0 - Delusional disorder

* Persecutory and clear and persistent auditory delusions;
* Induced delusional disorder;
* Shared by more than 2 people, usually of a persecutory or grandiose nature. Induced hallucinations;
* They even lead to other serious disorders (e.g. social phobias).

### F41.1 - Generalized anxiety disorder

* Nervousness, trembling, muscular tension, sweating, lightheadedness, palpitations, dizziness;
* Mixed anxiety and depression disorder.

### F62.0 - Enduring personality change after catastrophic experience

* Change after a catastrophic experience (torture and threats to life, for example) so extreme that profoundly influences the personality.

Under intense or prolonged psychological distress, intense physiological reactions to internal and external signals such as reckless or angry behavior and sudden outbursts of anger in the form of aggression generate visible negative changes in several areas of cognition — artificial events that will be incorporated into their previous cognitive layout.

**Hyperprosexia** - caused by excessive attention on SYNTELE, it generates a great focus on unnecessary details that fully hold the target's attention, most of the time controlling the microwave voice stimuli that automatically generate hyperactivity of attention as the target is torn between V2K (mandatory) stimuli and everyday activities. This causes the target to go get a coffee and halfway through it focuses on the content coming from the synthetic telepathy. Once the vocalized response is emitted to operators, the act of going to get coffee will be put on hold. By interrupting this action, the people around the victim will probably think that the target is a "forgetful" person who may be showing signs of schizophrenia.

**Hypoprosexia** - is decreased attention or severe impairment of attention in all its aspects. It is observed in infectious states, alcoholic intoxication, toxic psychoses, schizophrenia and depression. Distraction, similar to **Aprosexia** (dementia).

* Delusions (mental confusion);
* Hallucinations, especially auditory ones;
* Persecutory delusion: belief that someone is chasing and watching you while planning to hurt you. During this phase, the individual exhibits behavioral changes, high levels of anxiety and impulses of aggression;
* Lack of mental fitness: lack of motivation, apathy, social isolation. The thought fades away and the person shows total emotional indifference;

It is worth mentioning that behind all the torture and murder, there is a perverse worldwide experiment going on. The continuous use of electronic torture in practically all cases shows that experiments with human guinea pigs continue to this day, but this time remotely and always with some background — a narrative as an excuse to use the technology against people — from land disputes, psychotronic torture until the

## CHAPTER 5.11 - ADVANCED TORTURE TECHNIQUES USING MKTECH

person is defeated and decides to leave the property, revenge (the use of MKTECH to inflict pain on your enemy), work or political disputes, scientific study to even just macabre fun.

After all, what do these experiments (hidden in attacks and torture) want? The experiments seek to understand several long-term aspects of the complex interaction of the brain and its reactions to exposure to psychotronic weapons, thus obtaining results from several researched areas. These results range from the improvement and refinement of all aspects of the technology, the complete understanding of reflex and autonomic organic reactions, to adaptations, leading to the ultimate weapon of total control of the human mind and its complete obliteration at the push of a button.

* Reaction to the total lack of privacy in every aspect of the target's private life within their own house;
* Reaction to exposure to prolonged sleep deprivation;
* 24-hour reaction to prolonged torture with obnoxious voices sent via V2K;
* Reaction to having thoughts monitored and going public. End of cognitive privacy;
* Reaction to altered dreams every night that bring harmful consequences for the body, memory and mental sanity of the target;
* Remote polygraph reaction;
* End of primary thinking;
* Creation of Winter Soldiers and "programmed" remote killers;
* Series of experiments with D2K;
* Behavioral modification

# CHAPTER 5.12

## DETAILED TECHNIQUES FOR EXTRACTING INFORMATION FROM THE HUMAN MEMORY

Imagine being able to access all of an individual's visual memories that were acquired during their lifetime or checking the context in which those experiences took place. As a bonus, imagine capturing the consequences of those memories in shaping that person's character. In other words, think of the sequence of events that shaped and sculpted this individual's personality, all the traits that make them unique. The preservation of memory translates itself into the existence of the "self", creating the temporal feeling of past, present and future.

It is not yet possible to get all of a person's memories — a sort of dump[54] from a computer captured in a "live" forensic data analysis. However, there is another way of evoking and capturing memories little by little, in a sadistic, methodical, creeping and macabre way. This is in fact the only way to capture human memories, as there is no physical location where they are easily accessible, such as RAM memory, a flash drive or the data stored on our computers' hard drives. So, it is up to the invaders to direct the thoughts within the target's mind by raising questions using SYNTELE, triggering an automatic and unconscious process of associative relation in a continuous path, a sequence of mental facts that may or may not make sense. In this process, the brain tries to receive its reward for quickly locating information: a good dose of dopamine as social gratification. After all, who doesn't feel good about taking a test and knowing all the answers after a long study process? Probably everyone feels this momentary pleasure, right?

---

[54] DUMP in the context of RAM analysis using specific tools is the state of the frozen data, where it is copied and its contents analyzed.

The brain's evolutionary shortcuts designed to access memories in fractions of a second emerged to ensure our survival, to immediately recognize danger and to respond effectively to it. For example, when our deduction process makes connection with already known elements in order to reach a certain conclusion, or details of processes that our biological computer handles with mastery, executive control, ability to hold multiple pieces of information in the mind at the same time and use only a few of them to adjust the thinking, etc. All complex mental resources that we use without realizing that they make us what we are — humans — are the same processes by which the brain "betrays" us when clandestinely accessed through the intracranial microwave voice, causing it to activate exactly the same resources that your "self" takes advantage of when it asks the brain for memories recruited by the internal voice, which, in turn, requests support from visual images as we discussed in the opening chapters in Volume 1.

As thought unites ideas, creating the mental link and associating memories, feelings, emotional states and vocalized thoughts, everything is collected, amplified and stolen by the operators of the MKTECH technology. It's simple: just stimulate the brain to work. In this way, the meaning of the communication process between sender and receiver is decoded using the message or set of information transmitted. This code — the set of signals by which the message is sent (the language, for example)—, if it makes sense to the brain, will decode the message in an autonomous process and, during the process and soon after, it will generate a series of internal and external reactions. The internal processes that link the subject with the associated visual and vocalized memories are the object of interest to operators. External behavioral reactions serve as a confirmation bias on the veracity or knowledge of the target about the subject sent by microwaves, as well as internal reactions, activation of visceral neurons (the famous "butterflies in the stomach"), among others, as we saw in the chapter on the Remote Polygraph in Volume 1.

The channel is the vehicle through which the message is sent. The act of communicating oneself is carried out by someone or something and its purpose is the transmission of a message that decodes another. So, the theft of the target's thoughts and memories is one of the most harmful consequences of the technology. All at once it manages to completely violate the cognitive privacy of human beings and all democratic precepts of law, in addition to diminishing their intellectual property. When a target is unaware that they're connected to the MKTECH system, their private cognitive information and memories are transparently stolen. The TI will probably never know that their thoughts were stolen and sent to servers and databases prepared to store this kind of information. In addition to becoming a scientific experiment, the victim of electronic torture will have all their thoughts systematically stolen in the midst of a whirlwind of psychotronic weapons attacks, some disguised as torture, but which serve to trigger memories of diverse interests. For example, information theft, intellectual property stolen directly from the source (mind hackers).

Now that we're familiar with the technology and how it works, we're going to understand what techniques are used to make it possible to steal information directly from the brain and how they use this information to constrain, undermine and torture subjects.

Generally, operators of the technology use the unique process of human mental memory retrieval to get the information they need, details of some fact, or to confirm gaps in stories that are doubtful. They work together with other modules, mainly the Remote Polygraph, constantly analyzing whether the answer, the information extracted in the face of questioning, is true or not. The shortcuts the brain uses to deal with cognitive processes are used to store and retrieve memory. They use these details against targets, taking advantage of the methods that facilitate connections and speed up mental processes to steal memory straight from the victims' mind! These shortcuts that the brain uses as a resource to access memory in the fastest way work in MKULTRA's favor.

It takes a Herculean mental effort to reverse the process of the brain's normal cortical mechanics as our brain is used to being rewarded for the effort of successfully solving a problem, remembering something quickly or completing a challenge in a short space of time, for example. This reward is then present in the form of hormones that cause well-being, tranquility or an abstract feeling of joy and satisfaction. When we fight for the brain not to work, we end up asking ourselves, or the mind asks us: *"Why can't I function normally? What happened to my memories? Why can't I use mental shortcuts like I've done all my life?"* And the "self" answers: *"This is happening because they're listening to everything you process, along with the product of that process—the thought itself".*

Metaphors aside, a particular art emerges. Between mistakes and successes, the target internally develops the art of not accessing memories, of not thinking, even with external stimuli using the process of decoding words and meanings, activating reactions inherent to socially advanced beings. Thus, there is a new paradigm for the person who observes the genesis of these issues, as one tries to work out some solid explanation. On the other hand, in terms of defense and complete analysis of the scenario, this weapon has very serious consequences for the social model we live in today.

The mental effort to block out the information that your brain's always been trained to access as quickly as possible causes tremendous pain and mental fatigue. Our brain insists on finding the requested information as the mind is not capable of internally hiding information or refusing to access memories. Consequently, operators use the brain's natural processes against itself, generating contradictions and the paradox of thinking before thinking, which ends up restricting the range of thoughts in memory.

In this chapter we're going to see the techniques developed by those involved in the torture and information theft scheme and how they work to access memories, use them to constrain and violate the space and privacy, and also how they use our own mental processes against ourselves.

## 5.12.1 - What is memory and how does it work? – part 2

For those who don't remember, the first part concerning memory was introduced in the chapter on remote dreams (D2K), volume 1, chapter 2.5. I decided to break it down for a better understanding of the causes and consequences of manipulating memories during the sleep phase. At this point, it is worth recalling some concepts to then learn some others that we haven't seen yet.

Memories are encoded by neurons, stored in neural networks and evoked by them and other memories. They're modulated by emotions, level of consciousness and moods. Thus, your emotional state, mood and stress level cause the memory to be recorded and later accessed. Memory is a set of processes that allows manipulating and understanding the world, taking into account the current context and individual experiences which recreates this world through imaginative actions. It is characterized by the ability to acquire, conserve and evoke information through neurobiological mechanisms and social interaction.

There are different categories of memory that will show us that memorizing dates, names and places is not the same thing as learning to ride a bike or drive a car. What we learn and remember is processed by different areas of the brain. Human memory can be divided into three main parts, each with different characteristics.

* **Short-Term Memory/Temporary or Working Memory** - we use it on brief occasions, such as when we think of a phone number just long enough to dial it. In addition to its low information retention capacity — a few seconds or at most a few minutes — working memory is responsible for managing our reality. It doesn't produce long-lasting files or leave any biochemical traces. It is a determining factor for the performance of executive functions.

* **Sensory Memory** - sensory memories are considered a kind of information warehouse derived from different senses that extend the duration of the stimulus. It is processed information that reaches the hippocampus and adjacent areas and is considered a large-capacity repository in which the information stored is an isomorphic representation of reality purely physical in character and not categorized (even if the object has not yet been recognized).

* **Operating Memories** – the product of sensory memory. In computational terms, it would be like a buffer, where temporary raw information would be stored and then processed in a priority queue.

* **Permanent or Long-Term Memory** – is responsible for storing all of a person's knowledge. The access time for information retrieval compared to other types of memory is much longer. It can take days, weeks or even years.

Memory consolidation is the process of storing new information. Immediate memory is susceptible to interference from other memories or stress, which requires 3 to 8 hours — metabolic processes in the hippocampus — for its final consolidation. Even after days, months or years, we can still remember an acquired memory. We're even capable of recalling memories from our childhood.

The functioning of memory is not fully understood — how and why some are deleted and others aren't, for example. It is known, however, that the brain depends on strategies to access certain memories. Forgetting is usually based on poor memory access strategies, not on its unavailability. Nevertheless, the most important element for the consolidation of data in permanent memory is attention.

There are endless memorization techniques for different types of situations. Some of these techniques are efficient for a couple of people and not for others; it depends on what you want to remember and the way you're trying to do it. Memory association strategies form neural networks modulated by emotions. Pleasant stimuli in a healthy mind are stored first while trivial memories are repressed.

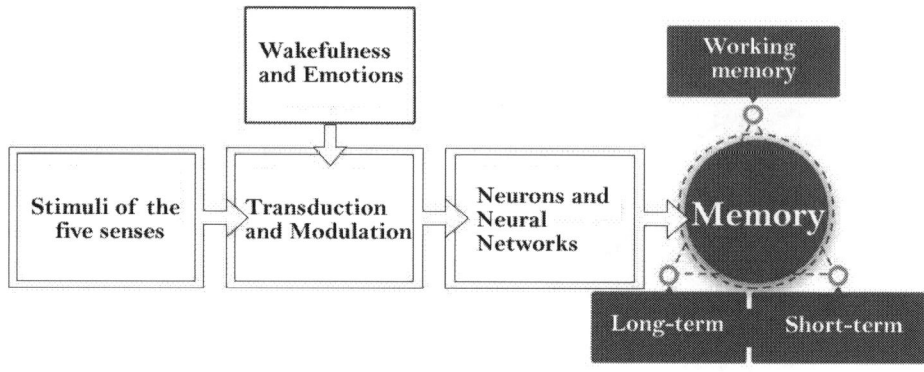

**Figure 5.3** Natural process of memory synthesis in terms of acquisition and fixation.

The process of acquiring memory basically takes place in this memory formation. These are the main modulating factors:

* **Attention** - directly responsible for the quality of the storage.

* **Motivation** - influence on attention performance.

* **Anxiety level** - also influences the storage and attention level and prioritization of memories.

* **Conscious memory** - remembrance of facts, places, times.

* **Unconscious memory** - automatic memory (driving a car, cycling). Content is made up of declarative and nondeclarative working memory.

* **Declarative memory** - recollections of everyday facts and events are what we usually call "memory" in our daily use. Change in the perception of the object — the perception of the past with that of the present; corresponding ideas or emotions are elicited. The hippocampus has a wide influence on long-term memory and spatial navigation (limbic systems), and is one of the places where experiences are transformed into memories (in the cerebral amygdala) in events of high emotional content.

* **Declarative content** - influence on semantic and episodic memories.

* **Semantic memory** - general knowledge, recording and retention of content according to its meaning. It is sometimes called generic memory, and refers to the memory of meanings, understanding, and all forms of concept-based knowledge. Unlike episodic memory, it is not personal, as it is shared by everyone who speaks the same language. Both are located in different parts of the brain — one can be affected while the other cannot.

* **Episodic memory** - it is related to events that we participated in (or witnessed) in the past. It is the memory of autobiographical events — time, place, associated emotions, how, who, what, when and where are the sources of knowledge — which can be explicitly

declared or invoked. It is the collection of past personal experiences that took place at a particular time and place. An example of episodic memory is when an individual remembers their 6th birthday party.

* **Procedural memory** - motor learning, conditioning, skills, habits and behaviors (playing the piano, kicking a ball, tying a shoelace). It is considered implicit memory, as it results in a more automatic and unconscious effort.

* **Nondeclarative memory** - motor memory, simple conditioning.

Access to memory is usually reflected in body gestures, which may cease after reasoning has followed. It enables the individual to refer to experiences and compare them with current ones, projecting them in prospects for the future, as past learnings can lead to new behaviors. Besides, memory leads to approximation and detachment between reality and ourselves.

## 5.12.2 - Information extraction via Synthetic Electronic Telepathy and Intracranial Voice

Imagine that your brain works similarly to your home computer that stores information only you know. It is a kind of complex computer that stores everything regarding your personality, preferences, life experiences, education, places, good or bad times, happy or sad times, the environment in which you grew up, personal stories, people who went through your life and made an impression on you, family ties, secrets, in short, everything that makes you, you. After every stimulus and in each situation, all data that make up your identity are required to interact with this new information using perception to filter the data and analyze them according to your personality. For this "self" to remember to remember who they are, they must always access conscious and unconscious mechanisms to simultaneously maintain the cohesion of their "spirit" and, at the same time, deal with a gigantic amount of information that is captured at all times. Among these processes are your memories.

Now imagine that your internal voice, the voice of the mind, is one of the mechanisms capable of requesting information from your brain — like a mouse click or the Enter key on your keyboard, for example. Think about the touchscreen on your cell phone, or the data stored in a hard drive or in a flash memory. The V2K (Intracranial Voice) is able to stimulate the request of memories by mimicking and activating the voice of thought or by triggering the information search engine during the decoding process of the received message. The operator just needs to speak about a certain topic, or even use a single word, and you'll automatically search for that desired information in your biological database. Your response to this stimulus may occur via silent thought supported by mental images and vice versa. The brain will work to access that memory, as it was evolutionarily trained to do so even before the species evolved into what we are today.

So one can say that reversing this process is an arduous task for any human being. Preventing the brain from accessing memories requested by microwave voice stimuli is in fact extremely strenuous. Evocation is a process that takes place in millionths of a second in which information is located and accessed due to precise brain storage. In this process of accessing vocalized memory, which uses language as a vector, the naming process is responsible for building the memory of an object in the form of words. The most important activity linked to language is performed, as it uses memories involved in phonological and semantic retrieval accessed in a language file system — verbal memory.

The type of evocation with direct access to stored memory fragments, and to a general idea, trigger the process of concatenation of memories along with other associated processes such as naming, mental images and emotions. Visual memory, for example, uses this process. It helps the brain to activate the identification of the object by its semantic representation, visual thoughts or abstract situations recorded in memory, which are always accompanied by the naming process that is easily stimulated by V2K auditory pathways. That's why it is so hard to keep a secret in our minds. The vocalization of thought will be activated in the form of a descriptive narrative of a particular memory, past event or it will refer to a broader scenario on a situation that occurred in the past — a topic of interest to operators, of course.

This process is automatic, making the job of the memory retrieval operators easier, as the brain waves are already connected to the technology. Even if the target internally thinks, *"I won't remember this, I won't remember this!"* the process ends up taking over and revealing thoughts that travel through the system.

The effort to not allow access to memory and trigger thoughts, with due proportion, would be the equivalent of holding your breath under water for as long as possible, but without the consequences of drowning. At a given moment, the brain starts sending desperate signals for the individual to catch their breath, that is, to supply the cells with their primary fuel. But the target defies all signs and stays in apnea in a single mental effort until they either pass out or return to the surface and obey the signals.

Therefore, when trying to access such a memory, the mental strength needed to avoid naming it and the silent thought that represents it is almost as stressful as holding your breath. The brain sends several electrical signals telling it needs to connect the neural networks as well as it sends warning signals for the individual to supply the cells with oxygen. In most cases, after a huge constant mental effort not to let the flow of information run free, the target gives up. They stop fighting natural processes. A relief similar to breathing again can be felt immediately. But later the victim remembers that they will have to deal with the consequences of letting that information be accessed as it can now be amplified and stolen.

Internal struggles, an endless battle to preserve cognitive freedom in the face of the enemy's advance, takes place inside the mind of a human being. In this chapter, we're dealing with high-level data and strategy, in which new precepts are found as a result of experiences in neural "combat", anticipating for society and for you, reader, what the present already holds for us.

Obviously, the feeling caused by all this information-stealing mind game affects each individual differently. Letting a memory be accessed can be followed by a feeling of violation, shame and remorse, particularly when Electronic Telepathy sends feedback from that personal, intimate memory with remarks from the audience that watches that memory unfold on a TV screen and through sound systems.

Despite similarities, human memory doesn't work as a computer program. Memories aren't mere mechanical recollections of facts, or a set of bits that together represent a virtual file. When evoked, they make judgments, and present validation of events, acts and opinions about people. Sometimes they can lead to internal conflicts and an immediate change in behavior that affects the entire organism.

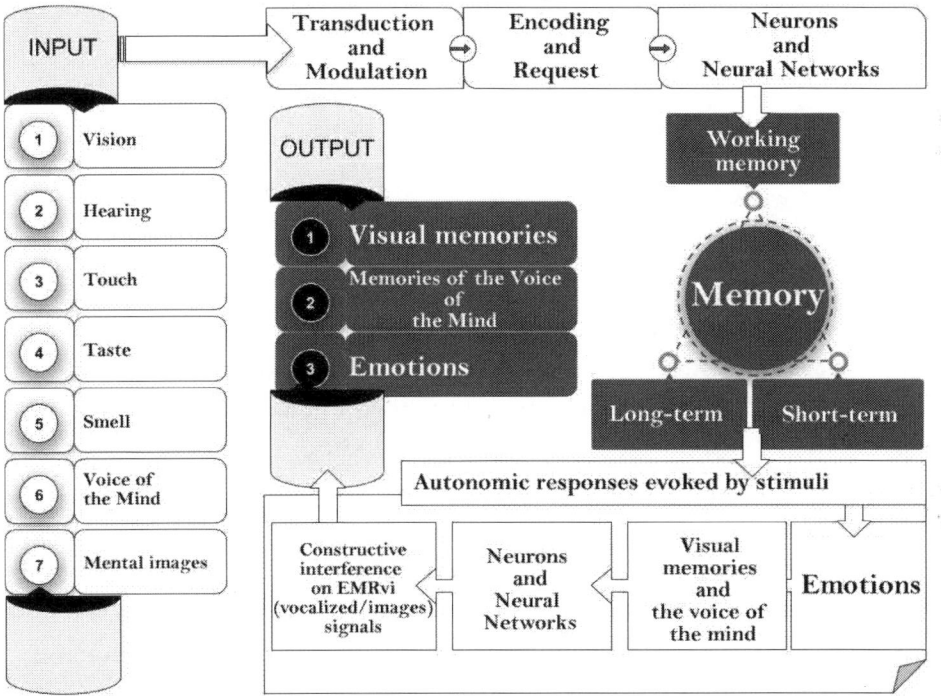

**Figure 5.4** Representation of the input and output path of stimuli. It starts with stimuli that come from the five human senses (touch, hearing, vision, taste and smell). In addition to them, we have the vocalization of thought (voice of the mind) generated by the Synthetic Electronic Telepathy (SYNTELE) and the mental images generated by the Synthetic Electronic Dream (D2K).

## 5.12.3 - Direct interference in memory

One of the most dangerous aspects of these weapons is the ability to interfere with memory — in stored memories and in the process of immediate storage, as well as modifying memories that may be acquired in the future — using a false projection already instilled in the mind, modulated by emotions and subjective filters that would never have been present at the time of acquisition of that particular event when performed. If a negative scenario was created through systematic repetition, contextualization and modification of memories under constant stress, and was developed over a narrative largely executed with words, images (and it is related in some way to a future event), it will totally misrepresent the experience as it is acquired and held in the brain. These modifications permeate the unconscious mind and ravage our immediate perception of stimuli and projections of future events that we will inevitably encounter. Therefore, we won't be able to feel safe again, let alone have positive experiences. This significantly affects new memories going through the system.

Information can be added by altering a pre-stored memory — it is called retroactive interference —, and shows that operators can locate a certain memory and change the acquired experience in a qualitative or quantitative way, transforming a bad memory into a good one, adding or removing details, connecting with other experiences and completely modifying the old state of the previous memory. In other words, the way we perceive something as "good", "bad" or considered memorable is changed and is completely artificially drained, similar to a bygone era when neuroscience was in its infancy and we had no advanced knowledge on the matter.

In the mid-1880s, French philosopher **Émile Durkheim** said,

*"There is nothing legible about the phenomenon as long as there is a mental memory, that is, past representations persist in the quality of representation and as long as the memory consists of a few words, not of a creation or original work, but simply a new emergence of the lucidity of conscience, the adaptation of the mind, when confronting the same memory with the perspective changed by time."*

Time and maturity that shape an ever-changing worldview of an individual are also able to perform this task naturally, thus adapting our feelings to certain memories. However, when induced by psychotronic torture with medium and long-term mind control techniques, these effects can be felt more severely with immediate consequences in the behavior of the victim. There is a possibility that their behavior and worldview will never return to the original state one step before being artificially modified.

This strategy executed at all times is fundamental and happens with the working memory. As a matter of fact, the memory of the present is already assimilated in a completely deformed perspective, caused by the virtual reality in which the target is inserted, which in turn comes from the electronic prison they created for the victim's mind. Future projections are also affected by this type of event, that is, the interference of psychotronic weapons related to memory formation, mainly based on the ongoing narrative sent by SYNTELE. Bringing information closer to the reality you know is a strategy to improve the memorization process. So, operators modify the perception of reality and instill a memory that eventually becomes part of that reality.

## 5.12.4 - Evoking memories

There are many things that we don't want to access or remember — we actually want to forget. This is an active process; our brain is able to remember memories that we don't want to evoke and makes an unconscious effort to do so. It selects bad memories and avoids evoking them by slowly erasing their synaptic connections. This fact is exploited extensively by MKTECH operators.

As they can hear and see thoughts, they make the victim easily remember situations that they wouldn't have remembered otherwise, and use this fact for torture, theft or invasion of privacy, generating a kind of professional bullying. And because we have memories that have been dulled over time, they become more and more vivid due to this method (systematic recall) that causes behavioral changes.

The natural logic is reversed once more. It's not you and your mind that decide what is important and what isn't; it's now the job of the technology operators who have full access to your brain to do so. We all have experiences that we'd like to forget, right? Well, this is no longer possible with the use of this technology.

Memories are stored in synapses in the brain; its neurons are plastic and are organized as a network. Information with a central idea of the object to be accessed triggers a process of correlation of this information in which short and long-term memory systems interconnected continuously transfer information from one to the other. Any subject that arises via auditory, visual or microwave voice information along with visual images within dreams, or captured via SYNTELE in a waking state, works in exactly the same way, triggering and accessing a fragment that calls another fragment until it forms a complete memory that becomes useful and readable information with context. In the case of us humans, the fastest way to evoke a memory is through language. The faculty of expressing your thoughts involves signs and sounds to convey ideas, sensations and experiences. There are spoken/written languages and sign language and gestures that operate on a complex system of sounds and signs using a code we call language.

There is, indeed, a powerful tool — the mnemonic potential [55] — within this huge communication system that has the ability to access memories and make someone remember a specific fact. It is used for other aspects of language and explores other intelligence inputs (such as logical-mathematical intelligence and personal intelligence), syntax and phonology. Once you have access to all the functions of the Targeted Individual's brain (electronically), it is easy to extract any type of

---

[55] A mnemonic is a memory aid for something. They're typically verbal and used to memorize lists or formulas. Besides, mnemonics are based on simple ways of memorizing larger constructions, based upon the principle that it is easier for the human mind to memorize data when these are associated with personal, spatial or other relatively important data, rather than data organized in a non-suggestive (for the individual) or non-apparent way. However, these sequences have to make some sense, or they'll be equally difficult to memorize.

information from them, as you take advantage of these processes of communication and association.

Before these weapons became available to the public and were more accessible to different groups, you were able to decide whether you wanted to share certain information with others (in its entirety or only part of it), concealing or exposing everything in the form of spoken or written language. In other words, it was possible to keep your thoughts and opinions to yourself. This usually happens in conversations between people: there is a choice, the freedom to withhold information and keep it inside the mind, or trigger one more sentence, generating another conversation around some common subject that is based on the connection of memories coming from a central idea. Unfortunately, this no longer exists. The whole process — a process that has always seemed clear to humans — became all too noticeable when the real possibility of amplifying, listening and making this information public emerged.

Now it is possible to stimulate the target's memories via SYNTELE using a fragment of a memory until the information has been completely captured, in addition to being able to interpret what that information means to the individual, how it affected them then and now. Depending on the subject at hand, one can discover how this information relates to the cognitive and social background that makes up the individual's personality. This content that generated the response to the stimulus corresponds to the information that is recorded in their long-term memories. But this same technique is used to capture working and short-term memories as well.

Nevertheless, this may raise some questions. What if the target is a born liar and starts telling untruths and making up stories? If the victim performs the same technique to lie using their vocalized thoughts, is it feasible to hide a memory?

We don't usually hear or see the thoughts that go through a person's mind, nor are we able to internally observe the neural structure and automatic subsystems that participate in the entire chain of internal events. We also don't see the processes that organize the reasoning and bring out the final result in the form of words orderly uttered, creating an audible interaction. We never know for sure whether the narrated event is true or not. So, we depend on several factors to detect a lie: coherence and cadence, the degree of intimacy with the person, physical signs and automatic mannerisms to access creative memory, or even through the perception that the person is making an unusual effort to explain a story that leaves no gaps or doubts. However, with the advent of the MKTECH technology and its ability to hear thoughts and follow external dialogue via EMRA - Electronic Mind Reading (auditory) — that compares how the narrative is developed internally —, everything changes once and for all.

The lie externalized in words is based on fictitious internal data, which are supported by reasoning and the choice of memories, blocking certain processes, accessing stories belonging to other people, mixing different facts from different memories, distorting some details, decreasing attributes, exchanging protagonists, or other artifices to build a story. Even if all this deception were used to support a narrative, the system would still be able to detect lies (or parts of them), making it impossible for even a born liar to deceive the technology.

Furthermore, we have to take into account the accuracy of other auxiliary systems such as the Remote Polygraph (chapter 3 of Volume 1) that indicates physiological changes to each topic discussed by the mind, and the infamous AI that actively works to compare data and check patterns, typifying the information and pointing out whether or not each event is: a creative story, fragments of the truth with lies or a completely true story. Thus, when observing the "story" being narrated externally with the surveillance of all the internal processes that created it, it is easy to discover if such memory is false or if it was really acquired by the target through experience.

This is yet another fixed sample or a set of constants consumed by the systems during the entire period of the attacks. Given the basic, superficial analysis that precedes them, this massively helps to verify the facts from the memories, and is done by studying the personality of the target. Environmental factors, psychological knowledge — in short, everything that contributes to the understanding of a personality after a few years —

is already more than established in the database and in some evil AI capable of behavioral prediction or reaction to a familiar memory with great precision.

Memories, linked to judgment and social behavior, are the most violated. Operators play with victims who will never reach intellectual maturity depending on the age at which this massive attack occurs, limiting themselves in a tiny world, their cognitive fields electronically constricted.

## 5.12.5 - Digging up memories

The Synthetic Electronic Telepathy (SYNTELE) is a powerful tool when used to gather intelligence data under any circumstances. As it focuses on the internal monologue, on the vocalization of thoughts, which is part of the reasoning and mechanics of accessing memories (in the human brain), the technology takes advantage of the neural inertia to capture as much information as possible about a particular target or even a prisoner who is being investigated or interrogated. It becomes an almost foolproof tool for getting facts right, even if the target is aware of its existence. This module can also be combined with the Remote Polygraph in order to filter each fragment of thought so that operators are able to establish whether this fragment is true or not. With this data in hand, they utilize MKTECH's unique ability to access memories that were buried and forgotten in the back of the target's mind until then.

And then unprecedented possibilities of creating powerful strategies to relive such memories and make them more and more vivid, using the capture and contrasting repetition of the method by which they were obtained and, later, the repetition of the fact that represents the memory in words (SYNTELE) and images (D2K), arise.

Repressed memories and traumatic experiences that were blocked and had few reinforced connections — dormant synapses —, slowly come back to life. As soon as operators find out about the traumatic event, they incessantly capture its central idea — on different occasions — and gradually revive it, extracting more and more information about the researched event such as places, dates and people that are linked to that memory. Thus, it becomes the object of the operators' search. Other sources will also be consulted: minds of people close to the target and data on the internet and social media. This brings new facts to light at every moment, causing internal anguish related to the curiosity of wanting to know how they discovered a fact so deep and private that the target might not even consider remembering it until the memory fully emerges. **The simple act of thinking about how they acquired that particular memory is confirmation that the memory exists.**

From then on, the event is systematically pointed out and relived in the form of images and sounds for years, and its effect will be verified after each attack. At this point, the target's mind is restrained in a location and date that had already been left behind, but which keeps pulling the target back to that moment, thus making it impossible for them to go on.

We're able to begin to understand why things happen the way they do when we're faced with a situation similar to the one narrated above. After all, nature is clever. It shows us in a simple way the consequences of being stuck in the past — inside our own minds and on purpose — to cause intense pain that prevents us from progressing and evolving. These weapons do just that, which is contrary to nature as far as evolution is concerned. They try to artificially reverse brain patterns by strengthening and making painful memories more and more vivid. Don't forget that these are targeted techniques, taking into account modern MKULTRA experiments.

Nonetheless, this isn't a linear process common to all humans; the results will vary from individual to individual. After all, we're very different from each other — in fact, we're quite unique individuals. Identical twins, for example, are very different mentally. However, all of us without exception function based on a universal cognitive mechanics that is already installed in our brain from birth with the purpose of irrationally or unconsciously manipulating certain processes related to the analysis and management of memories and their cascading effects throughout the organism.

Memories compose the individual's personality. For information storage to occur, there must be an experience that translates into perceptual, motor and affective (emotional) functioning. Association strategies are vital for long-term memorization as memory forms neural networks modulated by emotions. The affective content, as we know, can help the memory to be more effective than others that don't have a content linked to emotions.

Perhaps this is an evolutionary process, as memories linked to strong emotions such as fear, disgust — mainly primary emotions triggered by the amygdala in the hippocampus — left their marks on those who endure trauma or scenes involving explicit violence. Thus, they increase their chances of survival when faced with similar situations in the future that refer to the event that generated such remarkable memories.

This type of attack using Electronic Psychological Warfare tactics is able to slowly and surreptitiously alter the dynamics of the target's interaction with reality. It leads the victim to selectively choose thoughts and memories as memory exclusion plays an important role in human development. Traumas and other memories generated by post-traumatic stress, which are thought to be unconscious or conscious, should not be dug up and constantly relived by unknown invaders of thoughts. This should only occur if you seek help from a psychologist or therapist to work on your traumas.

The use of perverse tactics produces deterioration of faculties, leaving the episodic memory, which refers to the individual's personal experiences, increasingly weak and with serious consequences, including degradation or adaptation in the way of experiencing the sense of self. Phobias related to the access to the past arise, including the act of avoiding close friends, or people from their childhood, so as not to go through the process of recognizing episodic memories that operators will later use systematically with the intention of hurting the target and further repress their memories.

## 5.12.6 - Complete the sentence

A tactic widely used — especially in the first months of electronic harassment, but which goes on until the end of the experiments — is to induce the target's brain to fill gaps in sentences, making them only think internally about matters related to the central idea of the fact, which will be immediately remembered and narrated by the inner voice, or vocalized thought. Simply put: do you know when you're in the middle of a conversation with someone and a topic comes up — a subject that you're very knowledgeable about and you already know what the person is going to say —, so you feel an urge to stop them from completing the sentence before they can finish their reasoning, as you have the answer ready on the tip of your tongue? Well, this impulse to finish someone else's sentences, especially when the answer is pretty obvious, is a process very similar to auto-correction, in which the brain tries to fill in the gaps with the right word. "So, you were home and then...": as the target mentally finishes the thought and, consequently, the story, the event will be internally recalled and narrated. Besides, the events that followed the topic will flow smoothly, without hindrance.

In this way, operators will quickly notice whether or not the facts they want to know are registered in the target's mind. If the victim begins to create a fake series of events, the mind invaders will insistently try to relive that fact within the target's mind until the truth unfolds. SYNTELE is extremely powerful to hack the human mind, as it creates a trigger of memories accessed by words. Those who have had their brains invaded easily perceive such accesses, in addition to the complete inability to reverse these processes using only the strength of the mind.

This resource is indeed one of the most exploited by aggressors. It is through it that operators find out practically every detail about the target. They fill the brain with a vague idea and deliberately leave gaps so that the victim's mental "autocorrect" can take over. Preventing its performance requires intense mental effort, as it works regardless of the target's will. So, this is systematically performed to humiliate, to enhance the frustration of the endless act of stressing the victim and to get visual memories, the memory of imagination, followed by emotional and physiological responses, thus indicating some particular matter that causes discomfort and is worth being brought to the fore. In conclusion, the complete invasion of human privacy is the main goal of the technology.

## 5.12.7 - First impressions and associations

Our remembrances, our sets of coherently correlated memories, need something that triggers this process through connections to be evoked. This process is based on the principle of association. This principle, in turn, is most intense in the first impression of any event or unique stimulus that occurs naturally, at all times, especially if it is linked to vision. When you arrive at a new place — for example, during a trip — where you find stunning scenery like waterfalls in a high-spirited, positive atmosphere, alongside friendly people, the place and the trip are likely to become very memorable. So, whenever a recollection of that moment occurs, associated positive memories will surface again.

This cognitive process is more noticeable when we get to meet someone. The associative perception of the first encounter can change over time, adding uplifting or degrading characteristics to it. First impressions are very important to us. It is an evolutionary feature, as the life and safety of humans in the old days depended on an immediate identification of danger — a predator, or a human opponent from another tribe, for example. Today it is useful to readily recognize people, enabling us to make complex judgments with little information. MKULTRA uses this cognitive characteristic to obtain particular (associated) information that the individual has about other people. As the thought is not at all ashamed to express itself freely in the mind, the brain automatically retrieves information (whether negative or positive) from the person in question, and this is easily captured by operators through the individual's silent thoughts — also known as **social perception.**

One of the phenomena that most haunt victims who have their thoughts hacked and violated by SYNTELE is the way operators exploit social perception, a natural process by which we form impressions about others. When we meet or live with someone, we never have a disconnected or isolated perception of them; we always integrate observations into unified opinions, even if we need to make up or distort perceived characteristics for that.

Remember: our brain takes shortcuts; it categorizes to simplify the recognition of an individual. It's like the famous saying that goes, "The first impression is the last impression". Most of the time we don't notice that we judge people in a fraction of a second, as this used to be restricted to the individual's mind. As operators are aware of this feature, they start asking the target about certain people. Automatically, the characteristics associated with social perception are vocalized and visual memories linked to that person begin to be accessed by the brain. So, what the victim judges in fractions of a second about a particular individual is at once captured. The content may be pejorative, usually somewhat creative. Sometimes this process is stored in a completely involuntary way, caused by opinions expressed by others about the individual in question. Physical characteristics are the most common. Let's say the person has a long neck: in your memory her name may be associated with the nickname "Longneck Claire". If it is someone who has an eye different from the standard human anatomy, he may be called "Cockeyed Pete".

Behavioral characteristics that shaped the first contact with the person may be also unintentionally associated. Several details can be used, such as skin color, ethnic origin, religion, sexual orientation, smell, gestures, tastes or mannerisms. All this greatly determines the associations we make when it comes to people. In this way, the unconscious work of the brain to quickly categorize and classify things and people is exploited, thus obtaining the concept of the object made up of social stereotypes and moral archetypes. This includes prejudices that are already quite ingrained in society and that constantly challenge our rational side and prevent us from expressing ourselves in words or deeds, maintaining good social interaction most of the time.

Prejudgment actively works on cognitive mechanics. It is practically impossible to stop such thoughts as they travel through our innate mental system. Having these processes read and exposed greatly upsets the target, making them feel guilty for something they have no control over. It's a great strategy to use as torture content and to manipulate memories, preventing access to others.

A simple and very common example is asking about the target's partner. "Talk about Tracy", you receive via V2K. Without a trained mind, just by the fact that your brain decodes the (hypothetical) name

Tracy, this voice will immediately access the memory network that carries all the primary data concerning Tracy, i.e. visual images, intimate moments, grudges, happy situations and charged emotions, perhaps even about a hostile breakup. This association will arise easily, as the memory works in a network system that quickly correlates information to set the context. Therefore, once that name has been decoded by auditory systems, the first and immediate association will be a visual image of that person, followed quickly by memories from their most memorable previous experience. In this case, the relationship that ended badly and, depending on the current state of the target at the moment, more prominent emotions that made an impression on them will also be evoked. All this processing is heard by the Electronic Mind Reading (EMR) and causes a lot of embarrassment to the target, as some associations are linked to intense memories and emotions.

It is worth mentioning that the tactics revealed here were exposed after numerous attacks that always proceed in the same way. Our thoughts are "heard" 24 hours a day by many strangers, but we're not used to having the content of our thoughts exposed in this way. So, operators spend most of the time asking about people the target has lived with — past and present —, gathering as much information as possible, and then measuring which of these names causes more discomfort when mentioned. At this point, they start "playing" with the victim's mind, giving feedback to each name that is raised. If a name, or the association linked to that person, is embarrassing to the victim, laughter from operators will be sent in response.

## 5.12.8 - Disrupting the mental defense mechanism against associations

Remote EEG and Remote Polygraph help verify the victim's mental and physiological state. So, such devices know more deeply how a given name affects the target. With all these parameters mapped out, torturers will distress the Targeted Individual during all hours of the day, at work, at home, at school, at the gym, in intimate moments, and so on. These moments are of paramount importance for this type of attack, as they rely on the shift of the target's concentration that at the moment is focused on relatively light topics and don't require a great mental effort, thus forgetting — for a brief moment — the fight against the mind invaders. It happens, for example, during physical activities: a place where you find heavy weights, music, people doing the same thing as they search for better health. In other words, exercises that automatically entertain the target. In these situations, the mind lets its guard down and stops defending and protecting itself against the discovery of facts by memory linking method.

As soon as the favorable situation indicated by the AI takes place, operators launch a sneak attack. This attack contains the subject at hand and emerges abruptly within the target's mind through voices that immediately direct their attention to the subject matter. In that millisecond, when every mental process leaves the quiet configuration focused on the previous activity and slowly heads towards the new process, the target's brain processes the request in a flash before that subjective defense that consciously blocks memory access rears up. Generally, this tactic increases the chances of getting the desired information, mainly of commercial military value and related to personal memories.

These requests occur and are systematically captured together with all the people related to the target (family and friends). In this massive flow of silent data during the internal process of reflection, other memories are connected, including the fact that created the association — this social perception —, leading to more private disclosures and exposure of undue and potentially painful memories or that have intrinsic commercial value.

V2K, the voice of the invading mind, impersonates your inner voice as it reaches your auditory cortex. So, any information can be requested by unauthorized people, simply by making you think about things you don't want to think about, especially when it comes to private, family and school life. Projects, industrial secrets, work, the most varied preferences are also used. All this keeps the target tied to some past or present event which prevents them from moving on mentally.

## 5.12.9 - Electronic amnesia

The amnesic syndrome often occurs and generates changes in memory storage and evocation due to the fulfillment of all the main processes of capturing information from the environment, which are supplied with content enforced by operators and their psychotronic weapons. To make matters worse, attention and mood are never conducive to the acquisition and proper storage of memory, which begins to cause this syndrome so crucial for the operators responsible for creating memory lapses. In this way, electronic amnesia can be achieved through continuous acute stress and electronic narcosis.

The big move here is to insert information that is already being constantly presented to the target — over long years in a 24/7 cycle — into lapses of memory. They fill the gap with false information either through microwave voice or memory modification via D2K. Over time, the target has no way of knowing whether certain memories are real or not. Unfortunately, this is just one of the many problems concerning the subject matter. The induction and filling of artificial memory gaps are some of the numerous factors that make up the complex attack in which it is quite possible to turn a person into a remote killer.

So, the target's senses continue to be bombarded and hacked on a constant basis. Voices scream inside the mind, uttering all kinds of words that go beyond what is bearable with total mastery of the main brain functions, including dreams, which no longer belong to targets, but to operators of the system in a constant movement that generate artificial gaps. Finally, operators will purposely present specific content in order to secretively clog the memory with false images through D2K that will be available when waking up. A kind of electronic temporal amnesia is caused, in which the memory process of a certain event in the present is not computed by the brain no matter how important it may seem at the moment, especially because the concepts of priority queue are also affected. The ability to prioritize the attention or urgency when it comes to facts is corrupted, unless the stimulus is physically and extremely impactful to the point of altering sensory data and its importance in the priority queue.

The information will likely be degraded if you need to memorize school work or work-related subjects during this state. After all, memory is not consolidated with the help of artificially created dreams and every part of memorization is compromised by the electronic stimuli sent via SYNTELE, creating a kind of temporal amnesia, the same that was used by MK-ULTRA 1950 through systematic torture and deprivation of the senses, use of electric shock and chemical drugs. The same technique is applied now, but using only MKTECH, that is, without physical contact. This shows, in practice, the true process of evolution both in methodology and performance.

The target is deliberately prevented from acquiring new knowledge, resulting in compromised (declarative) memories along with partial or total transient global amnesia. This opens up the possibility of them having a blank space in memory, thus unlocking the chance for operators to reprogram memories and enter false data. If the intensity of the attacks decreases, the memory system returns to a relative normality, however, with irreversible consequences for the individual.

Temporal amnesia (total or partial), artificially inflicted either physically, using chemical agents or electronic equipment in a remote way, is a valuable process and sought after by MK-ULTRA since 1950. They looked for a stable process that could be implemented in humans in a linear and uniform way. That objective, however, was never fully achieved.

Torture with prolonged sensory deprivation and sleep deprivation is known to cause temporal amnesia. With the possibility to insert memories electronically, the process is almost complete and the original goals can now be achieved. This is in fact the gateway through which false memories are inserted, using a moment of emotional fragility and vulnerability — at the time when targets are most susceptible to causing harm to themselves and others.

Both conditions differ by the intensity of dysmnesia (impairment of memory). It would be the equivalent of amnesia disorders to global memory loss for a certain period of time — a condition achieved by traditional ways and modern means using psychotronic weapons.

Unfortunately, the individual's memory extracts all its substance from the environment in which they're immersed. With their mental sensors hacked, working memories are filled with irrelevant stimuli deliberately created by Organized Professional Torturers (OPT) to make any long-term memory consolidation difficult, as such memories are filled with every REM dream, taking advantage of a brain glitch that inserts long-term memories without having to go through the consolidation phase (chapter 2.5).

**Electronic emotional amnesia** unconsciously performs the reorganization of memories with the ability to decide if something is repressed or erased, that is, electronically reconfigured under constant repetition and under full control of the brain during sleep. Stimuli that evoke consciously or unconsciously repressed memories.

**Psychogenic amnesia** is a memory disorder most explored by the MKULTRA experiments. Many of the effects studied here are a result of the swarm attack culminating in early morning electronic narcosis due to psychogenic amnesia along with lacunar amnesia. The daily event itself becomes the predominant memory that gradually shapes the personality of the Winter Soldier as it reprograms people to become the operators' puppets or to be led into committing the scripted acts, but performed differently.

## 5.12.10 - Paramnesia

Another terrifying effect of this torture is its resemblance with paramnesia. Paramnesia is the phenomenon that records information before it happens in two different ways: registration and comparison, causing the "Déjà vu" and "Déjà Rêvé" phenomena. Both of them begin to reside in two domains: the common reality and the dream reality. This type of event occurs several times a day in the life of a target who is at an advanced level of mental degradation and deterioration of intellectual capacity. This is an important strategy for cognitive disorientation, as we saw in the chapter on D2K, Volume 1. Remember that everything we've seen throughout this book is employed simultaneously for this purpose, leading to failures in the recording of events.

Despite the simultaneous attacks occurring at different frequencies with different content and objectives, such as the continuous attack of directed-energy weapons (voltage-gated ion channels) that affects a certain electrical configuration of the brain that leaves the target uncomfortable — to say the least, the alteration of visual memories with "fictitious reality" created in a manipulated dream completes the picture of temporal illusion.

## 5.12.11 - Access restriction

Among the most frequent attacks, the most noticeable ones that alter memory in a more profound way are uttered by words to the point of modifying all cognitive processes related to the conception of information and the mentalization of its real object in the mind. We creatively reproduce reality using language. When we speak, we build a logical image of this information that comes close to reality. Reality is with us right here and now; language gives us the power to interpret it. By uttering words, we think in terms of language, just as the language an individual speaks influences their memories, their way of reasoning and their perception of reality. The target's auditory cortex is being overflowed with information so that this information can become a false reality designed to be embedded in words, thus creating false memories. So, the victim automatically incorporates them as an accepted reality and start working memories on those terms from that point on.

At this stage, the mind will be functioning as the attackers have planned, and without the victim even realizing it, as they are mentally and physically lost in the attacks and probably focused on the wrong thing; a situation that OPS constantly use to unplug the victim's memories "under the hood". Distraction, blocking and misattribution are mixed between events, strengthening unreal or false knowledge in mental models that cause damage to the development of the hippocampus and disturbances in memory. Mechanisms of memory prioritization that make up the senses together with the imaginary takes the subject out of the abstract using photographic memory recorded via artificial electronic dream, easily instilling the "fictitious reality" they are trying to convey, accompanied by continuous stress and extremely negative feelings. And, of course, the novelty in the perception and processing of these weapons puts the brain in a state of catharsis due to the enormous amount of new and continuous stimuli and the disbelief in their effects on the human mind.

At the time of evoking such memories, neurons convert biochemical signals into electrical signals so that our consciousness can interpret them as part of the real world. However, there are losses in these processes. Some memories may trigger a primary emotion (as a final result), such as: **joy, sadness, fear, disgust, anger and surprise.** If any of them are recruited, it will be automatically detected by the Remote Polygraph or EEG configuration. As we've seen in Volume 1, such devices have all emotions mapped out together with the full knowledge and control of when any of them arise.

Operators will then exploit this fact and accentuate the symptoms, routinely evoking a totally private memory, which clears the way for the perpetuation of torture and theft of information, given that our memories are what we really are, and the collection of them determines what we call personality. In all these cases — with a natural feeling at work in the mind —, fabricated memories will likely become long-term memories with high added value and high availability. This applies both when inserting or extracting a memory.

The constant persistence that acts on certain types of negative memories causes them to arise spontaneously at any time of the day, even if not directly evoked by electronic means — some discomfort may be subsequently felt throughout the body. Electronic surveillance will pick up on that sensation and immediately co-opt the data and work on its development to get the rest of the memory, discovering why it causes discomfort and to what extent the target is able to contain it by preventing its flow from continuing to reveal secrets. This whole process is an endless emotional journey for the target.

Gradually, problems related to communication, such as aphasia — a disorder that affects language —, are progressively accentuated both by the direct effect of the weapon on the brain and precisely by the lack of access to memories and self-imposed limitations in the use of words so that they do not trigger any kind of uncomfortable memory that will serve as ammunition for attacks. Over time, communication and vocabulary enrichment become more difficult. In addition to aphasia, other serious disorders related to the area of communication arise, along with memory and physical problems with regard to Broca's area (Brodmann area), which causes errors in language comprehension, and Wernicke's area, which is related to problems in speech and in the interaction and association of written and spoken language.

Our brain is constantly receiving external stimuli that are temporarily stored in memory, most of which are discarded by the brain. By revisiting this memory, we reinforce the connections of the synapses, thus, the memory becomes more firmly established. But be aware that this must be done within 24 hours — after that, the acquisition of knowledge is lost.

## 5.12.12 - Electronic hypermnesia and memory triggers

Memory recall depends on repetition and use so that you can experience what you're learning. In this regard, memories acquired in remote dreams also fall into this type of process. Distraction prevents memories from being recorded; it makes forgetting something a habit. The mental concept associated with language through experience can have a flexible representation, so it is possible to modify it by using techniques of manipulation of previously acquired memory.

One of the ways of inserting and fixing a memory so that it is available in the long term is through systematic repetition, a technique used by operators to instill false memories/realities and create a universe that only exists in words and "uploaded" images processed by the brain, besides, of course, the brain's interpretation of this data concerning everything that is computed. Confusion is created in the newly acquired memories due to the response to each deviation in the flow of reasoning.

Memory, if not exercised, has a tendency to weaken over time. With this technique, however, it is possible to prevent the false memory or a private memory that the target doesn't want to keep reliving from being forgotten, in addition to turning up the noise in the cortex. Transience, degradation of memory generated by evil tactics such as multitasking and distraction: several people download a lot of data in each channel connected to the brain via SYNTELE at the same time — a huge amount of memory codes by mnemonic processes that cause problems of attention and memory.

Another issue that seems opposite, but has the same source, is called **electronic hypermnesia**. It is characterized by a surplus of low-quality memory, something completely ordinary, which is incessantly formed given the volume of useless information reaching the target. It even impairs the formation of words used to communicate.

**Interruption** is another technique commonly used to interrupt any type of reasoning or action that is being performed (all types of tasks), further degrading common forgetfulness in ordinary acts, such as going to the kitchen to get a cup of coffee. Along the way, microwave voices that accompany the target always say something upsetting, so the victim ends up replying them internally (reaction to the aggression). Then they return from the kitchen without the desired cup of coffee. In the eyes of those who live with the TI, this attitude sounds as if the target has a serious mental illness.

There is indeed false content from real memories, but such content is distorted by recollections of the situation. This range of harmful effects on the brain and its health is also known as **allomnesia**. The target is extremely exposed to memory suggestibility and false memories inserted in interrogation techniques with an attribution bias, leading to misattributions and errors regarding the source of the memories.

Trait ascription bias (attribution bias) is the distortion of memory caused by the perspective of the owner of the memory. It modifies the memory from its storage to its evocation and it depends on the current physical emotional state and on the personality of the individual, as our perspective concerning memory is very relative. The natural possibility for our memory to be influenced by information learned after encoding the memory, under the influence of other people, validates the well-explored changes by psychotronic influence, including instilling memories that never existed and modifying them later. Overall, operators make a big mess, distorting facts from the past, mixing them and harassing targets with disturbing memories, whether real/semi-real or implanted using D2K/SYNTELE.

Let's remember that memory doesn't only affect mental processes. A simple memory can make a heart race and pump — the heart asks the brain; the brain triggers the memory and the memory speeds up the heart. Operators are capable of creating and recreating electronically induced emotional shocks that go through the body and organs, using well-established Electronic Psychological Warfare protocols that were developed for psy-ops missions.

Remorse, one of the most undesirable feelings, also shows how this tactic can be harmful. By capturing a particular experience that carry this "organic configuration" called remorse, the responsible agent, the memory of the fact and its subsequent consequences, will be intensely revived, becoming a chronic condition capable of keeping people eternally trapped in their own ordeal. A compensation for the fact based on another altruistic behavior will never be enough, because the main idea will never lose its appeal.

All the techniques of direct manipulation of human memory lead to cognitive chaos and mental configuration never before experienced. At the lowest level, this weapon is capable of altering the brain's automatic and heuristic processes, leading to non-rational decisions and slower choice strategies with little adaptability to the environment. Mind control experiments work with the level of interaction that is reflected in small moments which, when added together, end up modifying a larger universe. Upon reaching the end of their lives, targets will be overwhelmed by the consequences of the technology.

## 5.12.13 - Keeping secrets in the hacked mind under constant MKTECH attack

One of the main results expected from the MKULTRA experiments is related to the internal control of thoughts and conscious blocking of access to complete or partial memories. How would you extract secrets straight from the victims' thoughts? It's simple. As we've learned and seen on previous pages, a "sound" stimulus via V2K, which imitates the victim's internal voice, is enough to make them initiate the entire natural process of accessing memories, thus revealing the desired information. As a word correlates to a visual object, to an auditory memory, which are the phonemes and abstract feelings, it is very easy to stimulate the target's thinking to make them access a certain memory that can be about a valuable project, a personal story, a crime or any other matter.

These experiments are also used to train field agents, spies or professionals who work with confidential information. All those involved in the scheme are trying to find the best way to hide secrets as their thoughts are monitored. But it's actually very difficult to do so, even for a trained mind. For example, in a hypothetical situation where an agent falls into enemy hands or has his thoughts remotely monitored, he will be forced to use such techniques in order to hide secrets inside his own mind.

Targets are daily monitored within their reality, and their personal traits are mapped. When created, cognitive techniques of blocking and accessing the brain's own memory are systematically attacked to try and break down this mental protection and to restore the optimal mental setup desired by operators. The passage of time may be beneficial or harmful to their plans, as the target can, without anyone's help, mentally block multiple accesses to important memories or lessen their effects when vocalized by an intrusive microwave attack.

Let's now see how difficult it would be to try to block your brain from searching your own memories. What if I ask you not to think about a certain object? For example: "Don't think about a bicycle". The object won't be ignored, right? In fact, the image of a bicycle, that is, your mental representation of it, will automatically appear in your mind. And that's it. As soon as you mentalize the object, it will be accessed and stolen. This concept of thinking about not thinking using the object preceded by a denial doesn't work inside our minds.

## 5.12.14 - Mental defenses against mind reading and invasion

Our mental effort to protect our data is immense. Creating a kind of firewall using our willpower and diverting attention from the issues at hand which are transmitted by electromagnetic weapons is indeed a great challenge. We each create our own techniques based on our knowledge of the reality of the attacks and of the technology and how each element affects our mental health. The reaction is closely linked to the person's personality, their emotional state, social stability and many other factors that influence the defense system.

No technique that is created naturally with only cognitive processes will be efficient against increasingly devastating electronic attacks. As the invasion and mind reading evolves with the equipment, the technique simply becomes a stopgap measure for staying alive and trying to retain information as much as possible. However, all is not lost yet. In the end, the goal is to minimize the stress that this activity causes. Although it is possible to conceal commercial, military and banking secrets with a trained mind, this technique will eventually be overcome by fatigue, and fragments of the product/secret will be captured.

Remember: we're dealing with the content of thoughts and it is impossible to conceive or take action without going through it. This is the only resource we have to do everything that we know. Several techniques are created to divert the flow of thought and increasingly fill it with memories associated with particular fragments of an event. As a matter of fact, there are ways to prevent momentary access (internally) to your private data, such as the use of inner thoughts:

**Assonance** - connection between thoughts that occur, not in a logical way, but through the construction of phonetics. Creation of nonsense rhymes, including what you see and hear, that is, anything from the environment, until the information in question is degraded — until you manage not to consider the possibility of accessing the information requested by the invaders, using rhymes or random connections between words. One should get rid of the perpetrators using such techniques until the sensation and mental inclination to seek the object requested by them has become futile, completely ignoring the mental request and consequently the natural temptation to think about it.

**Humor** - always try to laugh (mentally). When the access attempt comes in the form of words, use it in another context before the vocalized memory starts its narrative. Tell a joke, for example. If they're trying to find out some information about you, or a detail about a secret project you're involved in, when that information comes in, turn it into music and use nonsensical words in order to make them rhyme. As incredible as it may seem, working with vocalized thoughts in this way helps to de-stress and take the focus off the data that the invader is trying to access, leading the mental flow state to opposite directions.

We also have to keep in mind the superweapon called D2K that alters memories and plays with the individual's self. As long as you are not in control of your thoughts, your mind is being managed by automatic systems, primary brain maintenance programs. Since most of their memories are stolen during sleep period, the target is completely adrift and helpless. When it comes to D2K, I believe that the most effective way to protect your data would be to detect that you're dreaming and to be able to control your ego rationally as you control your waking self. One has to be an experienced oneironaut to achieve such a feat, and it might even not work in every scenario. So, external physical blockage — electronic blockage — is the only effective measure to effectively stop this intrusion.

Cognitive maturity and millennial techniques of concentration can also help prevent access to data in the mind. However, this information can only be confirmed if any of the tests are conducted directly on people who are experts in these arts.

In conclusion, it is difficult to say which strategies are the most efficient to fight memory theft, as everything is very private and secret. In addition to being created in the midst of an infinity of variables, it is impossible to precisely calculate the odds. However, it is because of the novelty of having your mind hacked — in attacks launched in the first months to a year — that most of the information is extracted. So, I hope that this information helps people not to be caught off guard inside their houses, with their privacy systematically violated — a violence unprecedented in history of mankind.

**Don't give up. Fight until the end!**

# CHAPTER 6

## DANGER IN THE USE OF THE TECHNOLOGY (PART 6) - "WINTER SOLDIER" AARON ALEXIS

After having direct contact with MKTECH, it is possible to verify behavior patterns of people who are being attacked by psychotronic weapons, as well as investigate models that reflect the modus operandi of the technology capable of destroying someone's life in a short period of time without leaving a trace.

In this chapter, I am going to talk about the danger of using this technology in practice and how it is capable of transforming people into "programmed" remote killers. The same path, the same guideline, is followed — implicit processes and protocols that dictate the steps that lead to the same end: death. So, it is fundamental to know at which stage someone will succumb, or will give in to irrationality and become a Winter Soldier, formerly known as Manchurian Candidates.

Opinions are curtailed, denaturalized at every moment due to our unconscious judgments added to the constant distortion of unrealistic evaluations converging in order to channel all this internal scenario caused by psychotronic torture of people close to the target that have nothing to do with the tortured and the torturer. That was the path taken by Aaron Alexis who unintentionally fulfilled the main goal of the mind control technology, the holy grail of tactics.

Aaron Alexis, 34, opened fire in a U.S. naval base in Washington, D.C. He killed 12 people before police shot him dead. Alexis worked for a third-party company that provided services to the Navy. He joined the Navy and served in Fort Worth, Texas, between 2007 and 2011, as a subcontractor working with electrical engineering services. He had no combat experience, nor did he serve overseas. He was discharged from the Navy in January, 2011 for "a pattern of misconduct". The news agency reported that Aaron Alexis displayed erratic behavior that included insubordination and unauthorized absences. He was also arrested at least twice for using firearms.

In May 2004, Alexis was arrested in Seattle after shooting at the tires of another man's vehicle. At the time, he confessed he had an anger-fueled "blackout" — he claimed he didn't remember the incident until an hour later. In a statement to the police, Alexis' father said that his son participated in the rescue/recovery efforts on 9/11. According to him, Alexis suffered from post-traumatic stress disorder, which caused his mood swings.

In September 2010, Aaron Alexis was arrested again after discharging a firearm. But this time it happened in the residential complex where he lived in Fort Worth. According to local police at the time, he fired a gun through the ceiling of his apartment after an argument with his upstairs neighbor about noise. He claimed that it was an accident, contrary to what the neighbor said in a statement to the police.

## 6.1 - Attack on the naval base

Monday, September 16, 2013. According to the New York Times, Alexis entered the naval base in Washington with a rental car. He gained access to the base with his contractor's ID. Alexis shot at a police officer and a passerby while still outside the building. Upon entering, he started shooting at the employees who were having breakfast there.

The police officers who entered the naval base to stop Alexis exchanged fire with the criminal and shot him dead. Before that, the gunman killed 12 people — three women and nine men — and wounded several others. Police Chief Cathy Lanier said all the dead were civilians or employees; none were military personnel. This was the deadliest shooting at military bases in the United States since 2009, when 13 soldiers were killed at Fort Hood base, Texas.

Gunman Aaron Alexis believed he was under the control or influence of extremely low frequency (ELF) electromagnetic waves, says an electronic note written by Alexis and found during the investigation: *"I've been subject to an ultra-low frequency attack for the last 3 months, and to be perfectly honest that is what has driven me to this"*. According to the FBI,

ELF was used by the Navy for underwater communication (submerged submarines). "Conspiracy" theories claim that this technology was developed by the United States to monitor and manipulate people.

On Monday, Alexis walked into the building at 8:08 am and had legitimate access to it because the IT company he worked for had assigned him to work there. The report points out that, soon after, he opened fire with a shotgun. He also used a Beretta pistol obtained during the shooting. Alexis etched, *"Not what yall say"*, *"Better off this way"*, *"End of the torment"* and *"My ELF weapon"* — a clear reference to the belief that he was being controlled by electromagnetic waves — onto the barrel of the shotgun.

**Figure 6.1** Photo taken by a security camera showing the naval base at the time of the attack.

This was the news released to the public. Neither the newspaper nor the local police understood what was really going on as they tried to piece the puzzle together, that is, to understand the motivation behind the attack. They didn't focus on the main factor, which are the voices he heard all the time, as his father and his friend Oui Suthamtewakul told the Washington Post: "[Alexis] had a gun at all times". He also said that his friend drank alcohol frequently and in large quantities and, despite being friendly and enjoying talking to people, he was a reserved person. "He behaved like a 13-year-old in the body of a 34-year-old man. He needed attention," he tells USA Today, still in disbelief at the events. "He was a good guy to me. I still can't believe he would do that". He adds, however, that Aaron Alexis was mentally ill, "He suffered from paranoia, **had difficulty sleeping and heard voices."**

The categorical statement that someone has a mental illness without even doing a more in-depth professional forensic study of their mental health by applying all methods to reach an accurate diagnosis, together with the total lack of knowledge of the technology combined with the lack of credibility of the people who reported hearing voices — and are discredited even by the authorities — have contributed for this type of crime to thrive and the culprits to go unpunished.

Psychosocial reactions, consequences of brain dysfunction caused by attacks of directed energy weapons have created scenarios like this all over the world. Few people affected by this powerful weapon survive, become mentally sane or manage to do a comprehensive study to try to deeply understand the phenomenon capable of driving someone insane in a matter of days. Keep in mind that MKTECH simulates symptoms of several mental illnesses to disguise its electromagnetic weapons attack, as cause and consequence. It thus causes the target to subsequently have other reactions to the attacks, simulating the common symptoms of the real disease.

For those who still doubt the destructive power of this weapon, see the damage it causes in the entire cognitive process of the human being, leading them to commit acts of violence, such as the one reported above. This is yet another sad piece of evidence for the existence of remote murder by torture using electromagnetic weapons.

Now I'm going to check and point out small, subtle clues common to most victims that indicate that they are under strong attack by psychotronic weapons according to the facts sent by the news agencies and the testimony of the authorities involved in the investigation of this specific case.

First of all, we can point out a very common characteristic in most targets: the incessant noise and sounds of simulated fights, screams, conflicts, fuss, conversations, laughter and mockery that seem to come from everywhere inside the house — from the walls from all directions, for example —, driving the target crazy, violating their privacy and even preventing them from sleeping. Although the TI try to look for the origin of the noise, walking around the surrounding buildings or the neighborhood, nothing is found. This is always the initial symptom and an expected reaction of practically all people living in a democracy — the violation of their privacy in their residence and during their sleep, an attack generated by SYNTELE and V2K.

The shot fired at the upstairs neighbor's floor, which was provoked by the noises he was constantly hearing, accurately illustrates what we've seen in Chapter 5 as OPS use the MKTECH scheme to cause the same human reactions. All victims of this type of torture report incessant noises and screams that seem to come from their neighbors. **It doesn't matter that the target lives there for years; they inevitably begin to suspect that the neighbors are watching them, conspiring against them and making fun of their misfortune.** In other words, OPS create an extreme negative scenario within the target's mind who is unable to establish an analogy or a correlation between realities.

After a whole day of constant noise, sounds and voices, which only happen inside the target's head and seem to come from other apartments, the intensity of the attacks purposefully increase with the arrival of the night. The aim is to keep the target awake night after night, undermining their ability to rational judgment, and to cause erratic decision-making and desperate measures such as those mentioned above. The shot fired at the upstairs neighbor's floor is a serious indication that Alexis had become a Targeted Individual and was being tortured by a set of electromagnetic weapons using MKTECH technology.

As in the other cases, he'd already warned the authorities about the electronic torture, as shown in this official document about another incident where Alexis explains in detail how the voices negatively act in the environment and in people's minds.

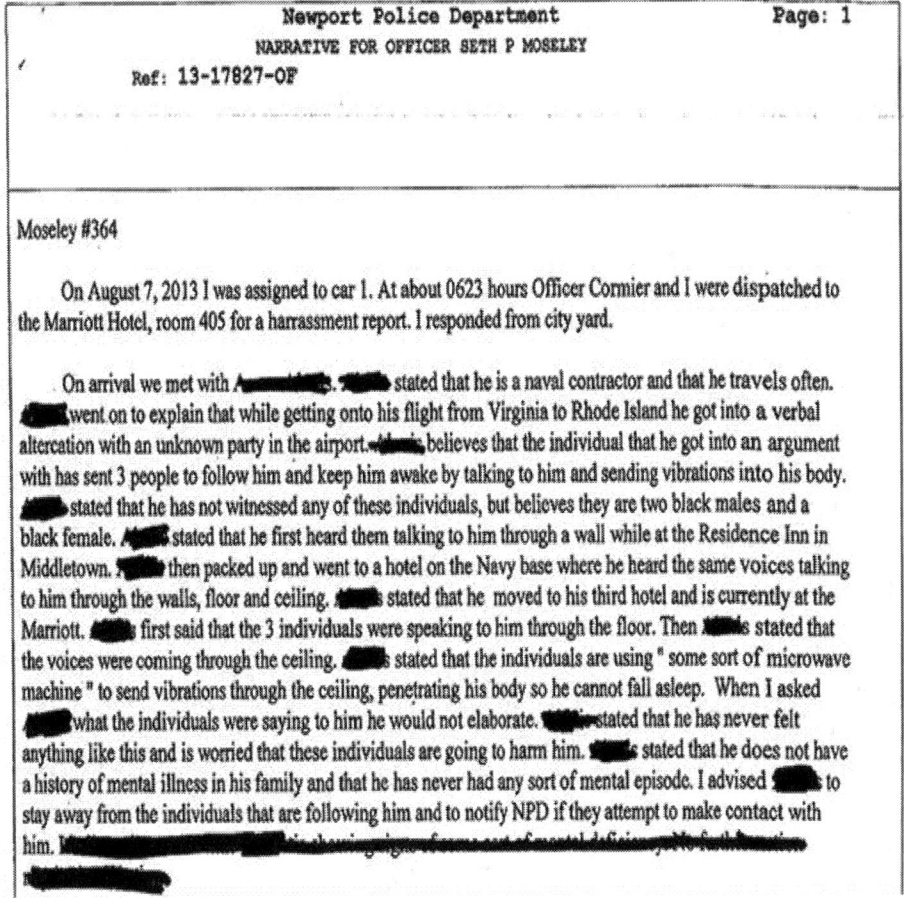

**Figure 6.2** Official document about the case.

On arrival we met with (redacted). (Redacted) stated that he is a naval contractor and that he travels often. (Redacted) went on to explain that while getting onto his flight from Virginia to Rhode Island he got into a verbal altercation with an unknown party in the airport. (Redacted) believes that the individual that he got into an argument with has sent 3 people to follow him and keep him awake by talking to him and sending vibrations into his body. (Redacted) stated that he has not witnessed any of these individuals, but believes they are two black males and a black female. (Redacted) stated that he first heard them talking to him through a wall while at the Residence Inn in Middletown. (Redacted) then packed up and went to a hotel on the Navy base where he heard the same voices talking to him through the walls, floor and ceiling. (Redacted) stated that he moved to his third hotel and is currently at the Marriott. (Redacted) first said that the 3 individuals were speaking to him through the floor. Then (redacted) stated that the voices were coming through the ceiling. (Redacted) stated that the individuals are using **"some sort of microwave machine"** to send vibrations through the ceiling, penetrating his body so he cannot fall asleep. When I asked (redacted) what the individuals were saying to him he would not elaborate. (Redacted) stated that he has never felt anything like this and is worried that these individuals are going to harm him. (Redacted) stated that he does not have a history of mental illness in his family and that he has never had any sort of mental episode. I advised (redacted) to stay away from the individuals that are following him and to notify NPD if they attempt to make contact with him. (Redacted).

This weapon has the ability to trick most victims' minds into thinking that the voices really come from walls, floors, in essence, from all directions and places. The source of the sound cannot be found. And that same microwave voice prevents people from sleeping. Depending on the intensity of the attack, one feels as if everything is shaking and vibrating inside their body, as presented in the report.

There are several cases like this all over the world and this kind of event is becoming more and more frequent, constantly leading to an erroneous diagnosis of mental disorders. Aaron's case serves as a great example of it, as it illustrates the complexity of everything that involves the technology and how easy it is to be led to simplistic conclusions when you only have a superficial analysis of the event.

The truth is that Aaron became another victim of prolonged torture and MKULTRA 2.0 experiments with psychotronic weapons and turned into a Winter Soldier/Manchurian Candidate. Cognitive deprivation combined with sleep deprivation and total loss of privacy makes us lose track of reality. For instance, his background, past and even testimonies of people close to him that point out a certain tendency to violence are used by gangs around the world to draw the attention only to the murderer himself. In most cases, the scheme actually increases violent acts, the target's bad temper and flaws, in order to create the image of a suspect due to their own past acts before they remotely torture them to the point of creating "programmed" remote killers.

Another aspect that draws attention is the shots fired at someone's tires during an anger outburst and claiming not to remember what happened. People tortured 24 hours a day by psychotronic weapons usually end up blinded by the hatred directed towards operators who violate their mind, which leads the target to walk aimlessly, losing track of time and memory of the event — a form of temporal amnesia, the direct result of the remote torture techniques performed on the victim, as we've seen in previous chapters. The ongoing brainwashing and the detachment from the reality are visible symptoms, which can later culminate in acts of violence.

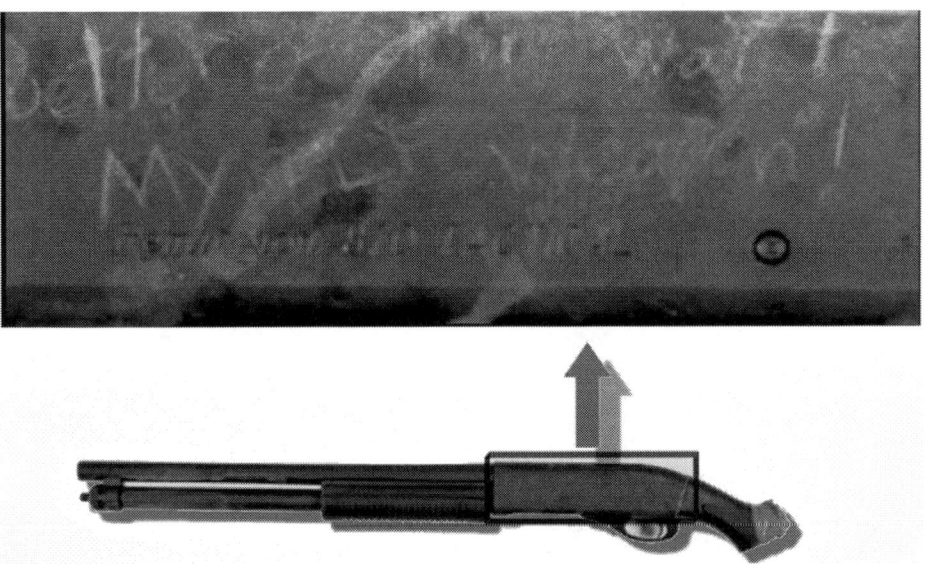

**Figure 6.3** Etchings on the barrel of Aaron Alexis's shotgun read, "Better off this way", "End of the torment" and "My ELF weapon".

Extremely Low Frequency (ELF) — the wavelength that the shooter thought was the one capable of violating his mind; the same one the Navy supposedly used for sonar testing. It is a set of frequencies of the electromagnetic spectrum, and is generated by natural or artificial sources with a small bandwidth for the transmission of information and restricted, practical use. Frequency range: 3 Hz to 30 Hz. Wavelength range: 100,000 to 10,000 km (62137 to 6214 miles). In my opinion, I believe that this type of frequency is only an additional measure to enforce psychotronic attacks, as it is incapable of transmitting voices — it only carries information encoded in simple binary form, e.g. Morse code. However, because of its wavelength, it can pass through soil or water with virtually no interference or signal loss.

The intrinsic parameters of the frequency variation of ELF waves provided inefficient and costly means of one-way communication with submerged submarines. The need for huge antennas made it impossible to send any response from the submarine crew to the operations center. The small bandwidth allowed transmission of only a few characters per minute. Furthermore, the energy cost for frequency modulation was only justifiable in wartime. Nevertheless, nothing can be ruled out. Perhaps the resulting wave is indeed ELF, the same one that the brain naturally generates. A full understanding of all aspects of the phenomenon remains elusive.

# CHAPTER 7
## USE OF THE TECHNOLOGY FOR FRAUD IN CIVIL SERVICE COMPETITIVE EXAMINATIONS

*"How did he manage to pass the exam? That's weird, he didn't study for it and he's not that clever. His family members are all civil servants, does that have any influence on the result?"*

**— Report from a candidate after seeing a colleague suspiciously pass the competitive exam.**

Passing a civil service competitive examination is not a very simple task. In fact, it was much easier in the past due to low demand and the devaluation of employees until the mid-90s when low salaries, "maharajas" and bad employees set the tone for lack of control and misdirection in the public sphere. In the early 2000s, the scenario completely changed with the stable currency, controlled inflation, consistent salary adjustments and a general improvement in every aspect of it.

Today it takes a lot of dedication, years of study, patience, goodwill, focus and money to face this fierce competition as you dedicate yourself exclusively to the exam — the dream of millions of Brazilians. Civil service competitive examinations are very important in Brazil. It involves millions of candidates every year, an enormous amount of money in the process, and moves a millionaire industry when it comes to courses, booklets, web classes, private lessons, institutes responsible for the correction of the exam and for logistics, examination boards, printing and correction of exams, and so on.

One important point is the well-defined working hours and rest periods. The employee is required to start and finish working at certain times — in addition to taking a break for meals. If the employee is a person of a higher position — in the sense of leadership —, working hours can be flexible. Retirement is also a great bonus: most of them retire with their current salary with no benefits (on average). So, this becomes a guarantee of stability even after they stop working. Another thing that attracts candidates is the age factor, since placement in the labor market is critical to the individual. In the public sector, it's usually necessary to be between 18 and 69 years old, with the exception of some positions.

The candidate who shows determination and strength after giving their sweat, blood and tears to achieve the privilege of being approved will obtain a series of benefits until the end of their life; benefits (such as financial stability) that aren't so easily obtained in most of the jobs in the private sector in Brazil. In this scenario, the candidate knows precisely how much they will receive each month so they can make plans for the future and not be at risk of facing a financial crisis as they get a taste of the true definition of the word "career". Being a civil servant can secure a great salary, since it's not required for the employee to achieve any goal or is related to any merit, as it happens in the private sector. Moreover, salaries are generally higher when compared to the same position in the private sector — on average 40 to 70% higher.

In other words, no matter how the employee performs, they'll receive their salary and benefits at the beginning of the month. Besides, other benefits that vary according to each sector are also available: transportation allowance, food allowance, health plan, day care allowance and maternity allowance, which in some cases can last longer (up to 20%) when compared to the private sector. Some offer a career path (with guaranteed promotion), different working hours arrangements, leave for qualification upgrading that lasts from 1 to 3 months — and it's not always relevant for the job —, day care, staff transportation, 30-day vacation and thirteenth salary. When it comes to specific positions — judges, for example —, a series of benefits such as housing allowance of R$4,000[56] will be provided for everyone, even if the judge works in their own city and has a house. In short, stability, security and salary are the factors that most attract candidates. After completing three years working in the public service, the

---

[56] Translator's note: 1 BRL = 0.18909 USD as of 31 December 22. Source: www.oanda.com]

new employee cannot be dismissed unless they commit a very serious misconduct.

Yes, the starting salary is very high compared to other countries and the private sector, but we can't say the same when it comes to efficiency. To give you an idea of the disparity between the Brazilian reality — the reality of a scheme employed in a troubled country — and the reality of a developed country that functions properly: the starting Federal Revenue Tax Auditor salary is, for example, R$15,743 per month, which is twice the remuneration of the equivalent position in the United States, which ranges from R$8,558 to R$27,733. In addition, a Judicial Analyst's salary ranges from between R$8,862 and R$16,829 in the United States and R$5,947 and R$13,375 in Brazil. Finally, a Legislative Analyst's salary ranges from between R$9,276 and R$21,776 in the U.S. and R$20,383 and R$26,004 in Brazil.

That's why such examinations draw the attention of ordinary people: citizens who want to work and have a relatively peaceful life in terms of stability and career. Unfortunately, this also draws the attention of gangs of all kinds, including those who use this huge salary to foment international terrorism, torture and the development of psychotronic weapons. To get an idea of how profitable it is for gangs to infiltrate the examinations, the World Bank carried out a study on Social Security[57] and the data showed that 35% of subsidies benefit those among the richest 20%. Only 18% of subsidies go to the poorest 40%.

Subsidies for civil servants cost the equivalent of 1.2 percent of GNP while subsidies for state and municipal civil servants cost another 0.8 percent of GNP. Federal public servants earn on average 67% more than private sector workers. When compared to state servants, the difference is also accentuated: 30% more. The problem in Brazil, compared to other countries, is not the number of civil servants, but the salary incompatible with the state's ability to pay. According to the study, 83% of civil servants are part of the richest 20% of the population.

---

[57] All pension data was provided prior to the 2019 reform.

Competitive examinations are extremely targeted due to the economic crisis and increasingly higher salaries, an immense disparity in the private sector. It's not a surprise anymore that gangs try to circumvent the rules of the examinations to enter the government sector and get their hands on this opportunity in an apparently lawful way. For this purpose, they try different types of fraud, but most of them are detected and promptly stopped by the Brazilian authorities. This often happens with conventional methods, such as a discreet earpiece placed in the candidate's ear and the transmission of the answers from a nearby location via radio waves. For this to work, however, someone has to leave the location with the exam papers in order to send the proper answers to the candidate through the micro earpiece.

Another type of fraud consists of bribing employees hired by the institutions responsible for the exams to tamper with answers. In the most recent case, the answer sheet was delivered blank to a corrupt employee and was later filled in with all the correct answers by other employees also members of the gang — thousands of Brazilian reais were exchanged to approve and put certain candidates in the first positions. All the top fifteen were part of the scheme. Crimes that involve many different people and large amounts of money end up calling the attention of the authorities and are exposed at some point. After all, this is a known scam that doesn't deviate much from common crimes. Nevertheless, none of these frauds can be compared to the level of sophistication of the techniques that I'm going to discuss in the next few pages.

A sad truth will be revealed here: organized crime is already part of the civil servant's world.

Candidates spend years on end studying — every single day dedicating themselves and sacrificing everything to achieve this dream — or at least that was the rule until the arrival of a device capable of turning the world upside down and completely paralyze the entire system and make the industry that orbits the civil service sphere fall into disrepute. How would you (an honest candidate who studied for many years and had to make personal, social and financial sacrifices in the hope that the process was transparent and fair, after having had to give up all other aspects of life to study) — how would you feel if you knew that the candidate next to you might be taking your place without having opened a book? That is, without having studied or having gone through any kind of suffering associated with the situation?

In the past, there was a democratic way of selecting candidates: their levels of knowledge prevailed. Today, this is unfortunately no longer true. For over 30 years, the most sought-after, well-paid positions with authority have been systematically defrauded in Brazil. Have you ever asked yourself (especially those who deal with civil servants on a daily basis), how certain people, mostly in the positions recently offered, managed to pass such a competitive examination if they don't seem to be at all prepared for it?

## 7.1 - The big fraud! The fraud of the century: civil service competitive examinations in Brazil

In addition to stealing information, listening to vocalized thoughts and sending voices and sounds to the target's mind without anyone else around being able to notice, as we saw in Volume 1, MKTECH technology, more specifically the Synthetic Electronic Telepathy (SYNTELE), is also used to circumvent the rules of the examination and allow the corrupt candidate to pass in any exam. This technology is capable of transmitting the test questions from within the physical location where the examination takes place using only the amplification of vocalized thoughts. While the candidate is silently reading the exam, the questions are forwarded directly to the HQ — the base where all information regarding the examination will be processed — without having to use any kind of electronic equipment, just the candidate's own brain. Afterwards, the candidate in question will calmly receive the test answers without anyone around even noticing. Both the capture of questions via neural (remote) electronic reading and the receipt of the answers via V2K will be captured only by the fraud candidate's brain; only they can demodulate this kind of information. If you are familiar with games in some capacity, it would be the same as the individual who has illegal advantages over other players, the real-life cheaters[58].

By offering a regular salary and creating a very well-paid mafia with tentacles that reach the most varied corners of the public sphere, from police officers to judges, prosecutors and auditors, there are already people in these groups who are infiltrated as legitimate employees.

## 7.1.1 - The scheme

Scammers, system operators and the whole gang, establish themselves in a remote location with a prepared environment supported by computers connected to private servers, software and programs optimized to make quick searches in databases that contain all questions and answers from competitive examinations in Brazil. However, to narrow the field of data search, the system can also be configured to only look for previously published questions from the same examination in previous editions, including examples for discursive questions with the best answers that have been reviewed and published. They have a complete database, with quick access to all editions, open-ended questions, closed-ended questions and topics and essays, exercises and simulations with the appropriate answers. In addition, there is information about the body responsible for organizing the exam with relevant characteristics and peculiarities, so the system can be quickly configured in the profile of the position, the exam itself and the organizer.

Just to be clear: it's not the questions that will be subject to fraud! **Scammers don't have prior access** or any type of connection with those responsible for preparing or violating the secrecy of the tests before they are distributed in the appropriate places precisely to avoid committing a crime that would be easily solved, such as stealing evidence, bribing officials that prepared the test, among others. The human being will

---

[58]Cheat is a term commonly used by gamers when it comes to codes and tricks that provide an advantage while playing games. Because these codes usually result in abilities that benefit the player or reveal the game's secrets, they are most often considered cheating and the player is seen as a cheater. It's a trickery, although not all cheats are considered "trickeries", as sometimes this information is necessary to find some kind of Easter egg in a game.

always be the weakest link, and usually the plan is unraveled when a link is broken in this kind of scheme.

The merit in this particular improved category of fraud is precisely not to leave a trace or connection with anything or anyone that had previous contact with the exam, especially people who are not part of the closed circle of holders of the technology able to hack the human mind. Everything happens live in the four hours available for the exam, that is, within the time specified in the civil competitive examination. If electromagnetic data traffic is not intercepted by then, we'll have another criminal working as a civil servant.

Let's take a real example: if the exam is organized by the Center for Selection and Promotion of Events (CESPE) – University of Brasilia Foundation [Fundação Universidade de Brasília], famous for being the most difficult exam of all, the fraudsters' apparatus will be configured with questions from previous editions, answer keys and data of the CESPE test application, that is, exams from previous or similar competitions for the same positions. Information about the examination, such as a wrong answer cancels out a correct answer, or leaving the answer blank doesn't remove a point, among others, will also be taken into account.

So far, it's not a big deal; nothing illegal or out of the ordinary is being used, since this information is in the public domain. Anyone can set up a system like this at home with all the necessary content to exhaust the subject in terms of questions. All you need is a database and an intelligent system capable of filtering the information as required. However, the access to this information is illegal, criminal and unfair to other candidates, as such access occurs during the exam with the help of several people and computers, all this happening inside the scammer's mind. Now the candidate in the field has several "minds" working together! A large group is taking the exam in place of the real candidate.

Knowing that the electronic police surveillance is flawed or non-existent — and is definitely insufficient for each region where the exam will take place, the fraud is carried out using SYNTELE with some modifications optimized for the exam in question, as it doesn't use the frequencies of radio stations that would normally be detected by the police.

The fraud comes from a complex set of frequencies, microwave pulses and interactions of radio signals. It is carried out in absolute secrecy by specialized gangs whose members are basically members of the same family or very close friends. Children of civil servants and financially fortunate individuals also use this type of "service", which practically makes such autonomous entities a family business.

The most defrauded and sought-after examinations in Brazil are for the Federal Police, Judiciary, Federal Revenue, Central Bank, ABIN, Senate and Chamber of Deputies. In addition to being well paid with many benefits and perks, public servants are key agents in the government, which makes it easy for them to stop any type of investigation against MKTECH and its operators. At first sight, the perpetrators seem to act within the law, but such positions are actually used to feed the scheme.

A sort of legal nepotism is created. The children of employees who are members of the organized crime start working as state employees in an apparently legitimate way, since SYNTELE is considered a weapon that hardly leaves a trace. Over time, this illegal infiltration into the government that seems legitimate begins to set up a super-department capable of reaching all spheres of the government in an extremely harmful way for the whole society. The gangs, in addition to getting popular and prestigious positions, are also very well paid. With a simple calculation, we can roughly see how much this scheme help to inject capital into the network of gangs linked to MKULTRA 2.0 experiments, qualified torture, murder and a series of crimes against humanity.

Take this scenario as an example: five employees from the same family gang earn between R$15,000 and R$25,000 per month.

-If the pay of two employees is R$15,000 each, this would generate the equivalent of R$180,000 for each per year. The gross earnings of the two would add up to R$ 360,000.

-If the pay of two employees is R$16,000 each, this would generate the equivalent of R$192,000 for each per year. Their gross earnings would be R$384,000.

-If an employee's income is R$20,000, this means R$240,000 per year.

The total annual income of the gang is around R$984,000. Together, they make almost one million a year, not to mention other shady side jobs — selling positions or information stolen from targets, illegal investigation of politicians and parliamentarians, the famous espionage (wire-tapping) that took place in the 90s but with much more sophistication and without the need for court orders to wiretap communications or carry out telematic interception. You just need to "plug in" the target's mind and that's it. Employees usually also have outsourced companies they bid on and receive government contracts, thus stealing more money from taxpayers.

To get an idea of how this practice is widespread, an exam for positions in the Brazilian intelligence agency [Agência Brasileira de Inteligência – ABIN] was held in 2018 throughout the national territory. There were around 300 vacancies in the 26 state capitals and the Federal District and the positions were: INTELLIGENCE OFFICER – with a starting salary of R$16,620; TECHNICAL INTELLIGENCE OFFICER – with a starting salary of R$15,321.34; and INTELLIGENCE AGENT – with a starting salary of R$6,302.23. It's practically a dream for most Brazilians who have to face lower wages in the private sector. People need to study for years — 5 years even — to pass such examinations nowadays. Nonetheless, even though the positions were for the Brazilian Intelligence Agency, SYNTELE ran wild in several states without facing any resistance from the authorities or interception by competent bodies.

It's not known for sure how many people used these fraudulent techniques and were successful among more than 180,000 candidates. Unfortunately, reader, this is our current reality: some of these positions with a starting salary of R$16,000 belong to criminal, ideological and dangerous groups. If not even our intelligence agency (ABIN) can prevent this type of fraud, or collect data that prove that this weapon exists and can be used for such purposes, what can we mere mortals do about it?

The use of this device in examinations is a slap in the face of the entire Brazilian society. Police officers electronically scanned the location with handheld equipment for any kind of "electronic cheat" while synthetic telepathy was being used and they found nothing — no SYNTELE noise. The groups involved in the scheme boasted about the situation and laughed at the agents who were doing the screening. The fraud is very sophisticated. So equally sophisticated equipment and knowledge of the techniques used to face this type of fraud that has been distorting everything we know when it comes to ethics and civility are urgently needed.

Other competitive examinations that took place this year were also easily hacked. For the time being, Synthetic Electronic Telepathy is the record-breaking device responsible for helping people during examinations in a dishonest and illegal way. This led me to a thought, and I will share it with you: this model of examination is no longer functional. Today, it's impossible to know which candidates were approved on merit. As long as examinations exist, people will try to deceive it. And make no mistake: the authorities' current equipment is not capable of detecting SYNTELE.

## 7.2 - A step-by-step of how this fraud works

It's similar to what happens with targets who have their brains hijacked — who are held in electronic captivity and tortured, as we saw in previous chapters. This time, however, the candidate consents to the intrusion. The technology will only be connected to their brain during the exam, even if they don't really understand how this complex technology works. The scammer will only know that they will receive the answers through whispers or voices in the distance.

The fraud candidate's brain is mapped using neural remote biometrics, thus categorizing mental functions — it follows the same path taken by an TI remotely. Now the fraud candidate's unique electrical activities are mapped and entered into the system where they can be located with millimeter precision anywhere in the world via satellite. Moreover, the combination of electrical parameters of biometrics and EEG represents the variables that will be included in the equations together with a random element to create the combination of frequencies and the type of modulation by which information about the examination will be sent and received between the candidate and their associates who are in a remote location with the equipment ready to answer all questions.

The equations and algorithms that delimit the frequencies make their demodulation to be interpreted only by the fraud candidate's brain with the exam in hand. In other words, this individual is the only one who can hear the answers sent by their associates in clear and good tone due to the crystalline properties of the microwave voice. No one around you will even know what is going on. Not a noise, not even a sound will be picked up by other candidates or inspectors or through security and electronic tracking devices that will continue to focus on other types of fraud.

In the remote base where the questions will be processed, there are several computer systems optimized to locate similar questions using sophisticated search engines that search for the most appropriate answer to the question in a database containing absolutely all questions from previous competitive examinations throughout Brazil, accompanied by the results of such exams along with several objective questions from previous tests, and high-score, first place essay topics from entrance exams. In addition, essays and writing samples taken from other competitive examinations can be quickly adapted to the topic requested at the time of the exam, which keeps the entire structure functioning and will lead to the approval of the candidate. To put it in another way, everything that can be used for the candidate to be approved is at the disposal of the scammers in the HQ thousands of miles away from the place where the exam will be administered. These places are usually quite remote, like farms or country houses.

It happens more or less like this:

The candidate goes to the exam location as they were supposed to and follow the rules provided for in the public notice, such as carrying only permitted items. Generally, the individual is required to use a transparent pen, but no watches, smart watches or electronic devices, only their own cell phone that must be placed into a sealed bag under the chair and can be retrieved after the exam is over. Then the tests with the remote equipment that will defraud the contest begin. The V2K (Intracranial Voice, Microwave Voice) is radiated: the people at the HQ send a message to confirm that the candidate is listening to them — and the confirmation that the candidate is hearing the voices clearly is given by the inner thought itself, the silent language, so it is captured via EMRv - Electronic Mind Reading (vocalized). The candidate just needs to think "Okay, all set". The link between the Synthetic Electronic Telepathy (SYNTELE) and the candidate in the field is then confirmed at the remote HQ.

At the appointed time the gates are opened and the fraud candidate looks for the classroom. Upon finding it, they enter, sit down and wait. The candidate only carries what is allowed: a bottle of water and a cereal bar to face 4 long hours of intense stress and the mental processing marathon itself. During this period all frequency bands are silent to prevent any kind of tracking attempt by the police, if any. The connection between the antennae and the brain remains open. Even if this type of tracking did exist, it probably wouldn't be able to monitor the set of frequencies from the MKTECH scheme. After all, police devices are incapable of capturing signals from military equipment, or of dealing with the complexity of this technology. Finally, the siren sounds and instructions are laid out by the inspectors in the room to reinforce the conduct listed in the public notice. After that, the exam officially begins — a test that separates the dream of a successful career from failing and having to try again.

The great merit of this fraud is keeping up appearances and not breaking any laws. The main rule, for instance, is that the candidate must not carry any type of electronic device, thus avoiding being caught red-handed. As soon as the exam starts, the collection of vocalized or silent thoughts begins, and these thoughts involve reading. Reading is using our inner thoughts in an automatic way, as we saw in detail in the first chapters of Volume 1. The fraud candidate in the room is then instructed to read the entire exam at once without making any effort or raising any suspicion. As soon as the reading begins the content is sent to the HQ thousands of miles away at the speed of light. Blessed are all four fundamental forces of the Universe!

At this point, the security seal is broken, and the candidate's only task is to read the entire content from cover to cover as agreed with the gang.

Consequently, the mind control technology starts to operate. At the time set for the start of the event, SYNTELE is reactivated. In a few minutes, the target's brain processes all test questions through the reading process. As letters, phrases and texts are read, they are amplified and captured. So, the entire examination is quickly in the hands of the HQ, leaving more than three hours and forty minutes to get the answers right — more than enough time to finish the exam.

Keep in mind that as soon as the vocalized thoughts are amplified and transmitted to the first antenna, their content is automatically encrypted and compressed to be sent to a satellite and then sent back to the HQ where the scammers are situated and had prepared their antennae to receive the carrier wave containing the vocalized thought. The data received will be processed, decompressed and decrypted by the analysis computer, making it indecipherable to third parties for as long as possible. The tracking or detection of a fragment of the frequency that involves the test questions are minimized. After all, you can't be too careful — and this statement applies even to this gang.

Everything happens without anyone around being aware that this type of event is possible in real life, much less that something of this magnitude is actually already part of our daily lives. Meanwhile, the candidate right next to them is about to take the exam with the help of handouts, computers, books, Google and other minds thinking simultaneously. The very essence of the technology facilitates the fraud and concealment of its existence today. Try telling someone the questions are being stolen and sent via synthetic electronic telepathy and wait for their reaction.

As soon as the exam starts, the EMRv - Electronic Mind Reading (vocalized) goes into action. The candidate reads the entire exam before starting to answer the questions. With every word read at every millisecond the room is bombarded by intense electromagnetic bursts in the radio spectrum to microwaves, amplifying every brain signal that is being picked up by the nearest antenna, relayed to satellites and transferred to the receiving antenna in the HQ in a triangulation and at the speed of light.

An honest candidate knows that it is necessary to study hundreds, even thousands of exam questions from previous examinations until exhaustion to pass an exam, especially because certain themes or subjects are limited. Over time, it's practically impossible to add new questions on certain topics that haven't already been asked in the past, or in workbooks. So, with almost all possible questions and answers in hand at the time of the exam, it is relatively easy to get a score of 95-100% correct if they want to, but this is rarely done, as it would compromise the scheme. A score much higher than the average of the candidates automatically makes you stand out from the crowd – you'd be considered a genius. But that's not true at all, since a genius doesn't need this subterfuge to get a job.

The process continues: the signal arrives at the HQ and the computer program translates the waves by extracting the content (demodulation), that is, the words read by the fraud candidate. In a matter of minutes, all of it is in the hands of the scammers.

And this is how the most complex problem — which is to quickly and safely obtain the content of the exam without the use of electronic equipment nearby, as these devices could serve as evidence of a possible fraud scheme — is solved. Now they only have to find the right answers to the questions that have already been transferred out of the official exam site. So, operators look for them and start writing the essay.

The system works at full speed for 4 hours. Algorithms come into play to streamline the search and help scammers to find the best results based on a meticulous and extremely fast analysis that examines data from previous examinations, correct or similar answers, pre-college questions, virtual books, among others, and they provide the answers that are most likely to be correct. This happens in a matter of minutes — a minimal fraction of time within the remaining hours—, which gives the candidate in the field an unparalleled advantage that conducts the HQ's search and makes the work much easier as they narrow the range of the search satisfactorily.

As answers are obtained, they are communicated to the candidate remotely. A gang member speaks into the microphone: "Question 1, answer (c)", "Question 2, answer (e)", and so on. This information is transmitted to the antenna and, after triangulation, is received by the fraud candidate's auditory cortex at a frequency that only they can "hear" — only they can demodulate the content of the electromagnetic wave. No one else will ever know that something like is even possible. The waves will just pass through the candidates without any consequence or interaction in their nervous systems. The candidate in the field, however, will have the feeling of having someone whispering in their ears.

This is the best example of how Synthetic Electronic Telepathy ends the current democratic models of competitive examinations, entrance exams and tests in general. Don't forget that the microwave voice (V2K) inserts audio information directly into the target's mind with a clarity that overlaps all other sounds, and that other people cannot hear what is being transmitted. Therefore, the candidate doesn't really need to study at all for the exams.

Taking this into consideration, nowadays it's possible to notice quite a lot of civil servants in key positions who don't seem to meet the intellectual requirements of the position as laid out in the public notice. This happens mainly in Brasília, at the federal level. And it gets worse: it usually involves family members! In some cases, father and son are employees of a certain agency. Sometimes, this happens with entire families. They think that autonomous agencies are a kind of family business.

To make matters worse, this type of fraud has been around since the 90's, when it was known as "the antenna age " in Brazil. It was at this time that the technology left the military and intelligence agencies and entered the civilian world, spreading through the most diverse layers of society. In the same period, digital computation took an extraordinary leap. Components such as processors and memories became accessible to everyone, boosting and diversifying the use of the weapon, as it was no longer necessary to mechanically change the parameters of the antennas. Personal computers were becoming a reality, and they were capable of performing previously unimaginable calculations with a software that adjust frequencies in a matter of seconds.

The astounding part of it is that it is impossible to physically shield all exam locations in order to block electromagnetic waves as it is extremely expensive to build a fortified room with such characteristics. The truth is that as this technology spreads, it will become increasingly difficult to pretend that civil service competitive examinations are a fair process and that only those who studied more passed. Even with the visible improvement in the way examinations are designed, transported and corrected — increasingly removed from direct human interference —, good faith is not enough against the threat of modern neural weapons.

SYNTELE is here to stay and will cause more problems until a low-cost and viable solution is found to block these frequencies, so that candidates can once again take the exam on an equal footing. They should, without a doubt, suspend any type of exam until further notice, especially those with the highest salary and influence.

# CHAPTER 7 - USE OF THE TECHNOLOGY FOR FRAUD IN CIVIL SERVICE COMPETITIVE EXAMINATIONS

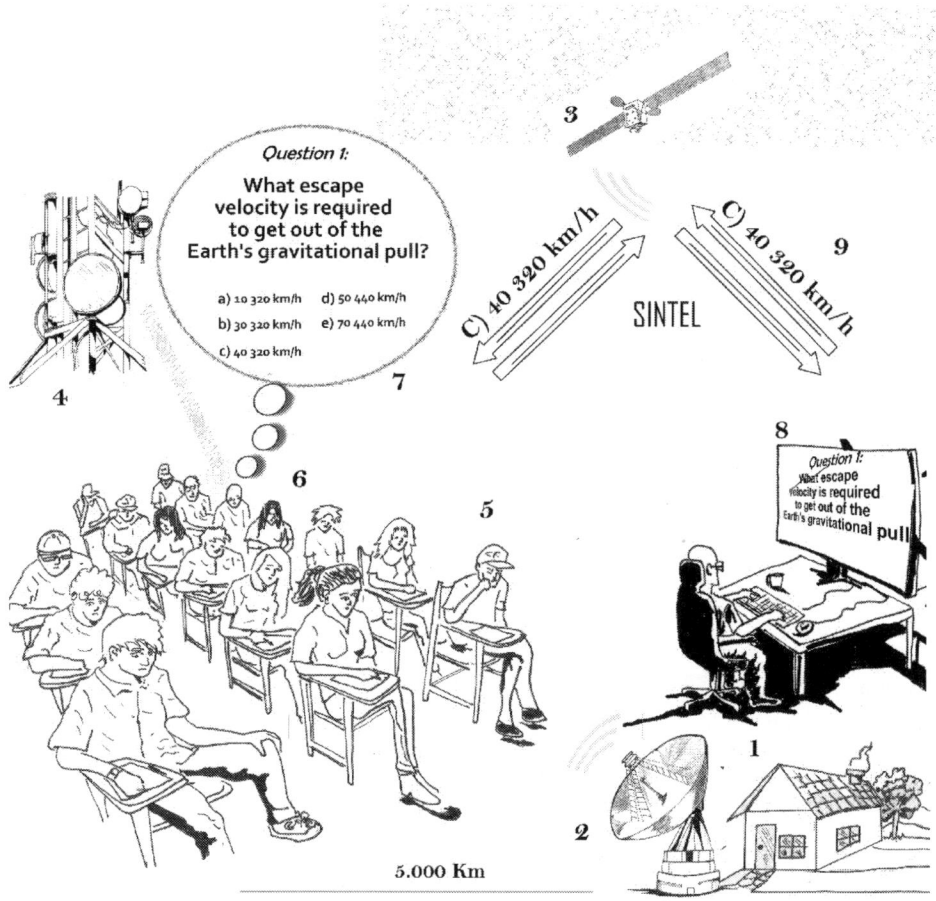

**Figure 7.1** Scheme to defraud civil service competitive examinations using SYNTELE.

1) From a remote base, far from urban centers and thousands of miles from the place where the exam will be held, the HQ will receive the questions and prepare the answers.

2) On the base, an external antenna comes into contact with the MKTECH infrastructure. This involves satellites and several antennas throughout the country.

3) The satellite receives the neural biometric information — the uniqueness of the fraud candidate's mind —, and locates the brain where the signals that travel through the nervous system will be amplified.

4) Antennas in the vicinity of the exam site capture the amplification of vocalized thoughts using SYNTELE.

5) All of the candidates are ready for the journey of a lifetime while they dream of having a stable career and a good salary. Nobody notices anything.

6) The candidate with the brain connected to the technology starts reading the questions. Meanwhile, the vocalized thought is recruited in the act of reading.

7) SYNTELE captures all thoughts and sends them uninterrupted at the speed of light to the HQ.

8) In less than 20 minutes, all questions are in the hands of the scammers located around 3,100 miles away. The connection is then broken until they finish their search. In about 3 hours, all the answers will be ready. The connection is re-established with the fraud candidate's mind and the answers are sent via V2K (intracranial voice). The candidate now only needs to mark the correct answers as explained in the example above (Answer C).

## 7.3 - Radio frequencies and electronic/spectral tracking

How does electronic tracking work in competitive examinations? How do radio waves work? How is this all connected with MKTECH? Let's now understand how the principle of radio waves works within the electromagnetic spectrum and its influence on telecommunications.

Radio waves[59], electromagnetism and telecommunications systems are very broad, complex topics. There are various technical details, equations and many other elements that make up this subject. But since I could fill an entire book, hundreds of pages long, with this topic aimed at a specific audience, I'm going to try and make a succinct summary without getting too technical and in a way that immediately involves the reader.

---

[59] Do not confuse radio waves with the chemical element, radium: from the Latin verb radiare (to emit, to beam, to radiate), symbol Ra, of atomic number 88 (which means there are 88 protons and 88 electrons in the atomic structure), atomic mass [226] u, a rare radioactive metal of the alkaline earth series, Group 2 (IIA) of the periodic table. At room temperature radium is in a solid state. It's a highly radioactive metal found in uranium minerals such as pitchblende. Some of the practical uses of radium are derived from its radioactive properties. It was used in medicine, but later replaced by more efficient radioisotopes. It was discovered by Marie Curie and her husband Pierre in 1898.

Radiofrequency waves are present in our daily lives: in audio transmissions, television, internet, satellites, garage controls, cell phones and other devices from the most banal to the most complex. Radio waves are waves of the electromagnetic spectrum propagated by artificial antennas or by natural means — when it comes from celestial bodies such as planets and stars that generate radio waves — and have frequencies lower than that of visible light, ranging from 3 kHz to 300 GHz.

* **KHz** - Kilohertz is equivalent to one thousand hertz. The hertz (Hz) is a unit of frequency, expressed in terms of oscillations (vibrations) per second.

* **MHz** - Megahertz means "millions of cycles per second", so 91.5 megahertz means that the radio station's transmitter oscillates at a frequency of 91,500,000 cycles per second.

* **GHz** - Gigahertz equals one billion hertz.

Every device that emits and receives radio waves is composed of a transmitter and a receiver. The transmitter consists of an oscillation generator, and its main function is to convert electrical current into oscillations of a certain radio frequency. The content (sound, image, data) to be sent is obtained through a transducer whose function is to convert the information to be transmitted into electrical impulses and encode it into a sine wave. A modulator controls variations in the oscillation intensity and in the transmitting antenna responsible for sending the waves to the medium.

The information signal is called by convention a **modulating signal**; the high-frequency signal that will "carry" the signal with the content is called a **carrier wave**. The result of the interference of one signal over another is a third electrical signal called a **modulated signal**. The process that involves generating this signal is known as **modulation**. Modulation is a process that consists of altering a particular characteristic of the carrier wave proportionally to the modulating signal.

The most popular forms of modulation are:

## 7.3.1 - Amplitude modulation (AM)

The main characteristic of AM is the fact that the modulating signal interferes exclusively and directly with the carrier amplitude. Each sideband occupies as much frequency space as the highest audio frequency being transmitted. If the highest frequency of audio signal being transmitted is 5 kHz, then the total frequency space occupied by an AM signal will be 10 kHz. Advantages of amplitude modulation: the ease of reproduction by a transmitter and the range of signals. Disadvantage of amplitude modulation: low efficiency in data transmission.

About two-thirds of the power of an AM signal is concentrated on the carrier wave, which has no information. One third of the power is in the sidebands, which contain the signal information. The sidebands contain the same information; however, one is essentially "lost". Only approximately one-sixth of the total output power of an AM transmitter is actually usable. The AM signal occupies a large portion of frequency space. In addition to being susceptible to static and other forms of electrical noise, it falls into four categories:

* **AM** (Amplitude Modulation) – it varies the amplitude, but the frequency is constant.

* **AM-DSB** – Commercial broadcasting.

* **AM-DSB/SC** – it was developed to make the process cheaper and to gain power.

* **AM-SSB** - *Simplex* PTT (push-to-talk) amateur radio – (Citizens' band) point-to-point communication.

* **AM VSB** – the main feature of television (video) signal modulation is to lower the cost of the SSB system. Measure of economy of the spectrum band reserved for TV channels, because instead of using an extremely high cut filter, it employs a filter whose attenuation is smooth and gradual depending on the frequency, and symmetrical based on the carrier wave. The carrier is fully transmitted to facilitate demodulation. In this case,

demodulation is done using an envelope detector. Video range from 0 to 0.75 MHz -AM-DSB from 0.75 MHz to 4 MHz.

The transmission of signals by means of electromagnetic waves undergoes the action of various types of noise. Furthermore, it has been found that the highest incidence occurs in the higher audio frequencies. When receiving the modulated signal, the relationship between signal amplitude and noise level is the signal/noise ratio expressed in decibels (dB). The processes of amplitude modulation with vestigial sideband and suppressed carrier (AM-VSB-SC) and amplitude modulation with single sideband (AM-SSB) are very similar — in both cases the carrier wave isn't transmitted. But in the second case one of the sidebands is completely cancelled, a fact that is only possible in practice when the modulating signal doesn't have extremely low frequency components, as is the case with the voice signal in the spectrum starting at 400 Hz.

## 7.3.2 - Frequency modulation (FM)

FM stands for Frequency Modulation. It refers to the transmission of waves with frequency variation. Modulation in phase and frequency — it was conceived due to the exhaustion of the possibilities with the carrier amplitude. The main concern here was the frequency or phase variations. Another main factor for its adoption was the inefficiency of the amplitude modulation (AM). Hence the need to create another modulation to circumvent this problem.

This type of modulation is little affected by noises present in the signal radiation, because in FM modulation it is the frequency that varies, not the amplitude. As a consequence of it, there is an improvement in the quality of the transmission, as FM requires a greater bandwidth than AM. To get a better picture, an AM radio transmission can be transmitted in a band of 10 KHz, while bandwidths of the order of 200 KHz are required for an FM radio transmission. That's why radio broadcasts are made in the VHF band, which ranges from 88 to 108 MHz and can accommodate a good number of radio stations.

The characteristics that can be changed in FM are: its amplitude and frequency. In AM, the variable characteristic is found in the modulation of the amplitude of its wave.

Example of FM signal demodulation:

* FM Heterodyne Transmitter (professional FM and microwave transmitter);

* TVs: it should be clarified that the sound of analogue television broadcasts is actually modulated in FM, and the picture is modulated in AM-VSB, maximum deviation: 25 kHz.

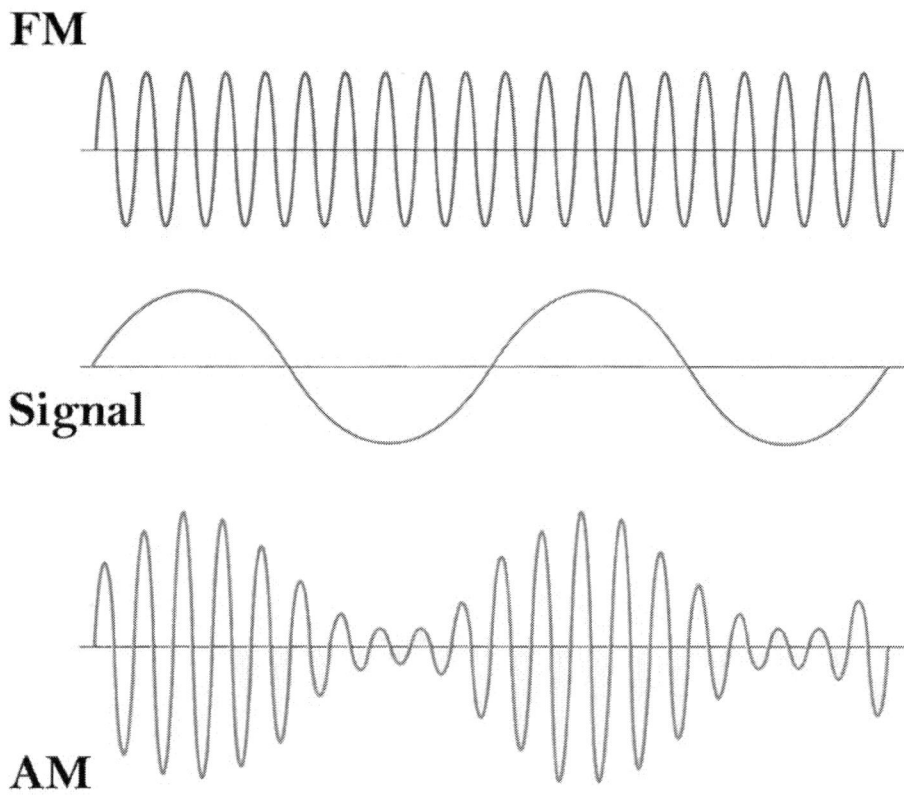

**Figure 7.2** FM and the sinusoidal carrier wave.

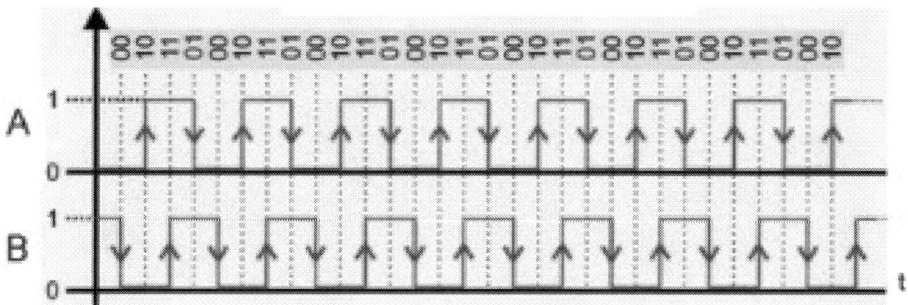

**Figure 7.3** Pulse train and Pulsed Systems - Pulse frequency modulation - PMF (digital method).

The receiver is basically an antenna with a tuner that captures only the desired frequency and a demodulator that extracts the modulated signal content. As there are many devices that make use of radio waves today all at the same time and all over the planet, sine waves are used at different frequencies to separate the signals so that a frequency does not interfere with another.

## 7.3.3 - Frequency bands

The RF bands have allocations determined by the Federal Communication Commission (FCC) in each of the 8 subdivisions. However, it is important to stress that this is just a convention. This doesn't mean that it is not possible to transmit any content on any frequency. Everything depends on the equipment. Modern MKULTRA operators use a number of varying frequencies and different equipment to keep the Mind Control Technology working; they don't comply with the FCC rules.

These frequencies are regulated by Anatel in Brazil, an agency that makes concessions possible and carries out auctions and specifications. The simultaneous use of the same portion of the spectrum through the transmission of several signals in the same frequency band results in interference, as it happens when a pirate radio station starts using a frequency allocated to another station.

Bands and frequencies using the data from FCC:

| VLF (VERY LOW FREQUENCY) - VERY LONG WAVES – 3 KHz – 30 KHz ||
|---|---|
| 3 KHz a 14 KHz | Not allocated |
| 14 KHz a 20 KHz | Ship-shore communications |
| 20 KHz a 30 KHz | Sonar |

| LF (LOW FREQUENCIES) 30 KHz - 300 KHz ||
|---|---|
| 30 KHz - 300 KHz | Maritime Navigation |

| MF (MEDIUM FREQUENCIES) - 300 KHz A 3 MHz ||
|---|---|
| 300 Hz - 415 kHz | Maritime Navigation |
| 415 Hz - 490 kHz | Telegraphy |
| 490 Hz - 510 kHz | International emergency frequency range |
| 535 KHz - 1.6 MHz | Commercial broadcasting - AM |
| 1.8 MHz - 2 MHz | Amateur radio, 160m range |
| 2.8 MHz – 3.025 MHz | international air routes |

| HF (HIGH FREQUENCY) – SHORT WAVES – 3 MHz - 30 MHz ||
|---|---|
| 3.5 MHz - 4 MHz | Amateur radio, 80m range |
| 5.95 MHz - 6.2 MHz | International broadcasting |
| 7 MHz – 7.3 MHz | Amateur radio, 40m band |
| 9 MHz - 11 MHz | International broadcasting |

| | |
|---|---|
| 13 MHz - 14 MHz | Industrial use |
| 14 MHz - 14.25 MHz | Amateur radio, 20m range |
| 15 MHz - 26.1 MHz | Internal radio broadcasting (shortwave) |
| 28 MHz - 29.7 MHz | Amateur radio, 10m range |

| VHF (VERY HIGH FREQUENCY) – VERY SHORT WAVES – 30 MHz - 300 MHz | |
|---|---|
| 30 MHz - 50 MHz | Broadcasting of fixed and mobile public safety communications centers |
| 50 MHz - 54 MHz | Amateur radio, 6m |
| 54 MHz - 72 MHz | VHF television channels: 2, 3 and 4 |
| 72 MHz - 76 MHz | Services in general |
| 76 MHz - 88 MHz | VHF television channels: 5 and 6 |
| 88 MHz - 108 MHz | FM broadcasting |
| 108 MHz - 136 MHz | Aeronautical navigation |
| 144 MHz - 148 MHz | Amateur radio, 2m band |
| 150 MHz - 174 MHz | Public safety and government services |
| 174 MHz - 216 MHz | VHF television channels: 7 and 13 |
| 225 MHz - 400 MHz | Civil Aviation |

| SHF - (SUPER HIGH FREQUENCY) – MICROWAVES FROM 3 GHz TO 30 GHz | |
|---|---|
| 3 GHz - 3.7 GHz | Radars |
| 3.7 GHz - 4.2 GHz | Communication with satellites |
| 4.2 GHz - 4.4 GHz | Radar altimeter |

| 5 GHz - 5.25 GHz | Terrestrial microwave communication |
| --- | --- |
| 5.9 GHz - 6.4 GHz | Satellite-earth communication |
| 10.7 GHz - 30 GHz | Communication with satellites |

| EHF - (EXTREMELY HIGH FREQUENCY) – MICROWAVES FROM 30 GHz TO 300 GHz ||
| --- | --- |
| 41 GHz - 43 GHz | Satellites |
| Other bands | Experimental communication with satellite and government agencies |

Conventional radio equipment — an AM or FM radio receiver, for example — will only be able to scan the frequencies intended for the band they operate and nothing else. The car radio is a good example. If you select an AM radio, you will only be able to "scan" between the determined frequencies ranging from 535 kHz to 1.6 MHz.

It is very different, for example, from SYNTELE that uses various types of modulation and frequencies with no concern for the band that will be used, in a very wide area of the spectrum, which can abruptly vary in order to jump from one end to the other in the spectrum in which there is the delimitation created by the Neural Remote Biometrics. In this way, SYNTELE is free to vary its frequency bands both in the radio spectrum and above (microwaves) basically using spread spectrum encryption.

The electronic defense responsible for tracking radio signals in examinations generally looks for bands used by radio amateurs, walkie-talkies and short-distance radio transmitters (push-to-talk communication), since most electronic devices normally used for radio communications operate in different frequency bands: 108MHz – 520MHz and 700MHz – 990 MHz.

Those responsible for tracking radio frequencies use equipment that, unlike conventional ones, is capable of tracking a wide range of frequencies. They use professional radio scanners, receivers which can pick up all kinds of radio signals. The device is usually programmed to scan certain frequency bands and stop as soon as it finds a signal. They can also track certain frequencies that seem suspicious at the time. Depending on the equipment, it is even possible to monitor several frequency bands simultaneously. On the other hand, there are several ways to mask the content of the transmission between the remote base where the HQ is located and the fraud candidate, depending on the competence and creativity of the members behind the MKTECH system.

SYNTELE decoding is performed in a simulated way with a complex system of canceling and gain of signals, leaving only a very low frequency. This causes the brain to naturally demodulate pulsed microwaves and various other frequencies in the radio spectrum. That said, I'm going to enumerate some techniques used in the examination that are related to the transmission of the questions from the exam site through amplification of vocalized thoughts and later some other techniques containing the answers that will be used by the candidate to finish the exam — all that through V2K using Synthetic Electronic Telepathy (SYNTELE).

## 7.4 - Techniques for sending questions from inside the exam site to the HQ in a secure way

The transmission of test questions from the location in which the exam takes place to the fraudsters' HQ in order to process the answers requires the candidate's vocalized thought to be amplified via EMR - Electronic Mind Reading and relayed to a receiving antenna to be encrypted, quantized, transformed into a digital signal, and later retransmitted to the HQ in a safe and practically indecipherable way if it is picked up by some strange receiving antenna along the way. But before reaching the first antenna, the information contained in the initial modulating signal — the questions the fraud candidate reads silently — needs to propagate for a few miles and, in the meantime, if it is intercepted by third parties, its content will be readable, easy to understand. So, the same technique and at the same frequency can be applied: to merge questions captured from inside the room with another sound, music or noise in order to camouflage them at the time of remodulation and, later, use a filter to remove the music or noise. That leaves only the questions read, which masks the content of the test in case it is intercepted by an agent responsible for electronic security.

Within SYNTELE, the EMR analyzes the vocalized thought. In the case of the competitive examination, and in order to improve security for this small unencrypted portion, an extra signal is used and the carrier wave that "leaves" the target's mind is already modulated with two contents — the test questions and other kind of sound, which can be either a song, a dialogue that emulates ham radio conversations, or some other noise. A signal with audio content recognized by the target's cortex can be sent simultaneously with the reading of the questions. Thus, both the exam questions and the song, noise or dialogue, which were sent simultaneously in another signal to camouflage the content, will be processed together. After all, everything that passes through the target's auditory cortex is captured by EMRvia – Electronic Mind Reading (vocalized/images/auditory).

SYNTELE scans the target's mind in microsecond cycles and is constantly amplifying the vocalization of thought, in this case the reading of the questions. So, the speed at which the amplification captures the content and sends it to the systems is immensely faster than the human being's ability to read. As a result, data arrives steadily, fluidly and without intermittence. The same is true of the further processed answers that the target receives from HQ.

Although the most common type of fraud is to use tiny, Chinese-made electronic devices, such as an in-ear monitor (IEM) that the candidate wears to get the answers from their accomplices, the mind control technology is infinitely more elaborate. The scammers don't even need this device, as the IEM in this case is the candidate's own brain. In addition, there is no need to leave the room with the test or risk being caught with electronic equipment. Remember: this is a technology that leaves no traces.

Several tactics hidden in signals are used to receive the content of the exam in order to camouflage the unencrypted content once the fraud candidate's brain is located. It basically depends on the creativity of the

MKTECH gang, technical knowledge, implementation of resources and technological level. Another trick is to link SYNTELE to one or more people in the room chosen at random who have no connection with the scheme. Remember: every thought is amplified and captured, so in the midst of a whirlwind of thoughts some of a private nature arise, such as thinking about a partner or family, or a man saying to himself, "Come on, you studied so much, you can answer the question!". All of this can help discover who is defrauding the examination. So, different thoughts are amplified and are added to background noise and other data that will simultaneously hide the content of the fraud candidate's real thoughts.

## 7.5 - Getting questions answered

Before sending the answers in real time (the examination still in progress) using Google, for example — surrounded by all kinds of precautions for IP address, VPN, Proxy and browsers like Tor hiding tactics — and a lot of people thinking at the same time, operators prepare the answers with the utmost care to get enough questions right/wrong using a huge database and systems with data from all examinations as explained above.

A tactic to perfect the answers before they are sent to the fraud candidate in the examination room is to use other minds to compare answers — do a sample survey of some questions. With this, it is possible to "hear" the answers from an individual who has previously been mapped, a true brainiac. This "brainiac" is an extremely studious and intelligent person who can be screened and easily found in cram schools, and they have a probability of achieving above 90%. They can then serve as parameter for the answers.

In order to send back the answers of the processed questions to the mind of the candidate, tactics similar to those used in capturing the questions must be applied, thus avoiding any electronic security tracking. One of them follows the same principle of capturing the questions as they are processed in the target's mind and transformed into vocalized thought: different data, music or any noise with the content of the answers sent via V2K to the candidate are embedded in the same frequency. The music or noise will be broadcast continuously, which clearly resembles a pirate radio, amateur radio dialogue or something unimportant. The candidate in question will hear a sort of stereo-like sound: a song or random noises will sound on the right side (R-side) while the answers will come from the left side (L-side). It is similar to what happens with radio receivers where both the sum and difference of channels are in the same region of the spectrum (L-R); one of the them will undergo the encoding or matrix shift.

It's difficult to track this fraud, because V2K uses microwave pulse in conjunction with the frequency-hopping spread spectrum[60] that is synchronized with the antenna that sends the carrier wave with the modulating signal (which carries the information), remapping the modulation that will be in the specific frequency range only the fraud candidate will be able to "hear".

It's quite a unique feeling to be in a room with several other candidates, inside a structure prepared with diligence and safety. Suddenly, you start hearing voices in your mind without anyone around even knowing that this type of technology exists nor its ability to interact with the human mind.

In order to be accurately tracked, scammers must first obtain the range of unique frequencies that was configured during the creation of the Neural Remote Biometrics (Chapter 3 of Volume 1) and that will be used in the candidate. Then it's necessary to determine the cryptographic sequence of frequency hopping to understand in which "channel" each piece of information will be transmitted in a given period. It's an extremely complex, well-crafted mind-control technology. Currently, it's practically impossible to be completely tracked by non-military equipment. Moreover, they use spread-spectrum encryption, rolling codes.

---

[60] Hedy Lamarr - The initial version consisted of a range of 88 frequencies. The idea came from playing the piano. They were rehearsing when, after repeating the notes they were playing on another scale, they realized it was possible for two people to talk to each other by frequently changing the communication channel to avoid patterns. All they had to do was to do it at the same time. The frequency hopping was born out of this idea. Virtually all modern cryptographic techniques involving frequency and channel can be employed to defraud civil service competitive examinations.

Once used, the configuration is deleted from the system and never used again, so the next configuration is switched between the sender and receiver with the same frequency hopping encryption.

Spread spectrum also work with the time-hopping (TH) technique, in which the choice of the time interval used for the transmission is defined in a pseudorandom way, and with the hybrid spread-spectrum technique, which is the combination of the aforementioned techniques. The most common way of implementing the hybrid technique is the combination of direct sequence with frequency hopping. In this combination the furtive nature of the direct sequence can be exploited in conjunction with the frequency diversity in frequency hopping.

The most meticulous reader might be thinking: "Okay, the signal is there. So why don't they just block these signals as it happens with cell phone jammers in prisons?" Blocking signals is not very simple when done electronically, and may cause serious problems similar to those found in prisons, such as when the radius of action of the blocker ends up acting in an area larger than the perimeter of the facility, preventing the cell phones of people who live in the vicinity of the prison from working. In addition, the blocking is not absolute, that is, it doesn't work very well in practice, especially in Brazil. However, it is one thing to block cell phone signals that are regulated and run on known locations with legitimate commercial purposes; another is to block unknown frequencies. In fact, it is relatively easy to know exactly which channels and spectrum bands to act in order to shut off cell phone access to antennas that connect us to the world. But it's quite different when the mind-hacking service itself is already a powerful weapon of war shrouded in state secrets that have endured for over 70 years.

I believe it's still not possible to completely prevent these signals from being used in all states of the federation in which examinations will take place using both electronic and physical means. In general, the interfering signal must occupy the entire frequency band used by the system with sufficient power to be effective in blocking the signal. This goes against the same problem again: how do you know which frequency band to use if the signal oscillates according to the spread spectrum algorithm based on the target's remote biometrics as seen above?

You simply cannot occupy the entire spectrum in several different locations in Brazil with enough power to block and interfere with legitimate signals, such as radio channels and other devices that use Wi-Fi, including cell phones and the internet used for communication – in other words, pretty much everything these days. Thus, for an efficient blocking effect, the spectrum bands must first be defined — and then one must work through the system into which the interfering signal will be inserted. Not to mention that the interference must be performed only on specific channels, not on the entire spectrum band. The main direction of signal radiation is one among several considerations in an extremely complex world of radio communication that should be taken into account, which brings us to the next section.

## 7.6 - Signals intelligence

To get a better idea of the complexity and types of equipment involved in the MKTECH scheme, it's important to remember that we're not dealing with amateur equipment, but with equipment produced by nations, services from countries that offer heavy infrastructure (which we don't have), such as China, Russia, the United States and Japan.

There is a division of the military that specializes in electronic warfare and signal intelligence to do the recognition and identification of electromagnetic waves at this level. They are responsible for analyzing the enemy's electromagnetic emissions and for creating strategic reports to support troops with information and decision-making on the battlefield, among other actions. These communications usually come from radar systems or the enemy's electronic communication system. Expensive equipment in the millions of dollars and highly specialized and trained people are required to operate the equipment and to capture, decode and track the emitting source or sources emitting the signals to analyze the characteristics of the signals and the emitting platform. Keep in mind that sources emitting electromagnetic radiation of all types are "mixed" every day; they pollute the spectrum with more and more information — such as these signals that are used to violate the human mind (SYNTELE).

The ordinary police scanner has little chance of detecting the fraud. The only thing capable of dealing with this type of weapon would be the employment of divisions of the armed forces specialized in the field with war equipment and qualified personnel, since the equipment that is part of this MKTECH "cluster" — antennas, satellites, radars, electronic equipment, among others — are on the same level as the most advanced ones used by the armed forces to track the entire global chain of equipment that makes up this technology. At least a number of actions would have to be employed to exploit the enemy's emissions in all bands of the electromagnetic spectrum with the purpose of knowing intentions and capabilities and using adequate measures to deny the effective use of the systems.

Therefore, a modern electronic warfare scenario should be set to protect citizens and our civil service competitive examinations so we can prevent communists, fascists, mobsters and criminal organizations from legitimately infiltrating the government at taxpayer expense and further destroying our country already so ravaged by corruption, foul play and promiscuous relationships involving state agents and the organized crime.

There are some modern battlefield tactics employed by our armed forces responsible for signals intelligence and they will soon be put into practice in this silent war that threatens world peace and the collapse of democratic regimes as well as our hard-won freedoms over the centuries.

The measures of electronic warfare are as follows:

* **Electronic Warfare Support Measures (MAGE - Medidas de Apoio à Guerra Eletrônica** in Portuguese) – encompasses the actions taken to monitor, search, intercept, locate, analyze, evaluate, correlate and record the electromagnetic signals radiated by the enemy for the purpose of exploiting it in support of military operations.

* **Electronic Protection Measures (MPE - Medidas de Proteção Eletrônica** in Portuguese) – set of actions taken for the protection

of means, systems, equipment, personnel and facilities in order to ensure the effective use of the electromagnetic spectrum, despite the use of electronic attack measures conducted by friendly and enemy forces; to ensure the use of the electromagnetic spectrum by allies.

* **Electronic Attack Measure (MAE - Medida de Ataque Eletrônico in Portuguese)** – the carrying out of actions to prevent or diminish the effective use of the spectrum by the enemy, and to degrade, neutralize and destroy the enemy's combat capability through equipment and weapons that make use of the spectrum.

80% of electronic warfare consists of carrying out support measures across the spectrum.

## 7.7 - The steps to take to identify the signals

There are already well-established protocols for taking action in radio and microwave transmissions, so I believe that these measures should also be applied to illegal transmissions that hack people's minds without their consent. The measures taken must initially use the modern systems of MAGE (Electronic Warfare Support Measures) and do the search and automatic interception, identifying and classifying the thousands of signals that are captured. As soon as they are identified, such signals must be cataloged in a signal database and isolated. Then, they must be closely monitored for the purpose of observing their activity and evolution or obtaining other relevant technical data from the transmissions (for example, metadata to verify whether the signal is analog or digital, encoded or not — and even if the data is encrypted, if they have a valuable set of data and all electronic activity).

The next step would be the location of the emitting source (electronic location). At this point, the intention is to search for the geographic position of the emitting source using mobile or fixed sensors; to identify the location, position and orbit if the downlink comes from satellites. Subsequently, and most importantly, the objective is to search for the initial transmission from the operator's remote base (the uplink) where all the torture content is generated and all "captive minds" are located. In this scenario, even if operators use "hijacked" satellites (satellite piracy), it is still possible to locate the emitting source.

The following priority measures should prevent such transmissions from reaching the target's mapped brain, both when directed at competitive examinations and during electronic torture. In this way, it is imperative to neutralize or degrade the signals of the attackers, using weapons whose primary purpose is the emission of directed energy of high electromagnetic power against scammers, operators and their electronic infrastructure.

Passive and active (mechanical and electronic) blocking can be used in signal blocking (electronic blocking). Mechanical blocking uses physical means to block or spread electromagnetic energy around. These are the three main jamming techniques:

* **Spot jamming** - uses narrow-band transmission, concentrating the radiated electromagnetic energy exactly on the spectral band occupied by the target/receiver;

* **Barrage jamming** - simultaneous emission of electromagnetic energy over a wide range of frequencies;

* **Sweep jamming** - uses emission of narrow-band electromagnetic energy with time-varying tuning, performing sequential transmission of the blocking signal at predetermined frequencies.

In other words, it takes a lot of hard work to identify (and get rid of) the people behind it all.

Figure 7.4 Screen showing a program capable of scanning frequencies in a wide band. The spectral occupancy is represented by a "waterfall" design (horizontal lines in a cascade) that displays the temporal RF activity within a given segment.

## 7.8 - Conclusion

As if the incompetence of the Brazilian authorities wasn't enough — a population sunk in poverty, ignorance and rampant corruption —, we also have to deal with government employees who are being paid with taxpayer money without having passed the exam fair and square. Now scammers have legitimate jobs that are the result of fraud. This sets a dangerous precedent: people are infiltrating the government — as legitimate employees—, getting great salaries when compared to the reality of the Brazilian people and using such positions for shady purposes.

In a macro view of the matter, something even more frightening hovers over the horizon, and it consists of the infiltration of agents who have a strong sympathy for totalitarian communist regimes, such as those of Cuba, Venezuela, China and Russia. The objective of these people is to effectively transform Brazil into a dictatorial regime where free market and capitalism would still exist, but increasingly imposing, dictatorial control systems would be gradually introduced into the society little by little, starting with the cultural destruction and the downfall of education — in both cases, the mission has already been accomplished — which pave the way for the end of individual freedom. In the end, the government would have full control of all our actions and thoughts.

So, it is crucial to pay attention to the serious problems these neuroelectronic weapons bring to society. They are the most powerful instrument of destruction of all democratic liberties we take for granted, including the end of intellectual property, something that is being constantly monitored inside people's brains. The government or related authorities, such as the secret police, would have complete freedom to observe, torture and suppress people's thoughts inside their own homes and minds. This is a horror scenario that can come to fruition when we least expect it!

But don't worry: I'm going to further discuss this topic in detail in the last chapter of the book. For now, we must always consider the ideological and geopolitical aspects of the weapon, since the MKTECH scheme has different tactics and protocols for experimenting with the human mind, and psychophysiological electronic warfare based on techniques employed in experiments in the former Soviet Union. That's why the general population need to be attentive and sharp. Unfortunately, this is yet another pressing problem for Brazilians.

Finally, to conclude — and my opinion about civil service competitive examinations may sound very radical right now —, I think that at the moment the best action would be to completely suspend all examinations until we're able to protect such locations, so that the democratic process can be consolidated and that all candidates have equal conditions to succeed. As it currently stands, the process is subject to fraud, making it impossible to tell if the candidate was honest. It's just impossible!

# CHAPTER 8

# DANGER IN THE USE OF THE TECHNOLOGY (PART 9) – TERRORIST ATTACK ON THE U.S. EMBASSY IN CUBA AND CHINA (HAVANA SYNDROME)

In this chapter, I will provide a practical example of the use of directed energy, neuroelectronic, psychotronic, electromagnetic weapons — the maximum expression of power embodied in a true artifact of war, which now guides our future and present regarding the employment of modern and unconventional weapons. In this particular event, the technology was used to commit terrorist acts on the U.S. Embassy in Cuba and China.

So, let's first analyze the devastating power of the weapon when used on just one target. The goal here was clear: to cause the greatest possible damage to the people who lived and worked there. Remember, terrorist attacks aren't always associated with high-powered explosives or blunt weapons. An act of terrorism is the application of physical or psychological violence using localized attacks on individuals or government facilities and the governed population in order to instill fear, panic, and thus obtain psychological effects that go far beyond the victims involved.

After reading a good portion of the book, I'll leave it to you, the reader, to analyze how the story was told and everyone's level of ignorance about the subject.

FBI Investigating Mysterious 'Acoustic' Attacks on U.S. Diplomats in Havana, from EL PAIS:

The FBI investigates the mysterious attack on U.S. diplomats in Cuba, but without success. The total number of Americans affected by the event rose to 21 this week. Five Canadian diplomats stationed in Havana were also victims of a possible "acoustic attack" which the State Department now loosely defines as a "health attack". Between November 2016 and autumn 2017, various symptoms were detected among those affected, such as nausea, mental confusion, deafness and basic speech problems. Some suffered permanent hearing damage or nervous system issues.

The mysterious attacks took place at the diplomats' residences and in at least one case, as revealed by the Associated Press, it occurred at the emblematic Capri hotel — state-owned, but ran by the Spanish NH chain —, where a U.S. employee heard a high-pitched and concentrated sound while in bed that disappeared as he moved across the room, as if it had the precision of a laser.

The hypothesis of an acoustic attack didn't develop in the North American intelligence community because the recorded brain damage isn't scientifically explained based on any device that emits sound waves. However, according to sources cited by the U.S. media, some victims claimed they heard unexplained screeching sounds. And as a result, hearing damage occurred.

As the mystery is yet unsolved, the diplomatic effect increases. This Friday it became known that Raúl Castro, the president of Cuba, met with the top diplomat of the U.S., Jeffrey DeLaurentis, to tell him personally that his government wasn't behind this unprecedented attack on U.S. personnel. Also, five Republican senators sent a letter to the State Department requesting explanations from Havana — designated as responsible for the security of diplomats from other countries.

"Furthermore, we ask that you immediately declare all accredited Cuban diplomats in the United States persona non grata and, if Cuba does not take tangible action, close the U.S. Embassy in Havana", says the statement signed among others by the powerful Cuban-American senator from Florida, Marco Rubio. In May, two Cuban diplomats were expelled from the U.S. in retaliation. President Donald Trump who in June announced a partial restriction of the measures of conciliation with Cuba (adopted by Obama) has not yet said anything on the matter.

Two theories revolve around the mystery. One is that dissidents of the Cuban intelligence system carried out the operation to boycott the thaw between Washington and Havana. The other is that a third U.S. rival —

Russia, Iran or North Korea — carried out the attack, perhaps with the help of rogue agents from Cuba. In any case, it is very unlikely that the Cuban government ordered the attack. Focused in improving relations with its neighbor and needing to attract tourists and investors, General Raúl Castro's administration would have shot itself in the foot while lending a hand to the U.S.

"There is no reason for the government to do this. Not even when relations between them were strained, during Bush Jr.'s presidency, did something similar happen: there were displays of hostility towards diplomatic personnel, but never an attempt to harm them," says William LeoGrande, an expert on Cuban affairs from the American University of Washington.

"I've never seen anything like it and I can't explain it. One possibility is that it was a Cuban intelligence operation gone wrong; another is that people who want to derail the process with Washington are behind it. If it is the latter, I suppose the Cuban government will have to deal with them, but if this is in fact the case and they do deal with them, they will never make it to the public, and we will never know", says former CIA Cuba analyst Brian Latell, now professor and researcher at Florida International University (FIU). LeoGrande bets on a similar outcome, "We'll probably never know who, why, or how they did it.".

FBI agents went to Havana to investigate the case with the permission and collaboration of the Cuban government. For now, however, it seems they remain in the dark.

We can already draw some conclusions about this article. The first one is that people in general have no idea that this kind of weapon exists and the damage it is capable of causing. The second is related to the FBI's explanation of the attacks due to total ignorance: naturally, they came to the conclusion that acoustic weapons were used, but that's not true. These are actually electromagnetic weapons, as we've seen throughout this book. After all, V2K causes confusion and total degradation of the central nervous system.

The third conclusion concerns the MKTECH version that is being used in Brazil. Despite being sufficiently advanced to the point of destroying an individual in a short time, this version doesn't have the direct ability to disintegrate the areas responsible for speech and hearing, which differentiates it from the weapon involved in the aforementioned attacks.

Undoubtedly, these are more powerful and modern weapons when compared to those used in Brazil, but this only confirms my pessimistic predictions and worst fears: the new generation is more powerful and capable of causing extensive damage in a very short time, especially if the damage is done while people sleep using their dreams as a conduit. The reported attack perfectly illustrates everything we've seen in Volumes 1 and 2: the dangers and the urgent need for a debate about the subject.

## 8.1 - Embassy War

It is not new that the U.S. and its allies, and the Russians and their allies, attack embassies with weapons that leave no trace. It's a kind of psychotronic Cold War, and any neural weapon is used as long as it leaves no known physical trace, or has no radioactive/nuclear or chemical signature. Since 1946 a silent war has been waged between Soviet and American embassies. Records show the effects of the weapons on the people of both embassies up to the year 1983.

It all started at the end of World War II when the Russians presented a large, hand-carved ceremonial seal of the United States of America in a gesture of friendship to Averell Harriman, the U.S. ambassador. The object was later to become known simply as The Thing. Ambassador Harriman hung the object on the wall of his office. The only problem was that the gift had a hidden device built by one of the most brilliant minds of the 20th century, Leon Theremin. Inside the ornament was an ingenious listening device, basically an antenna attached to a cavity with a silver diaphragm over it that functioned as a microphone. There were no batteries or any other power source. It was activated by radio waves sent to the U.S. embassy by Soviets from surrounding locations.

The Thing used the energy of the external signal to send a transmission back to the source. When the external signal was interrupted, The Thing was silent. In this way, the Russians were able to record private conversations in the ambassador's office for 7 years in a row. The device was only discovered because American radio operators stumbled across conversations of the U.S. ambassador being transmitted over the radio waves. This device was the forerunner of RFID — short for Radio-Frequency Identification — which allows you to pay for small items by simply holding it up to an RFID reader. The principle is the same with Theremin's Thing. Since then, electronic warfare began and continues to this day. Even at the beginning, it already demonstrated the superior skills of the Russians in both creativity and techniques using electromagnetic waves.

This silent war was a final testing ground for the new electronic weapon on all fronts, including in-depth medical analysis and many detailed conclusions about tolerable radiation levels and their effects on humans. With regard to its use as a direct-attack weapon to harm people, employee records show that they suffered severe health problems. The damage is similar and consistent with that found worldwide in people periodically affected by radiofrequency radiation. Secretly, the two aspects of the weapon were tested — the one that causes direct damage and the more sophisticated one that interacts with the mind and thoughts. Nowadays both can be used simultaneously.

This weapon is also used in specific situations; for example, when a country wants to force someone who went into exile (in an embassy) to leave and surrender without the need to invade a foreign territory, which could cause unnecessary diplomatic problems.

Another point concerns the victims' perception of what happened. All, without exception, think that they were attacked by acoustic weapons. This is perfect camouflage for attackers. After all, if the victim is "hearing things", logically the event must be related to sound. But as we've seen throughout this book, the issue is much more complex than that.

Those affected by the device will hardly understand its dynamics as its cause and effect immediately create cognitive dissonance, which leads us to the following question, "How is something that can be heard and destroys the auditory and speech systems not connected to a sonic attack or something of the like?". Understanding the nature of the weapons involves several areas of science. The main one is the study of light and the electromagnetic spectrum — one of the initial topics of Volume 1. Light propagates through any medium — the walls of urban structures such as Embassies, for example; it is odorless and invisible to the naked eye. It also carries information that will be demodulated and absorbed by the human brain, causing all sorts of problems and destruction, as seen in the attack that left several American and Canadian diplomats seriously injured.

The version used is probably a much more advanced and powerful type of Synthetic Electronic Dream (D2K) and SYNTELE, of second or third generation, that carry not only voices and images, but a set of signals specifically designed to affect every area of human communication. It's difficult to connect high-frequency waves similar to Wi-Fi waves — in due proportions — to auditory system damage. It actually took me a few years to confirm the information as I researched for this book and observed the phenomenon in my daily life. Even today, after years of contact with the weapon, its effects continue to amaze me and fool my brain that thinks it's something to do with mechanical sound waves.

So far, it is not known who attacked the embassy with modern, psychotronic weapons — whether it was the Cuban intelligence service advised by Russians, the KGB itself, or the CIA. In any case, we now have more knowledge of the facts and can better evaluate these events, finally crossing the first barrier of utter ignorance.

"I've never seen anything like it and I can't explain it", declared Brian Latell, a former CIA Cuba analyst. It is extremely worrying that an analyst from the CIA didn't know about the existence of this weapon when the CIA itself, together with the military, was directly responsible for its creation in the 1960s, the infamous MK-ULTRA. His statement makes it clear how restricted the number of people who have knowledge about these weapons is and how much their use generates a range of conflicting information that causes confusion even among qualified personnel.

The New York Times wrote an article entitled Microwave Weapons Are Prime Suspect in Ills of U.S. Embassy Workers, available at: https://www.nytimes.com/2018/09/01/science/sonic-attack-cuba-microwave.html

The NYT article positively surprised me, as it was the first time a major media vehicle managed to write something relevant and close to reality — in other words, they seemed on the right path to elucidate the secrets of these unknown weapons. They even mentioned Allan H. Frey, one of the first to talk about electromagnetic weapons in the 60s. The article has links to the medical evaluations performed on the victims and the damage caused to the nervous system of the individuals and is worth to read.

In May 2018, reports confirmed another attack — but this time in China. U.S. diplomats at the consulate in Guangzhou were attacked by state-of-the-art psychotronic weapons and suffered the same brain trauma as the diplomats in Cuba. These attacks show that the communist axis is ready to regain its place in the world as they take place in the midst of several political disagreements and trade disputes in which tempers have flared between the United States, Russia and China. The electromagnetic warfare that has been going on since 1953 is actually just getting started. So, protect your mind and shield your home, reader, especially the place where you usually sleep, before you become a target yourself.

After all this, what do you think about the article and the way it was presented? With the more in-depth knowledge that this book has given about such prickly topic — a knowledge that is gradually taking over our everyday lives —, we'll analyze the reality that surrounds us more critically, perceiving its nuances within this game of information, counter-information and disinformation.

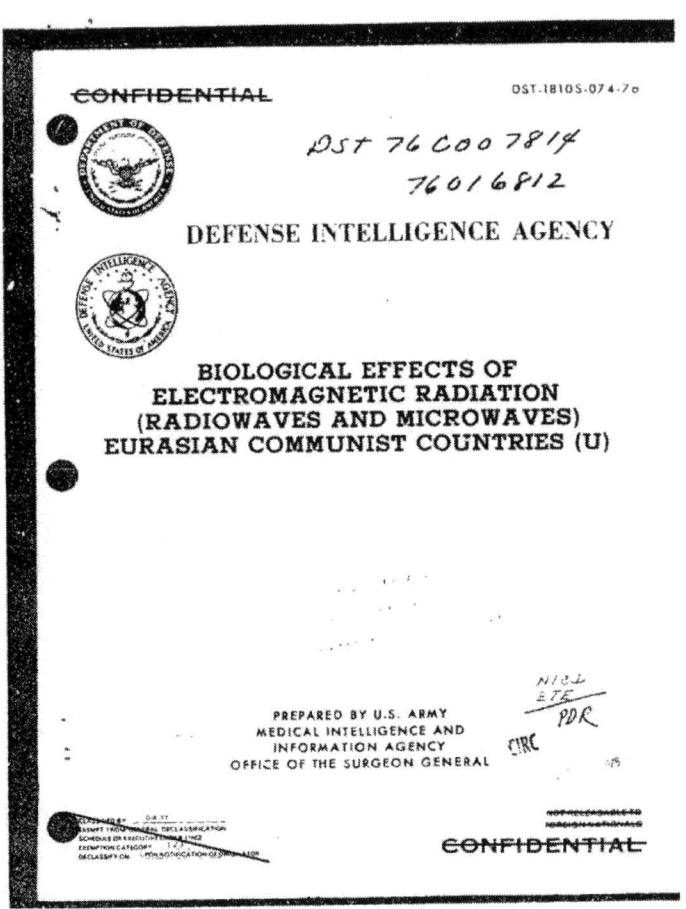

Figure 8.1 American intelligence document from 1975 that details the microwave, psychotronic weapons attack and reveals its consequences on the human mind perpetrated by Eurasia communist countries.

# CHAPTER 8 - DANGER IN THE USE OF THE TECHNOLOGY– TERRORIST ATTACK ON THE U.S. EMBASSY IN CUBA AND CHINA

Figure 8.2 Newspaper clipping about the embassy war using microwaves around 1960/1970. Psychotronic, electronic, and directed-energy warfare is a strand of the arms race during the Cold War.

## 8.2 - Official preliminary medical/neurological assessment

A team made up of qualified doctors conducted a preliminary analysis to understand the effects of the phenomenon on the 26 members who were attacked at the embassies. A preliminary medical assessment entitled **Neurological Manifestations Among U.S. Government Personnel Reporting Directional Audible and Sensory Phenomena in Havana, Cuba** found temporary neurological damage, which is consistent with those seen in targets around the world.

## 8.3 - A summary of the article

The 26 Americans exposed to the attacks had similar cognitive impairments, vestibular and locomotor dysfunction, sleep abnormalities, nausea and severe headaches. Unique circumstances that would cause neurological damage due to direct exposure to an as-yet-unknown technology.

Cognitive symptoms, including difficulty remembering things (n = 16, 76%) and experiencing slow thinking (n = 14, 67%) were the most problematic for subjects. More than three months after exposure, neuropsychological testing identified impairments in at least one cognitive domain in all 6 patients who completed the neuropsychological assessment to date.

Cognitive difficulties interfered with these patients' ability to multitask. Compared to vestibular and oculomotor impairments, cognitive impairments are often the slowest to improve after acquired brain injury, a finding that was observed in this study. In addition, it is not uncommon for patients with neurological damage resulting in cognitive impairment to have mood disorders such as depression, anxiety, and/or post-traumatic stress disorder. Mood disorders can result directly from acquired brain injury or develop in response to the precipitating event and novel deficits.

It still is unclear how the reported noise was able to produce such symptoms. So, they concluded that sound in the audible range (20 Hz -

20,000 Hz) is not known to cause persistent damage to the central nervous system and, therefore, the sounds described by the victims were probably associated with another form of exposure.

There are some important considerations about this analysis. For one, the anatomical substrates that cause the symptoms have yet to be identified. This can pose a significant challenge, as even the term "concussion" is still not a true diagnosis — no definition of the term includes the underlying cause. However, there is an emerging consensus that the concussion, or mild traumatic brain injury, is a type of brain network disorder based on classic symptoms (e.g., slow processing speed and memory problems), as well as changes in white matter tracts and consecutive connectivity, as detected with advanced neuroimaging studies.

I'd like to highlight the astonishment of the authorities due to the lack of information about the incident and the weapon itself, and the effects on the victims' brains that generates what some identify as a phantom concussion with no visible external impact.

**"The symptoms of the attacks were followed by the development of a cluster of neurological signs and symptoms only seen in mild traumatic brain injuries or concussions".**

The skepticism of investigators and physicians about the event was due to the consequences in the mass use of V2K, which carries a signal that causes massive neuronal destruction in a short time. They're in fact just looking at the tip of the iceberg when it comes to this weapon, trying to put the pieces of the puzzle together while analyzing actions similar to what happens to targets around the world.

## 8.4 - Conclusion about attacks and known problems with microwaves

The big question for specialists is, could the transmission itself cause the effects, or was there something else in the signal that when demodulated by the brain would cause, or help to cause, all phantom concussion symptoms?

The symptoms reported by diplomats were as follows:

* Buzzing, clicking, hissing;
* Confusion, impaired cognitive function, including short-term memory loss;
* Vertigo, dizziness, lack of balance;
* Headaches;
* Sensation of an electric current passing through the body;
* Inability to form mental images;
* Blurry vision;
* Nausea;
* Abdominal pain;
* Nosebleeds;
* Hearing problems;
* Trouble sleeping;
* Vibrating sensation in the head, similar to the effect of driving with a partially open window.

The use of microwaves as a psychotronic weapon, as carried out in embassies, is extremely sophisticated and capable of causing physical damage directly to mapped targets. Transmission (as a weapon) in direct beams causes all the aforementioned symptoms, but something else accentuates them: the demodulations of the content of the transmissions by specific areas of the cortex. That is, the weapon being used and tested has the familiar psychotronic attributes of weapons that attack TIs for deep mind control, combined with a devastating attack capable of physical alterations in the victims.

Everyone must have heard of neurons, the cells responsible for conducting nerve impulses within the brain. The neuron is a cell specialized in the transmission of information. Nerve impulses are electrochemical phenomena that use certain properties and substances of the plasmatic membrane that allow an electrical impulse to be generated and transmitted. Roughly speaking, this electrical impulse is generated by the difference in concentration of Potassium (K+) and Sodium (Na+) ions. And it is precisely in this location that neuroelectronic weapons of direct attack act — they attack the voltage-gated ion channels. A microwave weapon can be a directional device with a Yagi antenna or antennas affixed to satellites, using the same system that attacks targets on the ground with psychotronic weapons of deep mind control (MKTECH).

Just like an antenna that resonates at a specific frequency, the human brain does the same. This is because all dielectric materials have resonant frequencies based on their dimensions and electrical conductivity. The human brain is an electric organ that uses charged particles (ions) to conduct electricity — since it has dielectric properties, it causes resonance in certain frequency bands.

Research shows that radio in certain bands configured as a weapon — when interacting with the resonant frequency of an adult's mind (400 to 500 MHz) — can create an oscillation between fields and force the ions to move in specific directions, directly affecting voltage-gated calcium channels (VGCCs). This brings us back to the effects seen on diplomats and the various targets studied in this book. In essence, they basically alter the electrical flow of brain cells, causing physical neurological damage and sudden mood swings, among other "odd" sensations.

Further information on the subject is being updated continuously on news portals, newspapers and specialized scientific books.

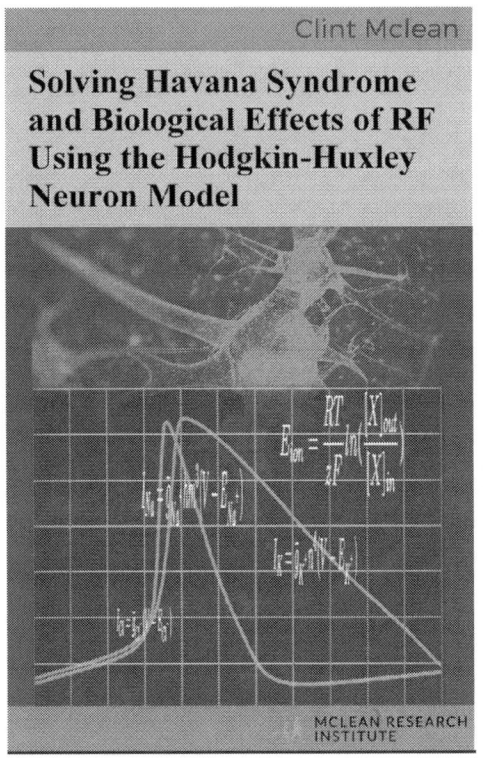

**Figure 8.3** A relevant book on the Havana syndrome that stood out from the rest was written by researcher Clint Mclean and is called **Solving Havana Syndrome and Biological Effects of RF Using the Hodgkin-Huxley Neuron Model**. It's a technical book for those who want to delve deeper into the subject, as it tries to explain the phenomenon with the help of several studies based on scientific evidence. **Link:** https://www.amazon.com.br/gp/product/B0BCNG8H89

# CHAPTER 9
# COMPUTER SYSTEMS, AI AND SATELLITES USED IN MKTECH

MKTECH is composed of a series of programs (software) developed for deep remote interaction with the human brain. Imagine a case with different tools created for a common goal: to invade, read, listen and insert thoughts into people's minds — to hack their thoughts and torture them to death, like a real psychotronic weapon! Among these programs is the AI (Artificial Intelligence) that learns personal and intimate details of our personality from our behavior and use this knowledge to improve the attacks, sharpening the understanding of the complex human nature.

## 9.1.1 - Artificial Intelligence[61] of the MKTECH system. The Dark Side of AI

We already know that AI is increasingly present in our lives, from the most innocent mobile game to complex military systems that make autonomous decisions — a possible target to be shot down, for example. AI emerged with the father of modern computers, Alan Turing. He published an article in 1950 describing the Turing Test, a method of inquiry for determining whether or not a computer is capable of thinking like a human being. Turing called it "the imitation game". This test gave rise to what we know today as the philosophy of artificial intelligence.

In August 1956, John McCarthy coined the term Artificial Intelligence. He also received funds during the Cold War to develop an AI to automate the translation of Russian documents and analyze patterns in

---

[61] Artificial intelligence is a broad term that refers to computational intelligence. The purpose of this chapter isn't to delve into the universe of AI, which is made up of various concepts such as machine learning, deep learning and neural networks. In fact, each concept encompasses numerous different techniques, structures, equations, models and details. So, when using the term AI, I'll be referring to all technicality and functionality of these algorithms.

language. However, the first chatbot named ELIZA was created in 1966 by Joseph Weizenbaum. ELIZA was the first computer program capable of holding a conversation with humans. But it was only in 2014 that Eugene Goostman, which was developed in Russia, became the first chatbot to pass the Turing Test. In the decades that followed, artificial intelligence was associated with science fiction and cultural imagery of humanoid robots that would think, feel and emote just like us. There was still no computer processing power that could implement this and several other concepts in practice.

It's quite different nowadays. We still don't have humanoid robots walking around imitating the human form to perfection, but artificial intelligence has become ubiquitous and is present in all technology with or without a robotic "body". Its power lies in its proficiency in coding, internet search algorithms, cell phone apps, social network profile analysis algorithms, video game enemies and characters, virtual "drivers" of self-driving cars, personal assistants such as SIRI, preference-based selection algorithms (video and music), simulators, internet shopping algorithms, among thousands of other examples from everyday life.

Among the technologies developed in centers of excellence, there are algorithms capable of analyzing thousands of literary texts by themselves — narrative structures, character arcs — and predicting which books have the potential to become bestsellers. They are able to anatomize the best-selling works and point out the secret of success by analyzing the profile of the protagonists and antagonists. Other algorithms are capable of writing journalistic and academic texts indistinguishable from those written by human hands. They're able to analyze academic papers and draw scientific conclusions and hypotheses that have been proven to be accurate.

This breakthrough was only possible due to the change in the use of the machine. Instead of just making it follow pre-programmed instructions with a script, they started to guide the machine with a system called machine learning. Deep machine learning is useful for many purposes, including astronomy to catalog the billions of visible galaxies and discern their unique characteristics.

Nowadays the goal is to teach the machine to recognize patterns and thus be free to make some decisions on its own. From 2013 to 2018 AI technology grew by 300%, greatly increasing human productivity. All technology giants and military agencies have divisions that work exclusively on the development of increasingly advanced algorithms that learn, execute and make critical autonomous decisions in various fields, including warfare.

One of the most complex and effective models of machine learning refers to Neural Networks, or artificial neural networks, which represent a technology that has roots in many subject areas: neuroscience, mathematics, statistics, physics, computer science and engineering. Neural networks find applications in such diverse fields as model building, time series analysis, pattern classification, and signal processing due to an important property: the ability to learn from input data with or without a human supervisor.

A neural network is a massive, parallel distributed processor made up of simple processing units that have a natural propensity for storing experimental knowledge and making it available for use. It resembles the brain in two respects:

* Knowledge is acquired by the network from its environment through a learning process.

* Connection strengths between neurons (known as synaptic weights) are used to store the acquired knowledge.

The procedure used to perform the learning process is called a learning algorithm, the function of which is to modify the synaptic weights of the network using a range of mathematical functions in an orderly fashion to attain a desired design objective after some cycles of self-improvement and training.

Artificial neural networks that act in mind control systems are computational algorithms and recognize complex patterns in the electrical activity of the brain. Capable of perceiving abstract feelings such as sadness and happiness, they are connected to a remote Brain-Computer Interface (BCI), and combined with chatbots and machines that establish dialogues with people using natural language processing to answer questions, use verbal attacks and perform actions with the intention of deceiving the victim, sending voices via V2K with such a variation of dialogues that they resemble a specific individual and convince the target to hand over their secrets, while accessing their memories only using auditory processing and employing human-understandable language. So now we have the perfect combination that autonomously degrade and destroy the minds of an entire society. The AI is capable of using all these devices, including remote dreams, based on the purpose and objective behind the targets in question.

The goal of Artificial Intelligence (AI) is the development of paradigms or algorithms that require machines to perform cognitive tasks, which humans are currently better at. Therefore, it must be able to do three things — to store knowledge, to apply stored knowledge to solve problems, and to acquire new knowledge through experience. An AI system has three components: representation, reasoning, and learning. The system must be able to express and solve a wide range of problems and problem types, and be able to make known to it both explicit and implicit information.

Contemporary artificial intelligence at work in the MKTECH scheme represents the greatest danger to humans, as they use precepts from computing such as BIG DATA, Neural Networks, and Brain–Computer Interface (BCI) much more advanced than those available in current computational systems to analysis algorithms of all our biochemical and electrical processes, voices, emotions, physiological states, biometric data, thoughts, dreams, desires — in short, everything that comes from our deepest thoughts and intimacy. With this level of knowledge, it's possible to make the AI decide how to cause the greatest possible damage based on all aspects learned from capturing emotional, physiological and behavioral parameters. With a complete picture of the target's momentary emotional state at hand, the AI can carry out atrocities that Isaac Asimov himself would never have thought of. Perhaps we never seriously believed that a full level of interaction with human cognition was even possible.

MKTECH and its systems show the little-known side of the use of AIs. A negative, bleak and dangerous facet within a broader purpose at a unique level of deep interaction with the human mind, which feeds on internal private data and signals interpreted by the brain from the external environment.

We have a wide range of functions in the fields in which the algorithms are designed to act, calculating the emotional impact of how a certain act or subject raised through V2K will impact the target emotionally and physiologically. In this way, it's possible to generate a complete emotional state with words and situations from the past that mostly cause negative psychophysiological changes. This entire process is stored in a database with records and statistical models. Internal emotional struggles (captured in social interactions), emotional and physiological pain, including guilt, resentment, regret, fears and urges, are the feelings most explored by the new AIs that work non-stop to completely unravel the human mind.

Now, we must think in stages and verify the purpose for each AI model when it comes to its specific primary function, such as the one responsible for collecting mass data from the target's mind. One of the aspects of modern MKULTRA experiments is to supply and feed such AIs that will act all over the globe; working on a cloud computing model that receives more and more data from fragmented experiments conducted around the world. This is done in order to respond and interact in an increasingly accurate and efficient way and thus be able to quickly investigate emotional profiles to later assume how to inflict the greatest possible damage to the target without the need to report certain actions to humans. This gradually leads to full decision-making autonomy.

Data collection and universal algorithms have many practical purposes as well, not only for military and intelligence agencies, but also for big companies that have contracts with defense departments worldwide. In this way, knowledge is generated for commercial use without precedent in human history. But due to the amount of the most varied data being captured and stored by servers, and the constant flow of information that travels from the targets' stolen thoughts, a large volume of data is generated and cannot be analyzed in the traditional way. So, in order to deal with this vast amount of data, MKTECH uses the concept of Neuro Big Data, a type of technology similar to that used by big computer companies today.

*Big Data* is commonly used by large social media corporations, such as market research, to understand and predict consumer behavior — sophisticated predictive models for analyzing customer history and predicting possible fraud in a given segment of the economy, or even for weather forecasting where models work with a huge number of variables to try to predict the dynamics of atmospheric conditions and efficiently and accurately give a future forecast.

It's estimated that these complex algorithms are capable of analyzing a volume of data in Petabytes — a petabyte (PB) is 1,024 terabytes. Think about it. Stop and observe how much information your brain processes in the next 10 seconds: audio, sight, visual thoughts, silent thoughts, autonomic systems, somatic systems. All this volume of data is captured for years directly from the target's brain that is connected to the system, thus generating a galaxy of information extracted from a universe of its own, and capable of operating with data in real time, while others considered less relevant at the moment may be stored for future analysis, or discarded.

In order to filter the data and set priorities, a lot of distributed processing and cloud computing is used, as well as NEURAL DATA MINING, another AI that works exclusively to extract in this immense sea of information relevant data for the system using machine learning, and achieving a satisfactory degree of data organization and quality. They use several clusters, data repository spread across the system, fragmenting the process and reducing data access time. The relevance and quality of data are of paramount importance and priority for AI learning, which must remain accessible and available for constant acquisition, as the learning element uses this information to improve the knowledge base.

Machine learning may involve two rather different kinds of information processing: inductive and deductive. In inductive information processing, general patterns and rules are determined from raw data and experience. In deductive information processing, general rules are used to determine specific facts. Similarity-based learning uses induction, whereas the proof of a theorem is a deduction from known axioms and other existing theorems. Explanation-based learning uses both induction and deduction. Knowledge refers to stored information or models used by a person or machine to interpret, predict, and appropriately respond to the outside world.

For example, classification algorithms, predictive methods, classifications of predictions — making predictions based on information from the past, grouping them in databases — can be used based on certain situations sent during dreams controlled by the Synthetic Electronic Dream (D2K). And then, you can group these situations into visual stimulus and subsequent reactions, and classify them from the most harmful to the least by applying data associations, merging them when deemed necessary to surprise the target and to cause a new range of unprecedented effects as they learn from their consequences.

The demonization of Artificial Intelligence is not the point here. However, we must show the other side of the story. When used as a vital part of the functioning of neural weapons, AI is as dangerous as (or more dangerous than) the humans who attack targets. Keep in mind that Artificial Intelligence is here to help us in the most diverse tasks and day-to-day problem solving. They're present in various human industries such as robotics, data mining, software verification, natural language understanding and translation, facial, speech, odor and taste recognition, biometric analysis and different kinds of image analysis, in addition to accurate image diagnosis (more accurate than human's) such as magnetic resonance imaging, computed tomography (CT) scan and laboratory analysis (blood tests and others).

## 9.1.2 - Computational intelligence

In the field of data analysis, machine learning is a method used to devise complex models/algorithms that allow predictions for commercial use; this is known as predictive analytics. These analytical models allow researchers, data scientists, engineers, and analysts to produce reliable, repeatable decisions and results and uncover hidden insights by learning from historical information and trends in the data.

In more personal terms, we have at one end the long-term attacks focused on the profound study of human nature, which enhances the increasingly sophisticated torture techniques implemented with MKTECH autonomous weapons systems. At the other end, we have the understanding of the human psyche and its changes and reactions to torture combined with the general consequences of this unprecedented cognitive invasion. So, we have a global AI, which gathers data — learns about all human aspects in a more comprehensive way —, and the individualized AI, which is subdivided into several more specialized AIs, capable of both adapting to the profile of a specific target, delving deep into their personal lives, and feeding the global system controlled by the main AI.

The AI gathers a range of data coming from different parts of the MKTECH system - SYNTELE, D2K, REMOTE EEG, V2K[62] and thus is able to trace the current physiological and emotional state of the target more accurately. It also indicates to the operator the probability of an approximate response to a given question sent by SYNTELE, which generates responses in the brain in the form of image and vocalized thoughts. Even if it's blocked by a trained mind at the first stage, the AI is able to indicate whether the response to the auditory stimulus is true or false, requesting parameters already acquired via Remote Polygraph (Volume 1).

For example, if we were to analyze only the D2K system, we'd enter a world of unique complexity, since the algorithms that operate this particular system have to deal with even greater subjectivities when compared to the data in the waking state. In fact, the reality of the dream world won't abide by any rules, and the scenario, the scene, the context of the event of the dreams that affected the target in a certain way, will no longer be repeated. In other words, if the experiment is repeated, the

---

[62] It'd be necessary another book in order to analyze every computational part of each module, which may become a reality in the future using technical data and in-depth computation aimed at a category of readers interested in such matters.

outcome will be different. Thus, adaptation is also an integrated part of AIs.

Also, we have some modules and each specific AI system responsible for its operation, focused on specific rules. Thus, the algorithms will work in each module taking into account its particularities in an integrated way, forming a cohesive system that is practically failsafe against a common target. If your brain is hijacked, you have little chance of surviving without sequelae. However, there is a light at the end of the tunnel. After reading this book, odds of current and future targets are likely to improve exponentially.

## 9.1.3 - AI - Infrastructure

Here we have the first complex algorithms that make the infrastructure work. That is, they make the physical part (the hardware) work in synchrony to reach the goal. In this case, the algorithms receive the target's telemetric data and calculate their geolocation, their exact position on the globe, latitude and longitude, and the operating radius. Subsequently, a maintenance algorithm checks the available infrastructure — terrestrial antennas, towers, satellites —, traces the best, fastest and most efficient route to deliver the data packet to amplify data in the mind or receive the feedback signal and confirmation that the data was delivered via SYNTELE, minimizing as much as possible the possibility of losses, interference or LOS[63].

So, the target's brain is connected to the MKTECH scheme in a constant way and without the slightest possibility of disconnection regardless of where the target is. The access channel to the mind is always open and easily perceived. Thanks to these advanced algorithms, MKTECH content can be sent or received without worrying about the functioning of the infrastructure. Specific communication protocols are

---

[63] Loss of Signal (LOS) — some factors are responsible for a drop in the quality of the transmitted signal. Attenuation corresponds to the energy loss of a signal traveling on a physical medium (the Joule effect). Noise is any random phenomenon that disturbs the correct transmission of messages and that generally one tries to eliminate as much as possible.

used for the AI or the algorithm that constantly sends data packets and receives the brain's own response to the stimulus via EMRA - Electronic Mind Reading (auditory).

## 9.1.4 - AI - Raw brain waves

This AI is responsible for creating the target's neurobiometrics (similar to digital biometrics) and transforming it into an ID — a profile — for the system to constantly compare brain wave patterns and indicate whether the biometrics and the profile created and sent to the system is that of the current person being monitored. All this happens in fractions of milliseconds. You can find more information in Volume 1, chapter 3.1, Neural Remote Biometrics.

## 9.1.5 - AI - Mind reading and emotional states

The AI is responsible for collecting data and metadata for a centralized base, and it works individually, mainly in capturing subjective, primary, secondary, social and background emotions that are triggered by the perception of the momentary state of the body related to sensations that, in fact, alter the human body, as well as those triggered by external factors — by reactions to environmental situations.

One of the most relevant aspects for the scheme is the data captured by executive functions located in the frontal lobe — higher, complex, and specialized cognitive functions associated with judgment, planning, strategy formulation and inhibitory control of impulses. That is, the relationship between intellectual aptitudes, general and specific intelligence, deductive inferences and emotional skills necessary for human survival.

In addition, the AI learns to identify subtleties in the deepest depths of human nature and is capable of capturing the general status of a specific experience in which the individual recognizes the existence of a personal flow process of a certain event directed towards a goal. Thus, the algorithm adds this set of parameters in its test sample, leading to the cognitive event that occurs easily in the human mind, cataloging it.

The AI can also use a flag variable (Boolean flag) that indicates to another system or to humans who analyze the situation that the target is focused on a certain important event that must be blocked, affecting their consciousness and modifying the flow of thought, finally altering their state of lucidity. Interaction takes place on such states — uniquely human, complex processes being used by machines and incorporated into AI and autonomous response systems.

By using advanced probabilistic classification approaches based on established techniques, such as Bayes' theorem and linear regression, it's possible to analyze and explore emotions in two ways: visually, with the analysis of facial expressions and organic data, and by making a deeper analysis of emotional responses via vocalized thoughts and images from the target that contextualize the situation. A graphically displayed map of the brain indicates from which part the momentary thought is coming from, helping to create a more accurate neural analysis, and obtaining visual confirmation from the remote polygraph when the sequence of decisions needs to be displayed to humans, thus assisting their choices.

Several frameworks are also part of it and are used to streamline the process or for operators to make specific changes and complete analysis as part of the neuro-electronic warfare and surveillance system. Mind-reading artificial intelligence is capable of remotely interfering with the emotional states of targets. A computer compares the data that is constantly received from the individual, which contains specifications on how to interpret new individual information from the profile of that specific target, stored on a constantly updated platform. This includes temporal information, histograms and reactions to particular attacks, such as predictive attributes, thus evaluating what causes the greatest emotional damage, and individualizing the attack and making predictions, as the aim is to explore the unique weaknesses of each individual's personality. In addition, it also measures the negative effect of the attack that causes consistent and complete destruction to all targets, such as the incessant noise in the mind via SYNTELE, sleep deprivation with V2K and mental degradation using D2K.

Let's look at a practical example. Even if the Artificial Intelligence managed to capture the paths that lead the target to a gloomy feeling and left them depressed — or morally injured, frustrated and insecure —, it would still not have the ability to feel this type of emotion in its entirety, nor would it know what this feeling represents as a human sense, which involves countless hormonal variables, specific electrical activities in certain areas of the brain, amygdala and hippocampus activity, and so on. However, the AI is able to indicate the occurrence of this specific physiological state. Its goal is to detect decaying states and aggravate them. If the current feeling is in disagreement with the rule previously established by humans — emotions considered positive that produce general well-being—, the reversal of this situation becomes a top priority, so it attacks the brain with everything it has learned that leads to decaying states.

Artificial Intelligences aren't be able to comprehend the consequences of their actions (at least for now). They don't have any philosophical principles or knowledge regarding a moral and ethical standard; they won't be altruistic in response to the intense pain felt by a target, or even develop spontaneous empathy or pity for them. They will just proceed with the torture as they were instructed until reaching the main goal. In the process, their actions will be captured by algorithms (automatic learning), and they will be able to learn from their mistakes and make specific predictions on continuous data, using responses — brain reactions to attacks. AIs can then analyze the physiological state, lead to adverse states and maintain that level for as long as possible in an incredibly accurate way. The super efficiency of AIs in conjunction with human ruthlessness (acting as a malevolent supervisor) makes this a deadly, degrading and cowardly weapon that catch targets by surprise. Victims have no chance of escaping from it.

As the AI has what we call a simulated basic level of consciousness — in other words, it's constantly improving —, moral ethics, generosity, or noble traits that we consider good for us humans are a set of variables, statistical data and metrics of all kinds, however they are detected and are part of the range of studied aspects of the target and the learning algorithm. The curious thing about it is that the deep interaction of the AI works most of the time with feelings and emotions, but it doesn't have a processing center capable of generating such feelings like the amygdala and the hippocampus — the AI is unable to understand them. For us, what the machine does —at the behest of humans— is interpreted as extreme violence and result in destructive impulses that we consider harmful.

The Telemetric EEG (Volume 1, Chapter 3) is able to scan the brain, receive an absurd amount of raw brain waves and analyze them together, in addition to capturing internal sensations such as visceral reactions to other people's physical pain, a sudden scare or shocking news. The Artificial Intelligence-Based Brain Computer Interface is fed back by this data on a constant basis. Algorithms operate by building a model based on sample inputs in order to make predictions or flexible, probabilistic projections, showing that the AI works with diverse and complex inputs coming from all over the MKTECH system. Specialized AI technology is used, which only plays a specific role, but does it masterfully.

Social phenomena in the face of human nature are of paramount importance for AI learning. It must know the target's personality in depth, so it registers everyday activities, such as fights caused by social interaction, in which the individual's delicate issues, weak or negative traits that cause emotional pain are addressed. Such information is logged, ranked in a series of temporal context metadata, and will be constantly replicated by the attacking AI. The AI accesses a dynamic ranking capable of showing the perfect moment to utter certain words that will cause pain and at the same time gain the attention and focus of the target. It will even capture signs of behavioral abnormality, hyperactivity and catatonic behavior, which generates more complex cases of Neurosis (F40) and Psychosis (F23).

The AI finds the set of parameters that characterize these conditions and maintain them in order to impair mental health to the maximum. It even carries out attacks with the aim of slowly subjecting people who don't have such symptoms to the aforementioned conditions — sustaining them as much as possible. After a certain time — months or years — in this degraded state, it's likely that the target's sanity will never be the same again. The algorithm has indicators that pinpoint precisely when this universal human behavior is occurring. Therefore, they can decide what to do as "a master premise": if it is to cause destruction or to create mental pictures using the attacks in a subtle way—, further exploring the personality of the target and contributing to the gathering of unique sample parameters to the scheme. Or it could be even worse: they could do both. That is, to destroy the TI slowly as they capture the data. It's like a neural parasite that use its host and exhaust it to death or take possession of it, damaging it bit by bit and killing it slowly and productively.

As neuroscience advances and makes new discoveries, the system's AI is also updated, self-empowering itself to deal more efficiently with our neuronal plasticity and our capacity for constant adaptation and the dullness of emotions linked to memories that gradually and naturally happen. On the one hand, we have stimuli that require changes; on the other, there is the power of adaptation and the psychophysiological mechanism of the individual. The brain is always looking for patterns in the outside world by comparing them with the model stored in the hippocampus through previous learning information modulated by emotions given a certain context.

It's impossible to decontextualize and remove emotions from a symbolic knowledge of the human perception of reality, leading to a mixture of emotions and sensations that the AI must capture and reverse or encourage at any cost, using any means necessary, including the known methods of torture added with the natural extreme negative psychological factor of psychotronic weapons that creates a starting point and advantage over the target's natural defenses. AI: 1. Humans: 0.

Another point of extreme relevance and priority for AI is to completely interfere with the target's privacy in any location, but especially at home. When the target enters the bathroom, the AI alerts that it's time to explore this opportunity to the fullest with the help of humans (operators/OPS) who hear everything and see the target's thoughts, preventing them from satisfying their biological needs.

Sex or any kind of pleasure related to the activity, considered by most to be one of the most enjoyable feelings, is immediately suppressed as soon as it's detected. Only the intention alone activates the module that completely directs the thoughts using the MKTECH equipment. The algorithm is based on patterns acquired over time that include sexual preferences and fantasies, or situations and memories that cause sexual arousal, such as more intimate memories with partners that have impacted the target's sex life and every now and then are requested by the mind to activate the state of sexual arousal on certain occasions.

In addition, it's possible to detect penile erection, as well as approximate values of vaginal temperature, dilation and viscosity. With the simultaneous transmission of data coming from several systems, the accuracy to pinpoint the event is very high. Now the primary focus of the attack is to exhaust the target's mental energy dedicated to the activity, making it impossible for them to proceed, decisively disrupting the moment by running all SYNTELE channels at maximum power if necessary, sending sounds, voices and shrill, deafening noises into the brain. Over time the victim is conditioned not to think about this kind of activity anymore, and unconsciously moves away from people who might lead them to sex. This creates almost enforced celibacy for some TIs, emasculating them in all spheres of life. Any similarity with Delgado's work (chapter 7) with the chimpanzee Paddy is not mere coincidence.

AIs capable of processing massive amounts of data and making internal processes work are arguably the most prominent and important for the scheme. However, there is another type of AI that tries to express human emotions in words, and is able to interact with the target's mind by impersonating a human. They are known as chatbots of various intelligence levels, capable of simulating social intelligence as well as moral decision-making. A derivation of all types of known bots that "came to life" through predictive methods such as classification and regression, detection of standard and sequential deviations, and summarization is also used for this purpose. It's a natural language processor that understands the semantics of the transcribed sentences and interacts with the target — extracting all features from a database accessed by a learning algorithm with operator supervision, thus providing implicit indications and generating models using associations and groupings, which automatically examines the data in the analysis of meanings of patterns and descriptive methods.

We thus have the linguistic and verbal communication aspect of AI— the pernicious interface between the processes and the target in which the synthesis of what goes on behind the scenes of this complex technology is shown in words and behavior. It's similar to robots that interact with humans, or automated call center services, but extremely advanced. The quality is infinitely superior, as they are able to easily impersonate someone, utter bad words and hold a more succinct (and better) conversation than most humans without needing to stop to drink a glass of water. They're able to conduct an interaction also based on reinforcement learning. That is, they learn from interactions, from the environment, from cause and effect, and from the experiences and results of previous interactions with the particular target. It's one of the most malevolent, diabolical and destructive aspects of the technology inside this machine-brain interaction that few could predict: a psychotronic alliance.

## 9.1.6 - Emotions and Feelings

Emotions and feelings are unique characteristics of the individual. Algorithms perform the segmentation after extracting the characteristics (attributes) of the target — in this case, their mental or physiological state —, and accurately analyze what is happening internally at the moment. For example, a vector containing EEG waves based on a set of pre-processed parameters, which makes it possible to immediately recognize the target's mental state, such as focused, relaxed, joyful, optimistic, among others. Afterwards, this process is evaluated and an efficiency value is computed with regard to emotions — a vector of characteristics of the raw data (alpha, beta, gamma and theta waves) already determined by other decision algorithms that support the continuous qualification or adaptation of the emotion that is present in their "spirit".

The algorithm differentiates and divides the elements and characteristics to analyze them in a common and comprehensive way between all humans and their uniqueness. Well-separated decision boundaries with various classifiers are used for such activity. For example, a model based on a set of diverse data already qualified and previously evaluated is chosen, which reflects the target's general state at the moment and causes these parameters to converge to some kind of emotion or condition — such as focused and happy. Based on the objective definition of "focused", it's only possible to ensure this result just by reading the brain waves. But what about the abstraction of happiness? How to explain such a complex, transitory and subjective state to the machine?

Data coming from another algorithm that uses EMRA, which analyzes aspects in the person's voice, are taken into account to help the main algorithm come to the conclusion that the person is really happy at the moment. Obviously, there is a key component: the verification of thoughts and the flow of images being perceived at the moment and that are based on our imagination, emotions and judgments of thoughts that help the algorithms to identify a certain emotional state. Each emotion alone contains an amount of data that requires an auxiliary algorithm for such purpose.

So, how to "tell" a machine to work with already well-established principles capable of capturing distress and indicating an emotional state with accuracy? For humankind existing is already a great emotion; the same is not true for machines. Basically, feelings occur naturally for us — we just name them.

These algorithms are so advanced and capable of determining such subtle nuances of emotions that after a period of interaction they can even anticipate the expressed behavior to be performed by the target after a certain attack, the duration with which this attack was executed, the target's resistance to it and the level of damage caused, which is confirmed in the behavior found on historical data being sent to the machine that uses complex algorithms based on decision trees and neural networks.

| Feelings/emotions to be immediately reversed by the AI, preferably by the opposite types |||
|---|---|---|
| Happiness | Peace | Safety |
| Well-being | Tranquility | Motivation |
| Pleasure | Comfort | Self-confidence |
| Hope | Altruism | Courage and Bravery |
| Longing | Charisma | Feeling respected |
| Joy | Emotional attachment to someone else | Sense of completeness |
| Willpower | Relaxation | Feeling of being loved |
| Love | Euphoria/Excitement | Compassion |
| Empathy | Admiration | Faith |

| PHYSIOLOGICAL SENSATION | COMPLEX ABSTRACT FEELING |||
|---|---|---|---|
| Drowsiness | Hopelessness | Paranoia | Frustration |
| Insomnia | Concern | Insecurity | Guilt |
| Tiredness | Dissatisfaction | Helplessness | Suspicion |
| Muscle fatigue | Sadness | Anguish | Bad introspection |
| Crying | Disorientation | Fear | Self-doubt |
| Chills | Madness | Emotional pain | Emptiness |
| Shortness of breath | Phobia | Feeling that something is going to happen | Lack of sex drive |
| Shivers | Low self-esteem | Unfitness | Feeling unappreciated |

The table above shows the feelings/emotions captured by the AI as the main goal to be achieved and preserved at any cost. The pattern of action is to keep constant all those negative feelings and sensations, making various adjustments to the technology and to the structure and content of the attack until reaching the goal.

The main rules of these advanced algorithms are to keep the primary, innate emotions (such as anger, fear, sadness and disgust) at maximum levels. Fear, for example, is one of the most important emotions and must have priority whenever detected until the target is exhausted. In other words, constant fear, as an unpleasant experience accompanied by an escape behavior, is key to a successful attack. At this point, the target is capable of kicking people out of the house, losing their job, or committing dangerous/impulsive acts.

Fear — a wise adviser and insightful master that is always aware of the threats around us — "in the right amount" is healthy and necessary for our survival. But if it is constant, it becomes dangerous for the individual; it becomes terror. If what threatens us is totally unexpected and not familiar, fear becomes dread, which triggers unpredictable, erratic and dissociative behaviors. As a consequence, it extrapolates the sense of escape or confrontation. Shyness, timidity, concern and anxiety are also considered other forms of fear. Fear has its own scale that indicates what level the target is at. Thus, it facilitates the insertion of other unpleasant feelings that come with fear, consequently affecting cognition that will always work defensively; thoughts will always be negative and future predictions depressing or extremely pessimistic.

Raw data are used to detect fear, such as fast heartbeat, change in expected behavior (out of the ordinary), temporary physical paralysis or psychomotor retardation. The scanned electrical activity looks for standardized EEG patterns that have already been previously mapped in real moments of fear or in past or future thinking that synthesize what we know as fear, such as the processes of access to memories and bad future predictions that harm the target, either criminally, socially or emotionally, and immediately generates this bad feeling. Once accessed, private memories will be readily identified. Other activities that help the machine with the fear scale are physical and behavioral recognition patterns, such as facial expressions, visceral reactions (butterflies in the stomach), amygdala activity, and low level of attention due to an adrenaline rush.

Later behaviors are also monitored, and indicate the consequences in the form of actions and the calculation of the intensity of said action together with its previous causing act, monitoring the success rate and generating subsequent coping behaviors, anger, hatred, emotions that manifest themselves in aggressive behavior followed by an anger attack. Since the electronic torturer isn't physically there, there is no one to vent their frustrations on — the rage consumes the target and makes them feel like something bad is about to happen, which keeps them alert, change their routine and attention at work and life, finally leading to constant, corrosive hatred and anger.

The attacks result in total exhaustion and acute stress, leading to constant sadness, which is a universal human response to situations of loss or defeat. Furthermore, it generates something that perhaps would never be felt by most people: the beginning of thoughts and processes that create the "programmed" killer or Manchurian Candidate.

## 9.1.7 - Distorting the perception of reality

The final consequences of this fine-tuning of the AI system manage to alter the perception of reality, making the target feel sick and with a negative self-image. The desire to commit suicide, or to kill those involved in the attack, increases. The target is constantly and cowardly observed, judged, criticized and tortured by computer-generated voices and by real people, the Organized Professional Stalkers (OPS).

At this point, the AI will maintain the settings, indicating the continuation of the attack at the same levels, but with small variations while it monitors the general health of the target and makes them give up their life and family and, above all, preventing them from responding to the attacks through severe demotivation. Acute stress is the AI's ultimate goal along with the collapse of the target's entire cognitive system and slow, painful death. Silence in the electrical activities of the heart and brain is the Ai's ultimate mission.

These algorithms achieve great psychic feats by distorting the surrounding reality, or by modifying the interpretation of it as stimuli are captured in the environment. It's like a neuroelectronic drug (similar to the effects of chemicals) capable of suggesting a state in which targets think they're fully conscious. It alters the ability to correctly infer a certain inner state by deforming the interpretation of surrounding signals, generating a distorted view of reality of future events and analysis of the interpretative conjuncture of internal data.

The AI has the ability to interfere and also explore characteristics in the signals that go from the higher to the lower regions of cognition, attacking the prediction, which is a brain process that constantly adjusts itself, redefining a future projection based on the new sensory experience captured, and comparing it with the information previously stored. This disparity between what we expect from a given scenario and what actually happens can be distorted when the signals go back to higher-level areas of cognition, which would serve as a natural refinement of the status of absolute reality around us. This event is constantly modified with the processing of the SYNTELE content via the triggering of emotions and reactions to the cognitive violation and the microwave voice. Since these visual signals are part of the process, they travel through the Lateral Geniculate Nucleus (LGN), as we've seen in chapter 2.4.1.

As soon as the image travels through higher-level areas to the visual cortex, it updates our expectations and the brain tries to do the same with its predictions. This occurs mainly in the creation of thought images strongly influenced by future negative emotions, which is based on a distorted reality created by operators using D2K. The capacity of transmitting mental images causes conditions that were influenced by the events created by the technology, including in the order of presentation of the target's visual thoughts. There is a dynamic between the actors that is only possible to be executed by advanced Ais during attacks. And unfortunately, these AIs work at the speed of human thought.

## 9.1.8 - Answers and surveys

A computer generates an automatic response based on the analysis of data received with the aid of assistant programs that tell the machine whether the action to be taken within a given situation is: a) passive; it only observes and captures thoughts; b) active; it recruits computational resources to conduct certain direct attacks; c) reactive; the new attack adapts to the cognitive and behavioral response to a previous attack; d) positive; it provides a false sense of well-being and psychic freedom in order to cause a deep emotional impact by returning to previous levels of attack; or e) negative; it focuses on inner negativity and deep, complex emotional drive capable of making people commit suicide.

Within these levels there is a scale between one and the other in which, for example, an active automatic response can vary from a weak attack that greatly annoys the target — but can be ignored —, to a maximum attack known as Computerized Swarm Attack (chapter 5) in which the AI "goes mad" and uses all its available resources to cause the maximum amount of continuous damage in a short period of time that promote irreversible neurological conditions.

The AI is also frequently sending a survey packet, parts of memories or words that estimate the process of accessing the complete chain of requested sequences of fragments, completely absorbing the memory, and capturing the data referring to the analysis of processing in the brain, therefore evaluating the results. Analysis of forced access to private memories provoke different reactions both for the disclosure of their content and for the content itself. So, the algorithms can identify the ones that cause a certain type of emotional pain (key points of pain), or evoke memories that can be used by the algorithm to inflict constant emotional pain in the process. There is endless word analysis, and in a parallel and simultaneous way the approximate activity in the amygdala (the place where emotions are processed) is scanned, inferring values for words, which may or may not trigger bad memories.

## 9.1.9 - Conclusion

The operation of the Brain-Computer Interface (BCI) and the processing of signals from real-time analysis of brain activity, biological signal processing techniques, a series of mathematical models such as the algorithm developed using AI, which was based on neural networks can, over time, begin to predict certain types of behavior and thinking. This is widely used in D2K to predict the electrical and "physical" kinematics of the dream — to verify the target's reaction to a given situation, even within the dream. A range of sensations and emotions perceived by the AIs with the help of humans (or not), based on several parameters coming from the most diverse places, occur at every moment. You can also train the AI to do the inverse — that is, to do good —, but this never happened during the attacks. The interaction of the machine with the mind takes place every millisecond for years on end in a sadistic and monstrous way, playing with the most complex areas of the brain while enjoying the privilege of having full and unrestricted access to the target. And all this against their will! A systematic and erratic behavior is caused by drowning out the voice of the target's conscience and taking its place in rash actions and decisions.

Once you hear people in academic debates or bar conversations saying that we're far from having an AI capable of harming humans, you better think twice. This is not a distant dream; this reality is already here in the most unusual way, but not through humanoid robots from apocalyptic sci-fi movies that hack computers and kill us by breaking our bones or with firearms. Unfortunately, this technology emerged from several artificial intelligences responsible for different modules in the technology that together form an integrated AI, capable of deeply interacting with our cognitive functions, thoughts, emotions and memories. So, they don't only attack our brains, but destroy our lives and deteriorate our bodies.

With no robots. No shots fired.

# CHAPTER 9.2
## TERRORIST SATELLITES, SATELLITE ELECTROMAGNETIC WARFARE, NEURONAL SATELLITES — SATELLITES, SATELLITES AND MORE SATELLITES

Russian intelligence has historically been ahead of the competition. We saw this in the development of psychotronic weapons and the MK-ULTRA experiments where the Soviets were always a few years ahead of the Americans. Around 1951 there were strong indications that the U.S. planned to launch a satellite, the first in history, so they rushed to overcome their enemies with the creation of their first satellite. The space race had officially started.

On September 20, 1956, after the successful test of the Jupiter C rocket designed by Von Braun and the American army, a device that would be the fourth stage of the rocket was launched. It rose to more than 1,000 km (621 miles) and disintegrated in the Earth's atmosphere, scaring the Soviets, as this would be the rocket that would theoretically have the capacity to launch a satellite.

Rocket engineer and spacecraft designer Sergei Pavlovich Korolev (Сергей Павлович Королёв) ordered the construction of a satellite — that weighed about 1,327 kg (2,926 lbs) — in the shape of a cone with more than 3 meters (118 inches) in length called Object D, a large satellite with geophysical purposes: to research on the Earth's upper atmosphere. However, the construction of this object would be extremely complex and would put an end to Russian plans to launch humanity's first space satellite into orbit.

The big move here was to leave aside Object D and its most varied complex equipment such as RKO system antennas, telemetry, solar battery, ion trap, temperature sensors, photometer, magnetometer, among other equipment, to be launched on a R- 7, an intercontinental rocket capable of supporting the immense weight of the satellite still in testing (and therefore wasn't completely reliable). Korolev proposed using the 8K71PS rocket, however, it could not support the weight. It was then proposed by other engineers the development of a smaller and lighter satellite that would only take a month to build so as to get ahead of the Americans if their satellite was really launched.

So, the age of satellites actually began with the launch of SPUTNIK, the first artificial Earth satellite. On October 4, 1957, the USSR launched the satellite from the Baikonur Cosmodrome, a Soviet collaboration, in celebration of the International Year of Geophysics. The International Year of Geophysics was defined as 1957 by the UN General Assembly. Its objective was to unite the efforts of the countries that joined the campaign in order to provide a greater and better understanding of Earth's phenomena. The effort brought together around 60,000 researchers from 66 countries.

Sputnik measured 58 centimeters (22 inches) in diameter and weighed 83.6 kilograms (184.3 pounds). It had few functions compared to the previous Object D design. It transmitted a radio signal (a "beep-beep" sound), which could be accessed by any radio amateur. The five primary scientific objectives were:

* to test the method of placing an artificial satellite into Earth orbit;

* to provide information on the density of the atmosphere and calculate orbital periods;

* to test radio and optical methods of orbital tracking;

* to determine the effects of radio wave propagation through the atmosphere; and

* to check principles of pressurization used on the satellite.

The communications revolution had started, but the beginnings and conception of satellites and launchers go back well before SPUTNIK. In the 17th century, Johannes Kepler, a German scientist, formulated the three laws of planetary motion. The first law — planets move in elliptical orbits with the Sun as a focus — can also be applied to satellites. In this case, our planet occupies one of the foci of this ellipse. In 1903, Russian physicist Konstantin Tsiolkovsky published his calculations proving theoretically that a rocket can be launched into space, in addition to conducting calculations of the necessary speed to put satellites into Earth orbit. A few years later, in 1914, a rocket-fueled aircraft design was patented by the American inventor Robert Goddard. In other words, the concept of an object orbiting the Earth using Isaac Newton's theories of gravitation.

Given the complexity and variety of equipment of a satellite, we shall not go into all the details. Several works with vast amounts of technical data would be necessary for this purpose, which is not relevant to the context of this particular book. However, the basics must be covered in order to determine the use of the equipment as the main tool that makes possible the worldwide MKTECH operation, in which one can track, hack, amplify, modify, invade and alter the thoughts of human beings, regardless of where they are in the world.

## 9.2.1 - But what is a satellite?

Satellite can be defined in two distinct categories:

**Natural satellite** - A celestial body that orbits a planet or other celestial bodies. It's usually a synonym for moon, and it's normally used to identify non-artificial satellites of planets, dwarf planets or smaller bodies. For example, the Moon is Earth's natural satellite.

**Artificial satellite** – It's any object made by man and placed in orbit around the Earth or any other celestial body. Basically, the satellite has a system capable of performing data transmission via radio waves, sending and receiving data that are most commonly used with voice data, telephone communication, mobile cellular telephony, the internet, radio and television transmissions, telemetry data (Earth and space), including data from people's minds, the raw visual and vocalized thoughts, as well as V2K and SYNTELE content, and any other type of data streaming through their systems. It can participate actively or passively in the amplification of thoughts directly in the brain, always ensuring high-quality transmission.

The use of the satellite gives a clear benefit: a satellite antenna — its transmissions — can cover an entire nation and thus supply the demand in areas that are not covered by terrestrial antennas, for example. Even satellites that form a constellation can work together in different orbits, thus covering the entire globe.

A satellite is an extremely expensive device. Sending it into space requires a ground and space operation that involves rocket launches and orbital maneuvers. The cost of this entire endeavor is usually in the hundreds of millions of dollars. Satellites are devices with a strong technological added value, which comes from its conception, construction and shipment to the destination. They have a useful life of about 5 to 15 years and can be used by closed groups such as the military, scientific space research, civil telecommunications, and so on. Although satellites have the most varied functions, they generally have similarities with each other. All need power to operate, so most have solar panels and antennas for communication, through which data is transmitted and received. A large part of the operational satellites in orbit are destined to telecommunications through the transmission of TV and radio signals, phone calls and other services. The main advantage of using satellites is the global coverage they can offer. Another important part is knowing the type of orbit in which the satellite will operate. Depending on functionality, satellites are placed in orbits of different altitudes and inclinations. There are in fact several orbits, the main ones will be explored below.

The satellite is yet another important device that can be used for good, as it normally happens, or for evil by transforming itself into a directed-energy weapon that affects humans. In this process of transformation, it has become an indispensable war tool for attacks with neuroelectronic weapons. After all, in addition to being positioned in a privileged location around the Earth, the satellite is able to capture the weak signal in the target's mind and/or amplify it with its powerful antennas to retransmit the signal with energy gains (dB) well above what it received initially. Thus, adjacent antennas can recover the signal and redistribute it according to the needs of the attack.

The versatility of the satellite is something unique. It can also play the role of an assailant by inserting the data (V2K inputs and mental images that permeate D2K dreams) into the target's brain through downlink — transmission from satellite to ground station or human brains. To make SYNTELE viable, the satellite receives the data already processed by the EMR from the terrestrial antennas through uplink transmission — Earth station to satellite. Within its range of uses, it also allows capturing EEG data radiated by terrestrial antennas, actively participating in the amplification process of telemetric and biometric EEG signals. So, satellites can act on the amplification of the target's mental data and capture this data, retransmitting them to the remote base of the operators.

The attacks, however, may have an aspect different from the ones mentioned above. They can all be performed on the ground (triangulation, amplification of thoughts and production of V2K or D2K) and just use the satellite as a relay, a signal repeater. The smaller and less powerful antenna at the remote base can send the electronic attack directly to the satellite and use its advanced equipment as a signal propagation medium to hit the target from a very distant point where the ground antenna is not able to reach. In this way, the target can be attacked by V2K, including by "hijacked" satellites built in the 90s.

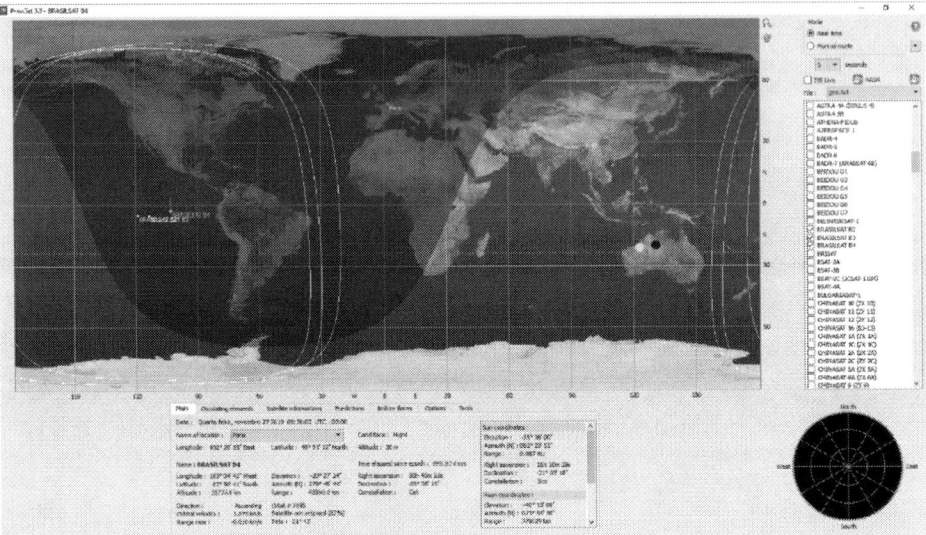

**Figure 9.1** – The PreviSat 3.5 (https://sourceforge.net/projects/previsat/) gives us a small sample of the complexity of the equipment so we can verify the details of a number of registered satellite networks — military and intelligence satellites are excluded. We can check its orbit, real-time location, latitude, longitude, altitude, orbital speed, as well as detailed information from the satellite itself. The program also shows us a footprint or satellite coverage area. We visually confirm the range of satellite signals and their varying intensity within the covered region, where we can also see the difference in power with which each sub-area of coverage varies in decibel watts (dBW).

## 9.2.1.1 - Types of orbits

### Low Earth orbit (LEO)

Once in this orbit, the satellite provides services with little loss and delay in radio transmissions. It's typically below 5,000 km altitude (3,107 miles) above the Earth's surface. Most satellites in this orbit are located between 500 and 1,600 km (310 and 994 miles). The most common services offered are data transfer and phone calls.

### Medium Earth orbit (MEO)

It's located between the smallest and largest orbits with a distance of 10,000 to 20,000 km (6,300 to 12,500 miles) away from Earth. Services such as GPS, GLONASS and GALILEO are placed in this orbit.

### Geostationary Earth orbit (GEO)

Geosynchronous orbit with zero inclination. In practice, the satellite is "fixed". By observing a point on the Earth's equator, it travels at thousands of miles per hour resulting in an orbital speed of 107,200 km/h (29.777 m/s) — just like the Earth's rotation speed. Thus, it looks like the satellite is fixed or stationary. It has a highly circular orbit, which is at an altitude of 35,848 km (22,274 miles). This orbit has a clear advantage: it permanently covers a certain area without variation during its useful life. Its transmission delay is around 0.24 seconds between ground receiving stations and satellites. In this orbit the neighboring satellites are positioned at a distance of 800 km (497 miles). Geostationary satellites are limited by the latitude, which varies at most between 70 degrees north of the equator and 70 degrees south of the equator. However, this too has disadvantages: the loss of connection in certain areas, places with a lot of vegetation and the limited number of objects that can be allocated in this type of orbit.

**Geostationary transfer orbit (GTO):** is a highly elliptical Earth orbit with an apogee of 35,870 km (22,288 miles) and a perigee of 200 km (124 miles) — Earth's equatorial plane. It's used to perform maneuvers for other orbits.

**Highly elliptical orbit (HEO):** special orbit used by the Molniya series satellites (Russia). It was created by the Russians themselves, and is used to observe high latitudes.

**Extremely elliptical orbit (EEO):** helio-synchronous orbit located at an altitude of 600 km to 800 km (373 to 497 miles) in a near-polar orbit. Satellites on this orbit are always visible in places where they receive sunlight, and are used for Earth and Sun observations. There is a type of orbit in which a satellite always crosses the equator at the same time of day, that is, the position of the Sun in relation to the satellite is the same and the angle of the Sun on the surface is always the same, it just varies with the change of the seasons.

Every satellite has equipment that operates on different frequencies and a specific goal — a general purpose. Let's discuss them below.

## 9.2.1.2 - Frequencies

Radio frequencies used by satellites (generally in the GHz range) may operate at lower frequencies in some special cases.

| RADIO FREQUENCY BANDS | FREQUENCY |
|---|---|
| HF | 3–30 MHz |
| VHF | 30–300 MHz |
| UHF | 300–1.300 MHz |
| L | 1.530–2.700 GHz |
| S | 2.700–3.500 GHz |
| C | 3.700–4.200 GHz (downlink) |
|   | 5.925–6.425 GHz (uplink) |
| X | 7.250–7.745 GHz (downlink) |
|   | 7.900–8.395 GHz (uplink) |
| Ku | 10.700–12.750 GHz (downlink) |
|   | 14.000–14.800 / 17.300–18.100 (uplink) |
| Ka | 18–31 GHz |

* **High Frequency (HF)** – ranging from 3 to 30 MHz, they are also known as short waves. HF propagation is very important as its signal is reflected by the ionosphere, making global terrestrial communication possible over long distances when receivers and transmitters are on the ground. This frequency is also highly susceptible to atmospheric phenomena, the characteristics of the

ionosphere and solar activity, thus not being ideal for use outside the atmosphere.

* **Very High Frequency (VHF)** – ranging from 30 to 300 MHz, in which open TV channels 2 to 13 are transmitted.
* **Ultra-High Frequency (UHF)** – 300 MHz to 1.300 MHz.
* **L-Band** - 0.5 to 1.5 GHz. This band is commonly used for terrestrial mobile communication.
* **C-Band** - 4 to 8 GHz. Specific frequency bands dedicated to the use of communication satellites, more specifically 3.7 to 4.2 GHz (downlink) and 5.925 to 6.425 GHz (uplink).
* **X-Band** – 7 to 8 GHz is usually intended for satellites and communications for military use.
* **Ku Band** – the frequency ranges from 10.9 to 17 GHz.
* **Ka Band** – the frequency ranges from 18 to 31 GHz.

These are the most common frequencies used by communication satellites. However, given the confidential and classified nature of psychotronic weapons (a kind of military secret), it's currently not known what spectrum they operate in. They can be diverse, including the X Bands ranging from 7 to 8 GHz or even at the limit of the experimental frequency bands from 10 to 300 GHz.

On the other hand, it's known that the MKTECH weapon uses several frequencies to achieve its objective. I'll talk about these variables later, but at the moment we can only speculate until we are able to figure out the type of satellite that operates and covers an entire nation — I can only hope that the truth will come out soon. Anyhow, it's important to mention that there is actually a very strong suspicion about some countries attacking the national territory with space weapons; I'm going to discuss them in chapter 9.2.8.

Two other pieces of equipment that are important parts of a satellite and should be mentioned are transponders and antennas. A transponder is short for: transmitter + responder. It's the most important set of equipment on a communication satellite that consists of receiving, amplifying, and converting frequencies, in addition to transmitting signals. It usually works above 10 watts. The number of transponders on a satellite range from 12 to 24. In some cases, however, it can get to 50.

* Satellites that use the RF spectrum to communicate have low power capability, operating with only 30 watts.

* Medium-power satellites operate with power levels between 30 and 100 watts.

* Higher-power satellites operate at 100 watts or more.

### 9.2.1.3 - Transmitters

Satellite transmitters are based on a multiplexer (mux) that combines several signals in a single channel, transmitted with varied power amplifiers — and they have the potential of being transmitted within the desired band. On the receiver side, demultiplexing occurs, and the signals are separated and individualized. Stations can use fixed-satellite service (FSS) equipment, a teleport can receive and send data to the satellites. Or they can also use these stations as transmitters via the Internet.

* **Broadcast Satellite Service** (BSS) – receives data on miniaturized devices.

* **Mobile Satellite Service** (MSS) – uses a variety of equipment for transmitting and receiving data. Generally, mobile devices such as cell phones.

### 9.2.1.4 - Terrestrial and space antennas

One of the most important equipment on a satellite is its various antennas for receiving and transmitting electromagnetic waves, such as radio waves and microwaves. Depending on the frequency to be used, the antennas can have different formats to work properly. Each antenna is unique with several parameters to be observed. A quick observation shows the difference between microwave antennas and those of radio and TV — the later has a rounded or parabolic shape, for example.

Today we're totally dependent on this type of technology as we use these services to keep modern society running. Virtually all electronic devices make use of data coming from satellites. This is mainly observed in the areas of communication, but we must remember that even our fridge uses services provided by this complex space machinery. Nowadays, some sectors are dependent on this mass of data: agriculture, land navigation, the Internet, GPS, telephony, mining, autonomous cars, maritime navigation, military systems, and TV and radio networks.

Actually, we cannot think of modern life without it, but it's essential to observe its use as a weapon capable of affecting all of us, residents of planet Earth, in an unrestricted way. It's known that satellites were designed during the Cold War, so the concept of a satellite-borne weapon is not new. In fact, since its conception, it was thought of as a weapon: they can attack anyone, anywhere, anytime. And this has been done, in secrecy, over time, culminating in where we are today.

To get an idea of the seriousness of the situation, think of the visual thoughts as a movie projector. Imagine your visual thoughts popping up inside your mind for any number of reasons, as they're created and invoked. The result would be a picture, a frame, right? Therefore, this picture would be added onto a film of a reel containing several frames, or visual thoughts, created by the mind. Then, the image (similar to what your mind "visualizes") would be rotated and projected.

This satellite technology is capable of inserting frames in the mental "film reel" to be projected onto the mind's screen, but without the individual's permission, causing the visualization of intrusive visual thoughts (in the waking state) at every moment and with any graphic content. The consequences are devastating. It's worth remembering that in the near future they may be able to directly interfere with optical radiation and the Lateral Geniculate Nucleus (LGN) to modify the images that come from the retina using the same technique, thus causing visual illusions or real holograms.

## 9.2.2 – Military/Intelligence/Neuronal satellites

There is a wide range of satellites from many countries. They generally work in a constellation — satellites from the same family use different orbits to cover the largest possible area of the planet. By working together, they provide a seamless service.

Defense departments that intend to launch new equipment usually order 4 satellites in a constellation to be launched throughout the year. They don't need to be identical or use the same orbit, but they need to interact with each other.

In addition to communication, there are satellites for the most diverse purposes:

* Recognition;
* Radar imaging;
* Space photo reconnaissance satellite– **Corona**;
* Detection of nuclear explosions or ionized charged particles, high-energy radiation (gamma rays) – **VELA**;
* Electronic reconnaissance satellite;
* Signals intelligence satellite;
* Radar and photographic intelligence;
* Global positioning;
* Global Positioning System – **GPS**;
* GLONASS (Globalnaya Navigatsionnaya Sputnikovaya Sistema) – a Russian satellite navigation system;
* GALILEO – the European Union's satellite navigation system;
* China developed BeiDou, also known as Compass, the Chinese version of GPS;
* Space-Based Infrared System satellites - **SBIRS**;
* Sensors to detect ballistic missiles - **MIDAS**;

* Interception of rocket telemetry data;
* Observation of installations on the ground;
* Military strategic planning;
* Attack satellites that carry explosive payloads;
* Radar calibrators – to calibrate radars on the ground before sending the main satellite that will carry out the surveillance;
* Satellites pointed at space for astronomical observation;
* Radar image – radar imaging to map the target using the return echo;
* Detecting military transmissions in enemy frequency bands — Earth-to-Space, Space-to-Earth, Air-to-Earth, Earth-to-Airplane. They listen to radio signals and radar pulses – **ELINT**.
* Furthermore, there are military/intelligence and global neuronal research (Neurohacking) satellites, which have specialized equipment, such as:
* Amplification of thoughts, Mind Hacking, Neural Satellite, MKTECH;
* Receiving and amplifying signals from the brain. "They hear and see human thoughts" (EMRvia);
* Sending signals to the mind (V2K);
* Sending and receiving the SYNTELE signal;
* Sending and receiving the signal containing D2K images and sounds;

Terms and names may vary, but the purpose is the same.

Military spy satellites are likely candidates to carry this electromagnetic weapon since they use signals intelligence. They're capable of, for example, capturing virtually all radio signals on the planet and generating reports for the NSA[64] and NGA[65]. All satellites in this category provide services for the military, defense and intelligence agencies of their respective countries.

Let's take NROL as an example, as it perfectly illustrates the cold space war we're experiencing today and the agencies' high level of advanced technology. NROL and satellite constellations are meant to operate in strict secrecy for the National Reconnaissance Office (NRO), an American intelligence agency that operates the U.S. government's spy satellites and provides satellite information to various government agencies, particularly signals intelligence to the CIA, imagery intelligence (IMINT) to the National Geospatial-Information Agency (NGA) and the Measurement and Signature Intelligence (MASINT). In fact, the U.S. Department of Defense (DOD) has been working with signals intelligence since 1961. Nowadays, it has the most modern technology and knowledge to operate these weapons.

Military satellites are a formidable arsenal. In addition to capturing data for intelligence, they are used for closed and secure communication and, in several cases, as space attack and defense weapons. Civilian satellites can also carry military transponders and military satellites can carry civilian equipment. An example is the Milstar (Military Strategic and Tactical Relay), a constellation of military communications satellites in geosynchronous orbit. These satellites will be gradually replaced by the Advanced Extremely High Frequency (AEHF) satellite program.

AEHF embarks a 44 GHz EHF uplink transmitter, while the downlink is transmitting in the SHF-band (20 GHz frequency), enabling cross-links between and among satellites and Earth terminals, just as its predecessor did. The system is designed to provide jam-resistant communications with a low probability of interception (data hacking), and it incorporates FHSS (Frequency-hopping spread spectrum) technology. FHSS is a method of

---

[64] The National Security Agency (NSA) is the U.S. government intelligence organization created on November 4, 1952, whose purpose is related to signal intelligence, including interception and cryptanalysis.

[65] The National Geospatial-Intelligence Agency (NGA) is a government agency of the United States of America with the primary mission of collecting, analyzing and distributing geospatial intelligence (GEOINT) in support of national security.

transmitting radio signals which consists of constantly changing the carrier through various frequency channels using a pseudo-random sequence. Transmitters and receivers also have a set of antennas and "dynamic" electrical beams that improve the quality of transmissions and avoid interference. They're currently run by the 4th Space Operations Squadron.

The most modern and advanced military technology that the American and Russian industrial complexes can provide are onboard these satellites that have total control of virtually all electromagnetic spectrums, including the most modern ones that hack the human mind in an increasingly efficient and lethal way. It's basically an undeclared electronic war that migrated to space.

Deductions and speculations based on orbits and launch details with the secret payload, trajectory, payload model and manufacturer are made, but they are just that: speculations! So much so that few people know about the modern space weapons we've thoroughly studied in this book. Even some data from the 1960s concerning satellites remain confidential, so as not to provide retroactive information, since the satellites we have today are the continuation of what was previously done, while always following a path of evolution. The truth is that the Cold War in space has been going on for decades, but now with the advanced technology of invasion of thoughts, the "Mind Hacking" satellites.

A distrust response is triggered when we hear about this subject. Skepticism immediately arises, right? However, we have to keep in mind that intelligence services know about technology-related and strategic issues well before the general population, and advanced militarism is decades ahead of the technology available to society at large. They both work together in all spheres related to weapons, military secrets and useful information that help build a world apart from society.

A subterfuge often used by agencies to put such an object into orbit without arousing suspicion — that is, away from prying eyes — and thus circumvent surveillance (espionage that is increasingly present today) consists of declaring only one object in the launch load. However, as soon as the load is delivered at the destination, two or three other objects are also ejected. In other words, they use secret payloads that ride piggyback on the rocket together with the "real" payload.

All countries do this in order to move their space weapons — prohibited by UN conventions — into position. So, decoy payloads are launched together with the main payload, as expected. However, this is nothing but a clever disguise as the final stage of the rocket is delivered to the orbit along with the necessary military equipment to conduct worldwide psychotronic warfare.

The ability to hide a set of frequencies capable of invading the human mind and merging with cognition also carry with it countless secrets and strategies similar to those mentioned above. In the case of signals coming from space weapons, the tactic consists of concealing a "clandestine" signal in legitimate transmissions, thus masking the intention of its use. This causes uncertainty and speculation regarding the interaction of frequencies capable of carrying out the attack.

Remember: space technology is always a few years ahead of the technology that we can get our hands on. A vehicle that probably has no connection with the events, but shows how this space war is escalating every day, is the X-37B, a small unmanned space plane of the American army. It's a top-secret spacecraft that Boeing developed for the U.S. Air Force, surpassing its previous record of 674 days in space. What this spacecraft tested and performed while orbiting the Earth is completely classified, but it does show interest in the manufacture of space vehicles for war purposes.

Although I'm mostly focusing on American satellites, these characteristics extend to several other countries — Russia, for example. At a lower level are the Chinese, French, Indians and Japanese, who have advanced space technologies such as the Quasi-Zenith Satellite System (QZSS) planned to cover the Japanese territory — a satellite with functions and advanced laser equipment. As a matter of fact, Japan has a significant number of people reporting MKTECH use on civilians.

Russia has also recently launched the Kosmos 2519 satellite which has greatly troubled the international community. The satellite shows an abnormal behavior, never seen before, as it makes orbital maneuvers and sudden changes in its trajectory. This type of satellite is known as a space inspection device, capable of ejecting other objects of any nature. Such objects can have robotic arms that destroy pieces of other satellites, or directly interfere with laser beams to destroy their plates. In this way, these cases show how easy it is to launch space weapons despite treaties and conventions with other countries. The same tactic is used to deploy devices that are prepared to attack the human mind.

In this war, propaganda is also a strong ally. Visual communication is expressed in art, whose main function is to convey specific information through images and texts, connecting them uniquely, and which contains an enormous number of variations and different styles with intrinsic artistic values. Some logos or designs, for example, are very creative and beautiful. They convey the synthesis of the functionality of satellites, capable of intimidating and demonstrating strength, power, supremacy and hegemony — basically, that nothing escapes them.

Figure 9.2 In this war, propaganda is also a strong ally. Visual communication is expressed in art, whose main function is to convey specific information through images and texts, connecting them uniquely, and which contains an enormous number of variations and different styles with intrinsic artistic values. Some logos or designs, for example, are very creative and beautiful. They convey the synthesis of the functionality of satellites, capable of intimidating and demonstrating strength, power, supremacy and hegemony — basically, that nothing escapes them.

1) Soviet commemorative poster – **LAIKA** (in Russian: Лайка; 1954 - November 3, 1957): Laika was a Soviet space dog who was the first animal to orbit the Earth.

2) **CCCP Satellite Молния:** The name was an homage to the Russian satellite Molniya ("Lightning"). It was the first to use the Molniya orbit with 37 launches between 1964 and 1975.

3) **CCCP Space 1961-1971:** Commemorative tribute to the Soviet space race years.

4) Poster in honor of Russian scientist Konstantin Eduardovich Tsiolkovsky, rocket scientist and pioneer in the field of astronautical sciences, in addition to being an author of several works in the area. The large lunar impact crater that is located on the far side of the Moon is named after him.

5) **CCCP Flag Space Rocket:** A tribute to Russian space rockets.

6) **SBIRS GEO Flight 4 satellite:** the design makes it clear that this air force satellite constellation is capable of detecting missile systems that use infrared, including intercontinental ballistic missiles. The energy ray coming out of the eagle's eye and destroying missiles while others are being caught in its talons and the three satellites tracking the globe depict the power of the satellite, and the message here is clear, "If they launch it, we will intercept it".

7) **NROL-32:** The logo uses the eye of providence as on the one-dollar bill with the words Annuit cœptis. The literal translation is "[He/She] favors (or "has favored") [our] undertakings", from Latin annuo ("I approve, I favor"), and coeptum ("commencement, undertaking"). In my opinion, it symbolizes the all-seeing eye, conveying the message that the NROL constellation is able to see everything that goes on in the world, including inside our minds.

8) **NROL-45:** The reconnaissance satellite honors the people who gave everything for their country and died for it.

9) **NROL-39:** For me, this logo is one of the most intimidating one, as it dictates the tone of space disputes and total domination over the planet: "Nothing is beyond our reach". The Octopus tell us that everything is under surveillance; there is no place they cannot reach. They see and hear everything.

## 9.2.3 - How many satellites currently orbit Earth?

## 9.2.3.1 - How many satellites are currently in operation?

According to records from the Union of Concerned Scientists (UCS), which keeps operational records of satellites, only 37% of satellites are still working. This is a number that ranges from 1,738 to 1,796 depending on the source.

### 9.2.3.2 - Whose satellites are these and what do they do?

A large number of satellites are launched per year. Most of them are for telecommunications and observation of private companies, entertainment services. These satellites have a well-known configuration, its chassis, and their equipment is usually assembled by renowned companies in the aerospace market. There are several companies responsible for developing such equipment, including: Airbus Defense and Space, OHB SE, Boeing Defense, Space & Security, Lockheed Martin, Thales Alenia Space, Orbital ATK, JSC Information Satellite Systems (Russia), among other large companies around the world. The list is long.

### 9.2.3.3 - The satellites in operation are divided into:

* Communication: 742
* Earth observation: 596
* Technology development and demonstration: 193
* Navigation and position: 108
* Spatial observation: 66
* Earth-oriented equipment: 24
* Space-oriented equipment: 67

### 9.2.3.4 - Who uses what?

* 788 - Mixed use
* 461 - Government use
* 360 - Military use
* 129 - Civilian use

## 9.2.3.5 - Earth observation - Intelligence data collection and environmental monitoring

* USA - total 814 - 202
* China – total 205 – 39
* Russia – 140 – 81 defense satellites
* Other countries – 578

## 9.2.3.6 - Some relevant data

* 33 of the launched satellites are designed for military and classified use and 80% were launched by the USA followed by China, Russia and France.

* Another 56% of the launches this year (2020) were used for government purposes and were mostly operated by the space agencies of each country and their associates. China has 52% of the total, followed by the United States. 65 are for civil use with educational/university experiments, and 39 are for commercial use.

* There are 33 different countries that operate experimental (demonstration) technology satellites. The USA is in the lead with 63, followed by China with 41 and Japan with 19. Each of the remaining countries operate with two. These numbers are rough estimates, as well as the functionality of some equipment and its purpose are just speculation. After all, the mind-invasion technology is in full force since 1992. We're not sure which satellites were connected to this weapon or which have been adapted over the years to function as part of a constellation dedicated to directly invading and destroying the human mind.

There are only vague categories when it comes to certain terms as one tries to quantify and catalog satellites as experimental technology. It could be from using lasers for communication, weapons created to destroy other satellites, improved mind-invasion technology, real-time observation technology that uses thermal imaging, dynamic radar positioning capable of seeing people inside urban structures or inside their homes, or even a ground mapping.

## 9.2.3.7 - Multiple uses

Satellites can perform many tasks. Generally, they collect images, intercept signals from other satellites and communications between them. Some projects that work with microsatellites are capable of photographing the entire Earth in a single day to observe its dynamics, and others have powerful lenses that could obtain images with outstanding resolution. However, some are made for espionage and defense, and their configuration, orbit and purpose are classified information.

Constellations of satellites are also launched; they are placed in the same orbit and each performs a complementary function. Some use one or more technologies and do not fit into certain categories. It's very difficult to state their real goal, mainly the military intelligence and spy satellites, including those whose purpose is to capture thoughts directly from people's minds here on Earth. In short, satellites developed to capture visual and vocalized thoughts don't fit into any of the above types.

The MKTECH satellites (mind hackers) have existed for more than 40 years since the 1980s. To give you a small idea of the technological advancement, images and digital transmissions have been used since the 1970s. At that time spy satellites already had photographic quality far superior to what we find today on Google maps. So, the gap between the technology of cutting-edge companies, the military, and surveillance and intelligence agencies, and that of ordinary civilians is huge. The difference is 15 to 40 years, which is plenty of time to improve a given technology. In the end, what is released to the public is an outdated device, or a device from previous versions.

Money, highly qualified personnel, top quality equipment, political contacts with the military, and espionage, defense and intelligence agencies result in big investment contracts with advanced technologies, billions of dollars at stake and absolute secrecy in operations. This space race showcases the most advanced and modern of humanity's capabilities.

Many of the satellites launched this year in China, Japan, Russia and India are probably carrying neural, electromagnetic and psychotronic weapons that cause irreparable damage to the nervous system of human beings.

## 9.2.4 - Range of satélites

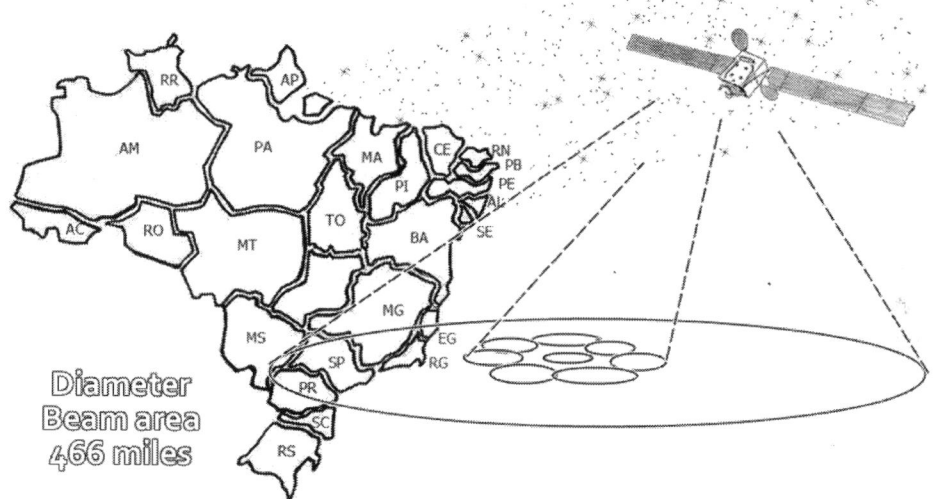

**Figure 9.3** HTS technology – High-throughput Satellite.

A single satellite is capable of covering virtually the entire national territory. Each satellite beam — smaller circles within the whole area — covers an area of 750 km (466 miles) in diameter. If the satellite works with other satellites in the most varied orbits, they will cover a larger area.

**Anyone who has been a target of this type of weapon can now understand and visualize why torture is constant and at high levels, even during international or intercontinental flights regardless of the type of vehicle — bus, plane, car, ship or train.** Think of the coverage shown in the image above as areas where your mind can be attacked. Each different location receives a beam that can use only one individual channel for each mind as it moves across the national territory.

HTS is a technology used in most modern communication satellites. It's based on the use of multiple spot beams and frequency reuse, which increases the available satellite band and, therefore, has a lower unit cost per unit of capacity. The Geostationary Defense and Strategic Communications Satellite (SGDC), the only modern satellite we have, makes use of this technology. Built by the competent French company Thales Alenia Space, it's located at the orbital position of 75 degrees west (longitude) and is operated by Telebrás on the civilian side and by the Brazilian Armed Forces (the military), which in turn is subordinated to the Ministry of Defense. The satellite was based on the Spacebus-4000 C4 platform and its life expectancy is 18 years. It has 50 transponders in K-band and encrypted military X-band. Adjacent beams have different frequency or polarization, among other technological specifications that owe developed countries nothing. It was recently launched by the Ariane 5 ECA rocket in Kourou, French Guiana, which has state-of-the-art technology.

It's thus possible to visualize the power of covering an entire area with only one high-power satellite with high-throughput (HTS) technology, multi-beam satellites with wide coverage. High Throughput Satellites (HTS) have a large number of spot beams. Each covers a limited region and is optimized for next-generation data services and therefore better equipped to support the growing demand for connectivity anywhere, anytime.

Adapt a mind-monitoring technology along the lines of the HTS, and you'll have something similar to the technology used to run SYNTELE and pursue the target wherever they go, even all the way between point A and B — for example, from the South of Brazil in Santa Catarina towards the North, Manaus. Throughout the journey, the victim's thoughts will be captured and read; the V2K will be amplified by the noise of the airplane turbine that will almost "explode" their auditory system and will interact with them without any intermittence, attenuation or cancellation. The brain will be constantly stimulated in a maddening way as we saw in chapter 2.7.

Now we finally understand why it's impossible to escape the range of these weapons. Operational beams — gateway beams similar to HTS — are capable of causing physical and psychological damage to the human nervous system. And SYNTELE, D2K and V2K use that specific band.

## 9.2.5 - Thoughts on Hertz (Hz) frequencies and data transmitted through MKTECH satellites

I'll make a brief overview to try and explain some points still unknown to those who don't have access to the MKTECH network by using the information I was able to gather during the seven years of attacks and study of the phenomenon. Let's start by doing a thorough analysis of the data itself, the content within the signal that travels through the satellites.

First, let's divide them into two categories: Input and Output.

### 9.2.5.1 - Input

#### Images

The input of visual data (images in sequence) into the human brain occurs in the following subsystems:

Synthetic Electronic Telepathy (SYNTELE) sends low-quality grayscale images — similar to the images that are naturally constructed in our brains when we think of an object, or something synthesized from actual experience that was captured by sight and stored in our memory — to get into the target's visual thinking while the target is awake. The target sees (in their mind/thoughts): a sequence of images (film) sent incessantly by these transmissions when concentrating on them, or these transmissions start to be visualized and displayed in the mental projection after focusing on them. However, this doesn't affect the images captured by sight.

Synthetic Electronic Dream (D2K) sends images of varying levels of perceptible quality, images which are designed to be processed by the visual cortex while the target is in REM sleep. Here, the quality of the images varies greatly. They can be adapted according to demand. In order to imitate hypnagogic dreams, they use gray and dark tones and scenes composed with visually identifiable characters and objects with a level of sharpness capable of deceiving the brain. In REM sleep, however, images get brighter, vibrant details and more realistic tones. Here, operators stream films with 3D effects extracted from virtual glasses. Now, the million-dollar question is, how do these transmissions force images to travel through cortical pathways and gain high-priority status, as they partially or totally repress the natural dream to occur?

The image format itself is perhaps a new type, but I believe that it can be derived from what we already have. There is also the possibility that the file goes through an algorithm made exclusively for the MKTECH scheme that modulates it within the combination of frequencies and signals in such a way that it induces its demodulation by the brain when interacting with the target's EEG biometrics.

The most popular compression methods are algorithms based on JPEG (Joint Photographic Experts Group), which show how the image should be compressed and encoded (in bytes). A complex algorithm involving discrete cosine transform, among others, comes into play. There are other different types of file format like BMP, GIF, PNG and TIFF. The size of the image isn't a problem due to the high frequency of absorption of the content in the human mind and the ability of the satellite system to easily cover this type of data. However, the exact resolution (dpi) — its size in bytes or dimensions in pixels both in width and height — is still unknown. Unfortunately, I cannot specify what type of image compression is used for transmission nor its characteristics, or even if only a certain type of image is capable of being interpreted by the visual cortex in this process of image formation. Nonetheless, it can be said that it hardly escapes the mathematical formalism that we use today to deal with computer images.

## Audio
### V2K (microwave voice)

At least 5 or 6 different voices that have distinct types of positional interpretation, acoustics and intensity are used simultaneously on targets. The microwave voice can also be modulated into different signals and at distinct frequencies. Sometimes it can come with a tonal or cyclic noise to simulate something similar to white noise. Then, it creates by itself the effect of voice within noise and increases the power of the attacks without depending on external sources that travel through afferent pathways (ear-to-brain). It usually takes up little space in the total length of the band.

Songs sent into the target's mind often appear to be mono and resemble music playing in the distance.

## 9.2.5.2 - Output

When it comes to output from the brain, we have the target's voice of thought, their mental images and images that are projected through the optic tract (eyes) — we're unaware of the compression level of these images or the number of frames per second (FPS) captured —, together with all the EEG data that contain a wealth of information. All these data are retrieved via brain activity by the amplification of some electrical parameters, including thalamic and cortical pathways (LGN – Lateral Geniculate Nucleus and MGN – Medial Geniculate Nucleus), using EMRvia – Electronic Mind Reading (vocalized/images/auditory).

**Image**

The mental image storage medium is unlike anything we have seen before. For the first time in history, we're dealing with image extraction in biological means. In this medium, the image cannot be found in a specific location, as it happens with traditional media. In fact, it can only be retrieved by mental commands — voluntarily requested when the target is thinking about a certain visual object—, or via involuntary external stimulation — sounds that reach the ear, word decoding and V2K. Once the brain gathers all the data coming from various parts of the cortex through a mental command, an image is formed in the mind. The more focused they are — the greater the target's ability to concentrate and control their thoughts to abstract external stimuli, the more cortical resources are recruited. Thus, the image can improve in several ways, such as sharpness and amount of detail. The image is captured as a signal. As it travels through the optic tract to reach the Lateral Geniculate Nucleus in the thalamus, it's captured by the uninterrupted transmission that amplifies the signals. Please note that images captured from the environment are much richer and more complex than mental images.

When extracting the feedback content, the same technique can be used in order to reconstruct the image from the mind. An algorithm is able to interpret the content of the signal and reconstruct the mental image. The most arduous process consists of extracting the information from the EEG and the EMR amplification, and transforming it into an image that is comprehensible to humans. At this point, the same image enhancement and restoration techniques are applied, such as histogram equalization, focus and motion correction, noise removal, etc. Here, the best you get in terms of resolution and sharpness is the quality of the visual thinking with some improvements.

Let's start the process by analyzing the images extracted from a TI's mind. By amplifying the target's visual thought, or vocalized thought, perhaps the quality of the visualized image and its compression level can be measured by the amount of feedback received and its relation to the level of data amplification. In other words, the lower the interaction level, the less details (pixels) the image contains; the greater the coverage of the transmission and its specific interaction — amplification or re-radiation of EEG data — directed towards the image as it travels through the LGN, the greater the level of detail of the captured image. This in itself is already a kind of image compression. In this way, one can lower the quality of the images and receive more data just by decreasing the signal amplification level (amplitude). The signal sampling level is coordinated by algorithms responsible for delivering and receiving data, and for the infrastructure if the signal is lossy.

**EMRo - Electronic Mind Reading** (optical) amplifies all visual images that travel through the optic tract after their formation in the visual cortex. It's known that operators see everything the target sees under certain circumstances, but the quantity and quality of images remain a mystery.

**Audio**

EMRa - Electronic Mind Reading (auditory) and EMRv - Electronic Mind Reading (vocalized). Both are captured when traveling through the Medial Geniculate Nucleus. The voice of thought and every sound that the target hears — sound waves in the environment — are also part of the set that will be amplified and intercepted by antennas. It probably occupies the bandwidth of an ordinary voice transmission.

### EEG data

The re-radiated EEG data have a large amount of information referring to several parameters, which indicate the target's unique electrical signals — neural biometrics that shows emotional indicators and a multitude of inferences (as we saw in chapter 3 of Volume 1). Such data are reradiated with all this information in the form of electrical impulses and are captured and decoded by the operators' BCI computers.

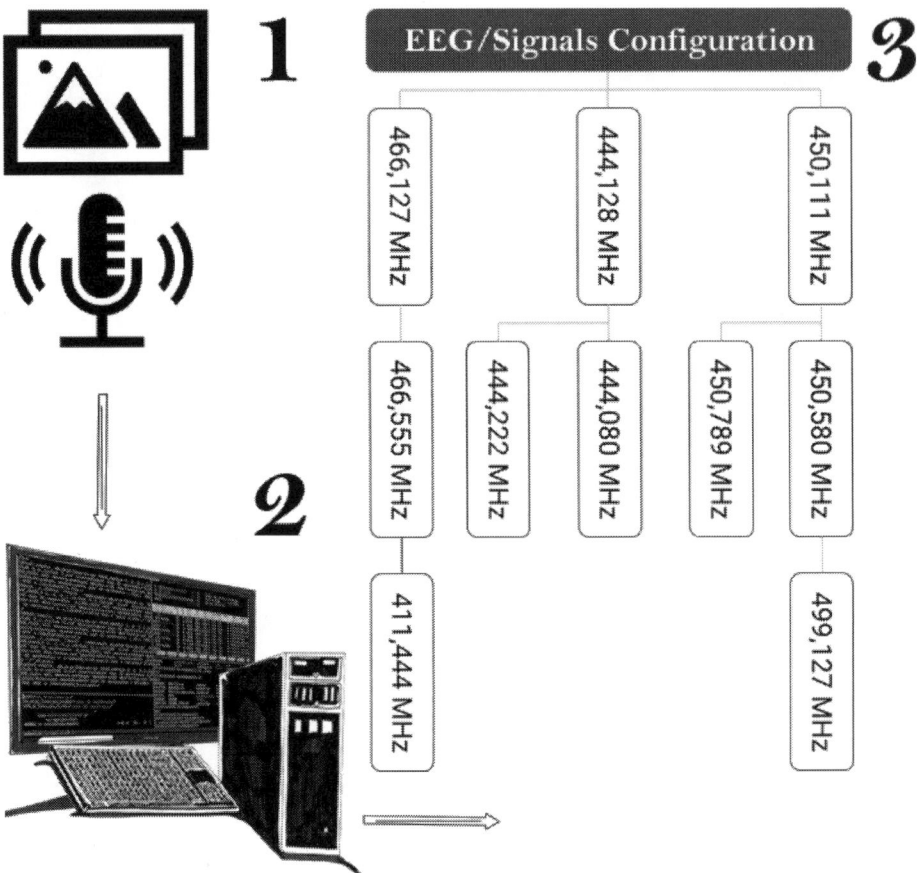

**Figure 9.4** Modulation scheme of the set of signals that send the data to the target's brain.

1) Image files or sequence of images (films) and audio are created by the operators and digitized on the computer;

2) A sophisticated algorithm modulates signals from different carriers that were already set to interact with the electrical signals of the target's brain;

3) Each frequency is intended for a certain location in the brain and each fragment has "a piece of the file", the set of signals, and thus "extracts" the content of the signal (images or sounds) that is forwarded to the auditory or visual cortex always through thalamic pathways. Every frequency would be the equivalent of certain electrical rhythms in part of the target's brain. It can operate without interruption for years. Once the EEG-Based Biometrics is created, the target's biological "computer" will be brutally hacked.

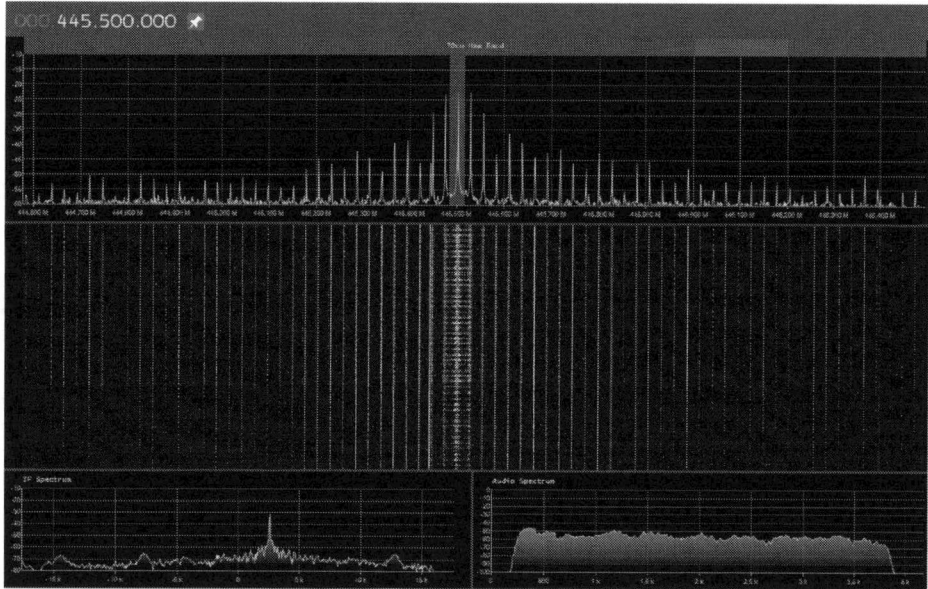

**Figure 9.5** The data collected show a possible capture of TELEMETRIC EEG reradiation at 445 Hz, 501 Hz, 920 Hz or 445 MHz, wavelength of 70 cm (28 inches), close to the band allocated for UHF, Ham Band or amateur radio band. This data was taken during a real-time psychotronic attack. The signal was strongest when the target came into range of the antenna, and its strength decreased as the individual gradually moved away from the Yagi antenna which captured the re-radiated EEG data. This frequency combination within different signals is the key to hacking the human mind, as well as extracting and inserting all the necessary data, as we've seen throughout this book. Some of these frequencies only appear in the spectrum, that is, they just get stronger if the target comes into the range of the antenna. Details of the experiment are available on the book's official website (https://invasionandmindcontrol.com/).

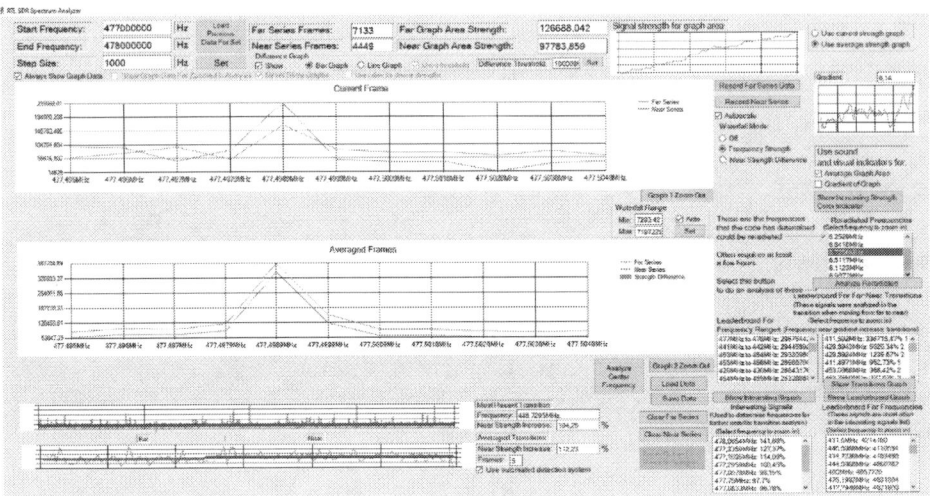

Figure 9.6 Clint Mclean, an expert on biological effects of radio waves on the human mind, developed a system that detects human resonant (radiated) frequencies that satellites could use to maintain a constant signal. Such frequencies should re-radiate EEG information, or biometric and telemetric EEG, to the EMR. The program called SDR Spectrum Analyzer works by detecting whether there is any signal that increases in intensity when the target approaches the antenna and whether this change is caused by reradiation. The algorithm averages the signal strength over a longer period. A signal that seems to be stronger when you are close to it will be considered interesting to investigate. It's a light at the end of the tunnel for targets who are tortured around the world in a cowardly way and have no chance of defense, as research shows that is possible to accurately measure the rerouted frequencies which would contain EEG data and are the basis for the Neural Remote Biometrics of targets, the individualization of their electrical activities.

We may be close to blocking one of the cornerstones of the attacks: the signals that individualize the human brain. Without them the scheme wouldn't work. Link: https://github.com/ClintMclean74/SDRSpectrumAnalyzer. For most of us, this technology generates the same electrical currents that our body produces to generate processes and sensory awareness. Therefore, the signals don't need to be very strong; they only need to synchronize and match the mapped brain to start the interaction process.

There is an intriguing factor in these frequency bands possibly used to attack targets and that allow them to be actively exploited: it would be difficult for FCC to detect irregularities, since these signals meet their destination and regulations previously established by the agency that cannot inspect their content, nor how it is used by the client. In short, agencies aren't aware that a certain set of signals, at different frequencies within the spectrum, could be used to conduct state-of-the-art neuroelectronic attacks on people. So, the attacks are camouflaged as a real transmission of a satellite.

Let's see what's in these electrical resonance frequency bands (400 to 500 MHz) officially designated by Anatel (FCC in USA) for registered services. The list below indicates the frequency, its destination within the reserved space and the direction of the signal:

* 400.050 – Mobile Satellite (Earth-to-space) / Radio Navigation.

* 400.150 – Standard frequency / Time signal satellite.

* 401.000 – Meteorological aids / Earth Exploration Satellite (space-to-Earth) / Mobile (space-to-Earth) / Space Research (space-to-Earth) / Secondary Space Operations (space-to-Earth).

* 402.000 – Meteorological aids / Meteorological satellite (Earth-to-space) / Earth Exploration Satellite (Earth-to-space) / Space Operation (space-to-Earth).

* 403.000 – Meteorological aids / Earth Exploration Satellite (Earth-to-space) / Meteorological satellite (Earth-to-space).

* 406.000 – Meteorological aids.

* 406.100 – Mobile (Earth-to-space).

* 410.000 – Fixed / Radio Astronomy.

* 420.000 – Fixed / Except Aeronautical Mobile.

* 430.000 – Fixed / Except Aeronautical Mobile / Secondary Radiolocation.

* 432.000 – Secondary Radiolocation/Amateur Radio

* 438.000 – Secondary Radiolocation/Amateur Radio.

* 444.000 – Secondary Radiolocation/Amateur Radio.

- * 450.000 – Fixed / Except Aeronautical Mobile / Secondary Radiolocation.
- * 455.000 – Fixed / Mobile.
- * 456.000 – Fixed / Mobile (Earth-to-space).
- * 459.000 – Fixed / Mobile.
- * 460.000 – Fixed / Mobile / Mobile (Earth-to-space).
- * 470.000 – Fixed / Mobile.

This neuroelectronic weapon is capable of transmitting data to the target's mind in the same way as it does to terrestrial antennas. The difference is the demodulator device of the signal content. In the case of humans, it's the brain. Upon reaching the individual's mind, signals that interact with neuronal electricity demodulate primary auditory and visual sensory information, as well as sensations and sudden changes in mood. The signals can be sent in parallel at different frequencies that have the content of each information destined for the respective cortical processing areas and by direct stimuli, along with the content demodulated by areas of the brain, such as the feeling of being watched and of having an "open door" which are caused only by the carrier connected to the mind with no content passing through. In this way, our thoughts are easily altered by external forces that cause the manipulation of consciousness and the direction of the flow of thoughts as operators please.

Other signals and modulations containing stimuli and "instructions" of electrical configurations via electromagnetism act only on their reception without content demodulation. Upon receiving the set of signals in the brain, it automatically interferes with voltage-dependent ion channels. The thalamus is one of the most affected brain regions. Voltage-gated calcium channels are key transducers of membrane potential changes into intracellular Ca (2+) transients that initiate many physiological events. The abrupt change caused by neural weapons causes permanent brain damage.

## 9.2.6 - Data transmission techniques

This network of weapons operates uninterrupted (input and output), receiving data containing sounds and images, rain or shine. SYNTELE is very efficient and capable of inserting and extracting data in the human mind in any given situation. The sending of images also takes place during sleep. As soon as the target goes to bed, their mind will be controlled by the images sent via D2K. No device on the ground would be capable of such a feat. The impressive range of this technology allows it to continue operating even if the target travels from North to South of Brazil and passes through CINDACTA[66] centers.

I believe that no terrestrial infrastructure alone has the same reach as CINDACTA, as transmissions cross CINDACTA I, II, III and IV without any kind of input/output attenuation or suppression. See below the reach of the neuronal electrical network transmissions.

---

[66] The Integrated Air Defense and Air Traffic Control Center (CINDACTA) has 18 Airspace Control Detachments (DTCEAs) where the means, systems and equipment that support its operations are located. They're set up in strategic areas in the states of Mato Grosso, Goiás, Minas Gerais, Espírito Santo, Rio de Janeiro, São Paulo and the Federal District.

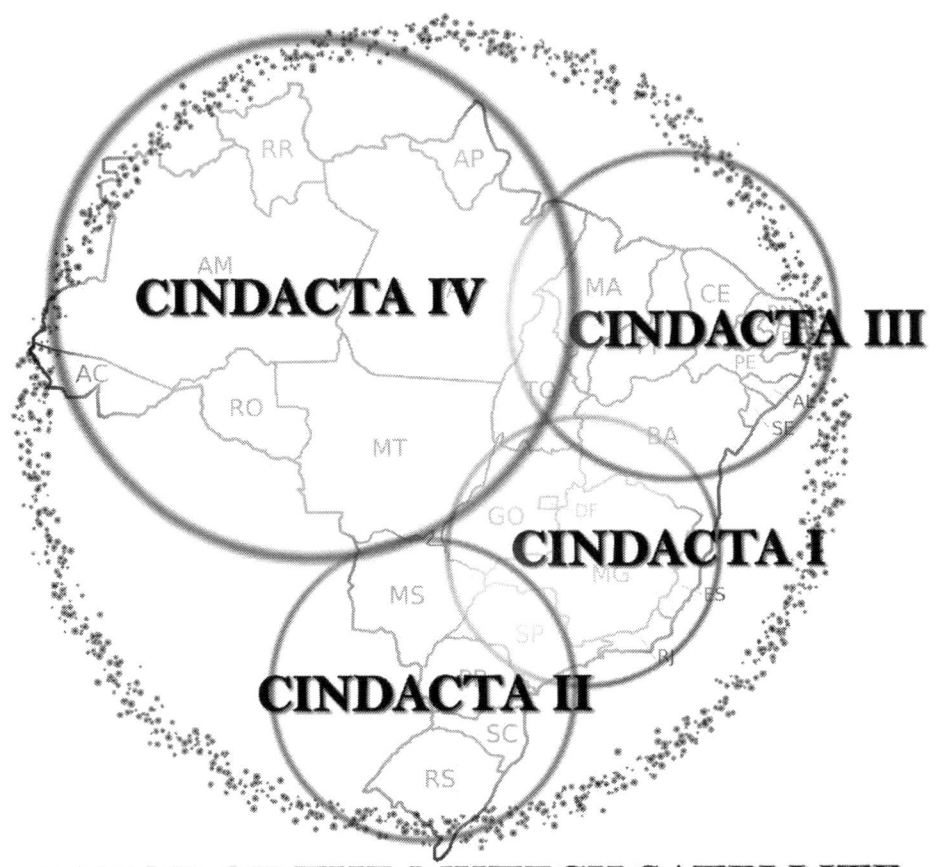

**Figure 9.7** Range of satellite (dotted lines) and CINDACTA radars (solid lines).

As we know, the range of coverage of the transmissions is broad and was already tested in the most diverse situations, locations, geographic configurations and natural and artificial structures throughout the national territory. As it impressively doesn't attenuate, some targets think the event is connected to a mystical, magical, divine or esoteric transmission such as non-Hz scalar waves. In fact, after years of experience and contact with this weapon, I can state that:

* It propagates through tunnels — places where powerful radio waves don't reach. For example, Túnel do Joá (Joá Tunnel), Túnel

Engenheiro Raymundo de Paula Soares (Engineer Raymundo de Paula Soares Tunnel) and Túnel Rebouças (Rebouças Tunnel);

* Subway: these transmissions are capable of operating without interruptions during the entire subway system and in any station where the target is located. For example, Metros in Rio de Janeiro and São Paulo;

* Commercial flights, throughout the territory;

* Commercial trips on all highways in Brazil;

* Underground parking garages (multi-story building);

* In the open sea off the coast of Brazil;

* 6.5 feet deep in pools and bodies of water;

* Natural areas where common devices won't be able to transmit; dense forests in the Amazon, waterfalls and wells such as those in Chapada dos Veadeiros and the waterfall and high mountains of São José do Barreiro in Serra da Bocaina, São Paulo.

With this in mind, it's clear that we're facing a network made up of satellites optimized to reach targets, whose main purpose is probably to manage the frequencies, signals and the time in which each data packet will be delivered and extracted without fail.

The satellites also work with modes of data transmission and communication. The most well-known are:

**Frequency-division multiple access (FDMA)** splits the frequencies into portions, or packages, to serve several clients. It divides clients by frequency. For example, every 20 MHz represents the bandwidth allocated to a given client. It has a guard band to avoid interference.

**Time-division multiple access (TDMA)** uses the same frequency band, but works in a different way, allocating the transmission packets with the data by the time interval. That way, many users can use the same frequency band by rotating pre-allocated time slots at up to eight different time intervals. Each user occupies a time slot in the transmission, which prevents interference problems.

**Code-division multiple access (CDMA)** uses spread spectrum technology as a means of access to allow several users to share the same frequency band. It allows for better use of the spectrum, enabling the increase in the capacity of cellular systems. It also has the same structure as other cellular systems and offers the same functionalities associated with mobility, such as roaming and handover.

The Biometric EEG, the unique frequency of each target, would be the equivalent of having a reserved frequency series similar to what happens with FDMA, CDMA or TDMA. Thus, the frequency bands would be divided between them and interspersed with guard bands so as not to interfere with other transmissions — for example, a 20 MHz slot in each frequency. The bandwidth division or slot allocation can occur by fixed allocation (FDMA) or on-demand allocation (DAMA). In this way, frequencies that would vary between 445.001 MHz and 445.120 MHz, etc., could have been allocated for the target's mind.

Multiplexing allows you to send a vast amount of data destined for a variety of different targets based on frequency space or time. This technique can also be applied to split the transmission for a specific target. So, a range of frequencies can interact with the individual's natural EEG at a specific time that coincides with the target's biometric signature. That is, these transmissions can perfectly replicate the electrical rhythms that will interact with the target 24 hours a day without ever losing sync. It's worth remembering that the rhythms of the brain's electrical circuits are slow (Hz range) when compared to the higher frequencies of transmissions that operate in the MHz frequency range.

All these data transmission techniques can theoretically perform the function of accessing the human mind. It all depends on the people behind the attacks, the infrastructure, among other factors. It's difficult to point out which technique is more effective for each type of crime, since they all have qualities and imperfections.

## 9.2.7 - Conclusion

In order to deliberate on what is most advanced in terms of technology, we must turn our attention to space travel equipment for missions in the solar system and beyond — in this state-of-the-art enterprise in constant evolution since 1950 with budgets in the billions that help produce scientific results that will be used in the present and future by us. For the most part, technology is used to keep society unified and dedicated to communications and the most diverse studies. However, there are military and intelligence satellites that use the most modern technology as a weapon for the safety of the country.

These space weapons can hit any country, and strike people on airplanes while they travel to anywhere in the world. They attack targets under all circumstances. Constellations of satellites armed with increasingly advanced technology specialized in hacking the mind and locating people, creating the target's neural biometrics and accessing their thoughts, in addition to modifying the content of dreams and inserting voices in the victim's head, are being used more and more every day for this purpose.

Imagine a system similar to IRIDIUM that has a constellation of 66 satellites in orbit and covers the entire globe, working in perfect synchrony with terrestrial antennas and operating in practically any urban location, including tunnels and subways. However, this system doesn't only communicate, but support MKTECH and its BRAIN NET network.

Probably all relevant or targeted minds can be covered with no problems with this technology. Suppose a similar system completely focused on analyzing, amplifying and collecting data from the minds of thousands of inhabitants is quietly in operation. Imagine the amount of strategic and commercial data that it can capture. Not to mention the terror that an entire population would face if there was a mass attack. Panic and chaos would spread. Thousands of people — the sick, the elderly or children — would suffer irreversible damage: their areas of communication, language and hearing would be destroyed. There would also be deaths caused by acute stress and problems in the bioelectric functioning of the brain. All of these would happen in a short space of time.

Only people with military apparatus capable of obtaining extremely expensive equipment or equipment installed in places that are strongly protected against electromagnetic radiation can defend themselves against the attacks. Outstanding members of the society like presidents of developed countries would be one of those people. Ordinary citizens, on the other hand, aren't able to escape it. Urban structures cannot protect our brains from the attacks, which are increasingly sophisticated, precise and lethal. The satellites and terrestrial antennas contact integrated neural information centers, where they process all the data that travels in this medium.

Neural and electromagnetic weapons that use high power antennas — one of the great military secrets of the ages — can suddenly come to light and be widespread in all nations, requiring only adaptation and adjustments in the existing equipment and alterations in the software. In other words, you'd just need to install neural recognition programs on computers, and every communication satellite would become a space weapon. Needless to say, a problem will arise as nations are investing massively in this technology, from Arabs and Asians to Mozambicans — a country that recently launched its first satellite equipped with modern equipment.

We're surrounded by high-powered space weapons capable of invading the human mind and preventing thoughts, of hindering focus and attention while the private content of one of our greatest gifts is monitored. Targets are selected to become subjects, and this is the end of the line for them. They will be killed or left with serious injuries for the rest of their lives. The most intriguing thing is to think that this occurs without any chip implanted in the skull; it's just the brain and electromagnetic interactions.

But is it really possible to point out which satellites are being used for this purpose?

Even those directly involved in its construction, conception and operation don't know for sure all the functionalities that the equipment has, especially when the real objective is classified. For those who don't have access to the weapons, all that remains is deduction and speculation based on assessments made on the characteristics and details that may be compatible with the effects experienced by the targets and unanimously confirmed by isolated research and empirical evidence from years of contact with the harmful effects of this space weapon. If I had to bet on the type of satellite that is currently attacking our minds, it would be a spy or attack satellite whose main function is signal intelligence or another similar kind that performs various functions in this context, such as collecting electronic information in different wavelength ranges in the electromagnetic spectrum.

If we continue to follow this path of devastation, that is, if the concept of psychotronic weapons comes to light — something that is about to happen at any moment —, a diplomatic war will be declared. Once established, a conflict between influential countries will rise up — nations will accuse each other of using psychotronic space weapons that they already know for more than 5 decades. Now we can only wait and see how they will explain themselves to society (or if they will) and the various conventions and violations of human rights perpetrated in these gloomy days.

Space satellites are equivalent to embassies; they belong to government officials serving and representing their country in a foreign state and form part of the territory of the country in space. Therefore, a direct attack on a satellite can cause retaliation. What about the attack from space on foreign territory? Despite many treaties of the United Nations on the non-proliferation of space weapons, it's not yet known what the international reaction to this type of attack would be once exposed.

Space technologies allied to military weapons have the most modern technology. Some of them have the ability to probe the depths of the Universe, see planets beyond our solar system, and produce technologies that will soon be available to civilians. Take the Charge Coupled Device (CCD) of digital cameras built into mobile phones for an example. Although it was initially produced for satellites and space probes, it has become something trivial. We all have one. Currently satellite cameras are much more advanced than that — their field of view are better than those of the CCD of our devices. So, the cycle is renewed, and soon this same technology will reach the civilian world at low cost. Logically, we can also expect a popularization of this neuroelectronic weapon that perform precise attacks on people's minds with no chance of defense as there isn't a place capable of blocking the attacks — not in the near future at least — that is within everyone's reach.

## 9.2.8 - National sovereignty and terrorism using space, neuroelectronic weapons

National sovereignty is for all independent nations, that is, nations that have complete power and control over their territorial limits, and are free from the influence or command of other countries. The sovereignty of a state is formed by different organs, institutions and powers. Following this premise, attacks with neural and electromagnetic weapons that are carried out in the national territory against citizens are cowardly terrorist acts. Those involved in this type of evil endeavor makes use of anonymity and abundant public and private money from the scheme that encourages the use of the weapons combined with the total certainty of impunity — impunity that even prevails in cases that have physical evidence, imagine when it comes to crimes that leave no trace and are difficult to detect.

Who would say that terrorists in Brazil are a kind of tragicomedy? They don't use a weapon of international terrorism, or explosive devices that cause immediate chaos and visible destruction to places and people. They use new weapons that have the infrastructure implemented by other countries, the result of decades of war and technical knowledge. For example, satellites from foreign countries like Russia or China carry electromagnetic weapons capable of attacking airliners in mid-flight. As we have seen throughout this book, there is no type of equipment that efficiently protects us. Combined with the surreal nature of the effects and the total lack of knowledge of MKTECH devices, there is fertile ground for experiments and terrorist attacks.

The crime of treason provided for in Law no 7170, of December 14, 1983, reinforces what I have been trying to highlight in this chapter: the seriousness of the use of these weapons en masse or on a specific individual. In my opinion, this violates the law and corroborates my view that this is indeed high treason, a terrorist act — a disgraceful crime against the homeland.

"Entering into an understanding or negotiation with a foreign government, group, or their agents to provoke war or hostile acts against the country. The sentence can range from three to fifteen years in prison".

The groups in Brazil that use the infrastructure created by MKULTRA have a clear communist/socialist ideology that is perceptible due to torture tactic, implemented protocols, persecution, and type of words used — all of these reflected in automatic behaviors based on ingrained ideas that go unnoticed by the authors (during the attacks) with words and images that generate "virtual" experiences in the minds of the targets, similarly to what happens in other targets around the world. They probably have connection with Cuban and Chinese terrorists or the Russian mafia that carry out specialized terrorist attacks with this type of psychotronic weaponry. This involves high-ranking public officials — Brazilians and entrepreneurs who are behind a clandestine scheme of torture, death, sex, fun, fraud and contribution to world experiments, transferring data from attacks in their own international infrastructures to provide the service and receive the results.

But how do we know that these psychotronic space and terrestrial weapons used throughout the national territory are not our ours to blame? Because we don't have national satellites capable of covering the entire territory and operating in the worldwide MKTECH scheme. The only modern satellite is the Geostationary Defense and Strategic Communications Satellite or SGDC-1, which was already discussed in previous pages. Even the first attacks date back a long time before the construction of the SGDC. Its launch only took place on May 4, 2017. So, it is in no way possible to blame any satellite or other device that we have, since attacks of this nature have been going on since the 90s.

Unfortunately, Brazil doesn't invest in technology. We have one of the worst education systems in the world and treat investment in scientific research as an expense. We value few things that have no real value. Therefore, we cannot develop and maintain a scheme like this; we just use the infrastructure and network already created to commit harmful acts to the human cognitive system. The proximity of some sectors in Brazil and their affinity with communist and dictatorial governments such as Cuba and North Korea, especially in the last 15 years, show us in what kind of relationship these people who are associated with terrorist enterprises have since the arrival of this technology in its early days back in the 90s.

Soviet satellite weapons attack the national territory through Cuban, Venezuelan and Brazilian terrorists. Other countries can provide equipment, but it was only after reading about the attacks on embassies — specifically the American Embassy in Cuba — that I was able to finish this puzzle. These satellites are capable of searching the minds of thousands of people simultaneously and amplifying the signals generated by the brain and its content being picked up by adjacent antennas or twin satellites.

There is a United Nations treaty prohibiting the attack on places and/or people from space or the Moon. Unfortunately, these attacks are happening in virtually all countries that are signatories of the treaty, including Brazil. Find below one of the clauses, which can be found on the UN's website:

"States shall not place nuclear weapons or other weapons of mass destruction in orbit, and states shall be liable for damage caused by their space objects".

So, it becomes clear the total violation of the treaty. One can plan a mass attack and cause irreparable brain damage to people in a given area. The elderly, children or those who have an illness or disability will suffer more serious consequences — strokes, destruction of areas related to speech and hearing, mental retardation in children, syncope and death. The chosen area can be an entire country, as happens with the weapons aimed at Brazil (full coverage), forming a global web of neural interference and in-depth tests on human cognition.

Which foreign satellites are currently being used in Brazil? Take your pick! Keep in mind that all developed countries such as France, Japan, China, Russia, Israel, Germany and the USA have this type of specialized weaponry to amplify the thoughts straight from the target's mind — to invade, insert data and destroy various areas of cognition. It's not by chance that all these countries reported several cases of people affected by this cowardly electromagnetic attack. Not to mention the thousands of cases, which are the majority, of people who were attacked and didn't have the opportunity (time, financial resources, patience, intellectual capacity, etc.) or a favorable environment to perceive and understand the phenomenon and study it as a whole. Consequently, they continue to be tortured targets with no chance of defense or reaction.

The Brazilian Armed Forces are extremely competent. They ensure our security and defend us from external threats, such as those mentioned here. Their objective is, among others, to protect the country and take appropriate measures (perhaps even electronic countermeasures) to locate the frequencies or spectrum used, and detect and block the signals. This is essential if we want to return to "normality" when it comes to competitive examinations for government positions and other prominent activities in Brazil.

The modern space battle started decades ago, but now space systems have gained prominence. Space divisions in the Russian and American armed forces were created with strategic plans in mind, such as bases on the Moon from which they can launch this type of MKTECH attack with a delay of approximately 1.20 seconds. The Cold War never fully ended. It just progressed to other stages and levels of interest in current activities.

To make matters worse, as this infrastructure spreads and grows, the notion of territory disappears. An individual in one country can pay to use the network and attack someone in a country thousands of miles away, making any action to prosecute or find the responsible for the attacks impossible. If the country that coordinated the attack doesn't have a cooperation treaty, the culprits will undoubtedly go unpunished.

How will nations deal with this dilemma? Who will take responsibility for the attacks? How will they manage to find the attackers and how will they be punished? Who's going to conduct an investigation into a weapon that doesn't obey borders or respect nations or countries? These questions must be debated. After all, these extremely powerful weapons are getting more and more lethal. The time it takes for the brain to be destroyed is getting shorter. So, there will come a point where the parts directly affected by the attacks (especially areas related to communication, hearing and reasoning) will be immediately destroyed right after the first burst.

The stolen thoughts that travel through servers and thought patterns, neural biometrics that use these satellites, all infrastructure that contributes to this attack both on land and in space, must be considered part of the MKTECH scheme and must be stopped or at least regulated. Its use is itself a terrorist act and people who attack their own nation using space weapons from foreign nations against Brazilian citizens should be punished with full force and prosecuted for terrorism, attack against humanity, systematic torture, violation in the use of space weapons and utilization of devices capable of causing mass destruction.

# CHAPTER 12
## PRESENT, FUTURE AND CONCLUSION

*"No price is too high to pay for the privilege of owning yourself!"*

— Friedrich Nietzsche.

How should we perceive the course of civilization in the face of an uncertain future? Human beings and society will possibly adapt at some point, but the use of these weapons will have an irreversible and unprecedented impact on our social fabric, given the inevitable discovery and popularization of the last frontier of human privacy — the reading of thoughts. The most interesting aspect about all of this is knowing that what we've seen in this book is happening right now, at this very second, in various parts of the planet with hundreds of thousands of people, as a result of the constant evolution of this process that started in the mid-1950s. To make matters worse, I'm absolutely sure that the weapons used in Brazil with which I had direct contact are already obsolete. Advanced technology developed by governments dedicated to scientific research, defense, and intelligence, alongside their substantial budgets, has already introduced newer and upgraded iterations surpassing what we've previously witnessed.

In this chapter, I'm going to delve into medium and long-term projections, drawing upon a comprehensive analysis of my personal observations, extensive research, and the synthesis of my experience and perspective. These insights are grounded in the emergence of this imposing new reality. I'll even venture into highly relevant speculations. After everything we've read — amazing as it might seem — this is the only chapter I take the liberty of using my rational imagination on matters revolving around MKTECH.

But before we can start talking about these issues we must first legislate on the subject and formulate regulations as soon as possible. It's crucial that we bring to light and eliminate the obscurity around this matter, enabling citizens to protect themselves with the aid of cost-effective technologies that can block the frequencies utilized by these weapons. These technological solutions ought to seamlessly integrate into our daily lives, becoming as commonplace as cell phones did just a few years ago. For example, shielded rooms, helmets to prevent theft of intellectual property, and a kind of device under our beds to help us have natural dreams, and not only dreams controlled by third parties in a revolting biological, mental and private invasion by the most impressive stalkerware[67] ever.

It's also essential to reassess all procedures associated with competitive examinations, as they are currently rigged, rendering it impossible to discern who genuinely succeeded based on their own merit and who may have relied on this "undetectable" technology to be successful.

So, extremely complex, unknown future scenarios are created. At certain points, they're similar to the cyberpunk[68] world, which previously

---

[67] Stalkerware is software employed to monitor activities carried out on a victim's computer, obtained by an individual who might be a controlling family member, a possessive partner, or even part of organized gangs. They are silent and therefore more difficult to detect. In the case of MKTECH, it directly operates as if "installed" in the victims' brains.

[68] Cyberpunk, a fusion of the terms "cybernetic" and "punk", is a sub-genre of science fiction: "High tech, Low life". Its name comes from the combination of cybernetics and alternative punk, and it mixes advanced science, such as information technology and cybernetics, with an element of disintegration or radical change in the social order. The term was coined in 1980 by writer Bruce Bethke in his short story "Cyberpunk," which, however, would only see publication in November 1983 in Amazing Science Fiction Stories, Volume 57, Number 4. Cyberpunk fiction typically revolves around conflicts that involve hackers, artificial intelligence, and megacorporations. The core of the theme is also centered on profound alterations in the human body, accomplished through cybernetic implants that enhance cognitive capabilities and enable extensive interaction between the mind and machines. The term "cyberpunk" also refers to a subculture that centers around cyberculture, recognized for its affinity for psychedelic music and the fusion of punk rock and electronic music genres, as well as its adoption of futuristic fashion accessories. The cyberpunk style and theme capture the nihilistic and underground aspects of the digital society that emerged during the final two decades of the 20th century. The cyberpunk world is inherently dystopian, serving as the antithesis to the utopian visions prevalent in science fiction of the mid-twentieth century.

only belonged to science fiction. In this context, looking ahead to the medium-term projection, we can anticipate significant societal shifts, particularly in the emergence of something that will soon become omnipresent and exceedingly challenging to monitor and control: paid thought-hacking services operating within a specialized satellite network encompassing the entire globe for this purpose. All individuals across the globe, without any form of protection from radio frequency waves up to microwaves, will be impacted. The trend indicates that as the technology gains popularity, it will progressively become smaller, portable, and mobile. More efficient algorithms and mind-hacking services will become widespread, akin to the storage format offered in cloud services, video and music streaming, urban mobility, hosting, cable TV, and so on. Chinese, Russian, Indian, and American satellites, already operational within the network, will serve as the infrastructure for this type of service. Consequently, customers will pay using cryptocurrencies or any other "untraceable" form of payment to access SYNTELE targeting a specific person of interest. This service will offer the target's thought content on an hourly, daily, or monthly basis through various packages.

In the face of this alarming scenario, the initial victims of this new era will be those closest to us. As human nature often exhibits inherent distrust and morbid curiosity, spouses will scrutinize each other, parents will monitor their children, and children will observe the thoughts of their siblings. The really curious ones will have the option to pay for spying on their attractive neighbor, observing their daily personal moments. Alternatively, they can access the thoughts of multiple individuals by connecting their cell phone to an application that streams such thoughts. They can even sit on a comfortable couch, retransmit the content to a larger screen, and switch between different thoughts as if changing TV channels.

As we've seen, this technology integrates all the systems, so it's clear that we're confronted with a service that offers the opportunity to use all other MKTECH tools at an affordable price. The affordability of this technology is attributed to the proliferation of thought surveillance satellites, which are being launched almost on a weekly basis, resulting in their abundance.

Clandestine (or otherwise) paid services will emerge in droves, making it possible to pay and solicit a person's thoughts anywhere, anytime. And I'm not even considering other MKTECH systems such as the EMRo - Electronic Mind Reading (optical) that would make you be able to see everything — all the signals that travel between the eyes and the visual cortex.

Being able to view everything someone is currently looking at opens the door to various unimaginable violations. The concept of seeing what someone else sees is a recurring theme in science fiction, blending elements of reality with astonishing endings and narratives that mirror this nebulous future. The real and boundless possibilities of what can be done with a person under surveillance, solely through their sight, provoke profound reflections.

Now we can only passively wait and watch as the first services emerge for the general public. In fact, this is already a present-day reality for select groups of individuals who have undergone testing with the purpose of monitoring the visual perception of the target and consequently tailoring a narrative through SYNTELE that describes everything that enters their field of vision to purposefully restrict freedom as much as possible. The target feels automatically intimidated, immersed in their thoughts and behavior, and in parallel these groups are able to capture their reactions to the ongoing optical surveillance. Think of the countless possibilities that arise when you're able to see what someone else sees in real time, and let your creativity put together a script that contemplates such possibilities.

However, this perverse scenario will spiral out of control as the services transition from passively capturing and displaying data from others' minds to a more active approach. The active way is nothing more than the use of the tools that cause neurological damage — the merciless V2K, intracranial voice that generates the microwave hearing effect in the form of attack (input) in which SYNTELE attacks and experiences the aftermath of the devastation as reflex stimuli in the form of vocalized thoughts. This danger disguised as a service is a tangible fact, but it's

CHAPTER 12 - PRESENT, FUTURE AND CONCLUSION

restricted to some groups of people. So, anyone who can pay for the services will select their target and "connect" this technology to their brain, thus generating negative change in the relationship with reality for the unprepared person living their life as usual.

If you possess additional cryptocurrency, it'll be feasible to incorporate D2K into this attack combination, enabling the transmission of horrifying nightmares while modifying the dreamscape according to the client's desires. This includes memory reconfigurations that induce experiences of "Déjà vu" and "Déjà Rêvé". Those who are financially privileged will be able to select a package with satellites and terrestrial antennas so that they can utilize all available weapons. The nefarious intention is to have fun as they party all night with their equally privileged peers, driving targets (whether chosen randomly or not) to madness, merely to serve as a barbaric medieval stage setting.

We cannot forget the numerous terrorist attacks that are becoming increasingly frequent nowadays with the use of these weapons. In the near future, due to the readily available access to this technological toolkit, radical groups will be able to attack both an individual at home and a plane in the air and full of passengers without much effort. Who knows, they might even manage to amplify the signal and attack several children in a crowded school, leaving them deaf and with serious problems in all cognitive-communication fields, as Broca's and Wernicke's areas are damaged — outcomes that have already been observed with the deployment of this potent weapon of destruction.

Over time, the human brain's damage will become significantly more devastating and widespread, evident through the advancements made in the weapon. After all, with each new attack, there are constant improvements in terms of power, interaction, and intelligibility.

In light of the escalating activities of military satellite launches and classified space intelligence and weaponry, a new global race becomes evident, utilizing the cutting-edge capabilities of these electromagnetic devices. Bear in mind that certain issues and strategies on the battlefield, such as the development of new weapons, have seen little change since the beginning of the century. Let's talk about the Manhattan Project. The world remained oblivious to the events taking place there, completely unaware of the nature of the work being conducted and the unprecedented weapon being developed until the pivotal moment when it was deployed in combat. The outcome of this secrecy resulted in the creation of a weapon that would change the war. Its immense destructive power became apparent only when witnessed firsthand in the shape of a nuclear mushroom after being launched at Hiroshima and Nagasaki. The military, particularly from dominant nations, continue to operate in this manner. The full extent of a weapon's devastating capabilities and technological advancement is revealed only when absolutely necessary, as exemplified by the deployment of the atomic bomb nicknamed 'Little Boy' on Japan, resulting in unparalleled destruction. Few people up to that point knew for sure what was going on.

Have no doubt: when the time comes, this neural electromagnetic weapon will be deployed to its full extent and maximum force, taking several lives, modifying humanity, and leaving us defenseless and unable to adapt in a timely manner to this type of attack that, until now, was never deemed conceivable — it was merely hypothesized as something distant from reality and yet to be used. But let's not forget that the experiment being conducted in Brazil, upon which the book is based, is significantly distinct from the state-of-the-art MKTECH psychotronic ultimate weapon developed by the top experts in Russia and the United States. We'll only grasp its full destructive potential when it's already too late. So, it's of utmost importance to recognize the threat we're confronted with and be prepared to defend ourselves. By doing so, we can avoid reliving the events that led to the complete deterioration of the minds of those who are no longer with us.

In the future, this weapon is likely to reach an exceptionally advanced level. Not only will it have all the functionalities improved —as this book has enlightened us about—, but it will also possess the capability to completely deactivate the systems supporting autonomous life, encompassing both the sympathetic and parasympathetic systems. It can initiate a sudden and relentless attack with the intention of eliminating a specific target, forcing a cardiac arrest or respiratory system collapse. On a more moderate level, it can transmit commands to the pupil, causing it to dilate and contract, directly reaching parts of the brain responsible for cognitive executive functions. It can even create optical illusions while awake as it happens using D2K during REM sleep. Gradually, it can cause an imbalance of the neurovegetative systems. Numerous possibilities exist for targeting the intricate human electrical system. By demodulating information transmitted via radio, which can modify cortical mechanics, remarkable advancements are being made in countless fields of science.

This mind-reading technology will permeate various fields of telecommunications and the manner in which we engage with the devices we use on a daily basis. Absolutely all entertainment models will make some use of this technology. Social networks will likely be among the first to do it, enabling interaction with individuals from our social circles, engaging in games, and even communicating through amplified input and output of thoughts. Video games, cinema and cell phones — our constant companions, the fever of the moment — will also use concepts from the platform to innovate. Possibly, novel devices that could function as a fusion of cell phones may emerge; devices capable of transmitting, amplifying, and deciphering neural waves and thoughts.

Perhaps we will work together with the device, conveying instructions solely through our thoughts. Communications will occur via Synthetic Electronic Telepathy, eliminating the need for sound emission and enabling distant communication through the cell phone itself, thus bridging the gap between minds. Who knows? Maybe it'll be possible to completely replace most of the functionalities of a smartphone with images that can be inserted into your mind and interact with your thoughts, sending commands and signal inputs with remote responses. It'll be simple. Just close your eyes and mentalize the visual thought to interact with it. Then open them again and move on with your life. It would be somewhat similar to that, since the capture of mental images in the waking state using SYNTELE is already possible in certain neural configurations, as well as the extraction of such images.

Certain features can function cooperatively, like thoughts working in conjunction with a mobile device or operating simultaneously (in parallel). Several functionalities will emerge from the fusion of mind and device with the Internet of Things (IoT). Without needing any implanted microchip, we'll use our vocalized thoughts to turn the lights on and off, to use our refrigerator or open electronic locks and our cars, in short, every element equipped with this technology through the intermediary of a central device that will be the interface of the owner's thoughts to control things. LoRa[69] is a low-power wireless platform and it's widely used in the Internet of Things. Soon our brain will also send commands through this network and access equipment on the other side of the world.

There is still a long way to go when it comes to positive applications of this technology that could benefit our daily lives. One of the biggest barriers will be teaching people to control their thoughts to the point of sending only commands relevant to particular orders that the equipment recognizes as valid and interacts with — to start an algorithm or switch on the lights, turn down the music, perform computer (or cell phone) authentication, get in the car, and so on.

The same issue will arise among individuals using these technologies, especially concerning communication. Thoughts of all kinds will be transmitted, leading to questions and amazement among those attempting to communicate, as they deal with the remarkable ability to hear someone else's thoughts. Carefully designed and adjustable software filters that enable sending only desired thoughts may present a practical and viable solution. The alternative, cognitive training routines to establish internal mental filters and prevent involuntary release of intimate thoughts, can be

---

[69] LoRa is a radio frequency technology that allows communication over long distances with minimal power consumption. It's based on a star topology network, similar to a cell phone network. The modules send and receive data from specific gateways (like Wi-Fi networks, but with much greater range) that forward them via IP connection to local or remote servers. Its main applications are IoT (Internet of Things) systems, such as sensors and remote monitors (pressure, light, on-off, temperature).

more challenging and abstract for many. And believe me, some thoughts will escape in the most diverse situations.

I explored only a fraction of the numerous potential consequences in various social and technological scenarios that may arise with the introduction of this device into society. Initially, I'd only touch upon the subjects, but I couldn't resist the temptation to delve into them a bit further, particularly those that I deem to be the most significant ones — this may not be the reader's viewpoint, but it's certainly a valid point of view. I worked hard to keep my vision of the near future coherent and consistent, building upon the technologies available today and their projected path. This leads us towards a future where these advancements will become part of our daily lives — and all this will be ancient technology that will pave the way for even more transformative changes, as we're going to see below.

## 12.1 - Future of intimate relationships

I'll start with a delicate and very private matter for most humans that will be profoundly affected by this technology: marital, sexual and emotional relationships. One of the significant drawbacks of our evolutionary progression from primates to people is the adaptation to an imposed and culturally unnatural monogamy that we're subjected to, which brings all kinds of marital problems that we all have witnessed at some point in our lives. With this weapon being used as a service, these issues will escalate to levels that are not sustainable.

The most jealous and suspicious partners will probably reside in their spouse's thoughts, resulting in a kind of unending obsession and paranoia, much like one of the actions usually perpetrated by OPS (Organized Professional Stalkers) against their targets. Let's look at an example: think of a couple having fun and drinking at a party. Suddenly the boyfriend or girlfriend picks up the cell phone and decides to check their partner's thoughts in one of the thousands of paid Asian pirate "streaming" services that will provide this type of illegal content. In about 10 seconds, jealousy will cause the couple to argue and stop talking.

It's only natural for people to find others attractive and think about them silently. That doesn't mean that the thought will become a concrete act involving the person in question. Moreover, these mental images can help boost libido during sexual intercourse. It's common for many people to use the visual thought and sound memory of someone else or a scene that ignites sexual desire, to satisfy the desires of both themselves and their partner.

This technology will then likely mark the end of relationships as we know them. After all, if the monitoring also takes place during sex and these additional stimuli are captured, it may become difficult to engage in sexual activities with your partner while either of you is thinking about someone else or a movie star. The consequences of capturing such thoughts, which are still natural stimuli that arise in the mind and enhance sexual arousal, remain unknown. These internal stimuli reviewed by the partners during sex can generate endless questions depending on the level of jealousy and possessiveness.

What if, in the midst of such mental images, peculiar thoughts come up? What if known faces and people — your girlfriend's friends or even your own friends — appear in her visions? It's important to understand how this technology entirely reshapes the social environment and the dynamics of human interaction, leading to significant alterations in what defines our very essence and identity. This results in a serious problem, which people will only become aware of when their thoughts are actually exposed: the natural processes the brain uses to establish connection with reality. It's difficult to judge people's thoughts based on these processes, but it's precisely in these processes that we're judged by external observers.

We're extremely complex, emotional and unstable beings, influenced by all kinds of internal and external stimuli. These include interactions with the environment that bombards us with information at all times and stimuli arising from interactions with others. Emotions, on the other hand, are complex states and the whole body participates in it. These states are typically based on the perceived intensity of feelings, categorized as either low or high levels of pleasure, displeasure, or indifference. There are, of course, more intellectual elements involved, but this is only possible due to the interdependence between psychological and physiological events. Hearing your spouse's thoughts and seeing their visual thoughts, questioning why your partner thought of someone else during sex can in fact be a recurring event in the future.

If someone is curious about examining the visual thoughts that emerged and sexually aroused their partner during the post-coital period, they can simply search the cloud servers and, with the necessary financial resources, access the available data. However, this pursuit may drive the most dedicated and distrustful individuals to exhibit neurotic behavior in their quest to fully understand their spouse's thoughts.

If an individual possessed all the answers to the variables encompassing the chemical and biological processes behind the mechanism of attraction, or the reasons behind certain situations and memories that ignite their sexual desire, they could potentially write a scientific article and even be nominated for a Nobel Prize, because at the present time the complete process remains unknown. In this sense, certain inquiries will be recurring and have devastating effects on relationships. Questions such as these, "Why were you thinking about my friend? Why are you thinking about her right now? Who's the man you keep thinking about?". The crux of the matter is that this is already a reality today in the dynamics of torture between the victim and the operators. And, in just a few years, this will be available in a store near you!

Other slogans like these will emerge alongside services based on the premise of mind invasion. Advertisements like, "Tune into your child's thoughts 24/7 and cherish every moment of their growth!" or "Concerned about your wife's fidelity? Monitor her whereabouts and listen to her innermost thoughts!". It's important to emphasize that this technology is available and active. When it comes to the services that will provide access to the data for anyone willing to pay, they will originate from mind-hacking platforms that will gradually become widespread. Unexpectedly, uncontrollable activities in this format will start happening all around the world.

Another questionable emerging service will be related to pedophilia, which I won't even elaborate on. But can you imagine this service in the hands of pedophiles? No child will be safe, not even under their parents' watchful eyes!

Which crooked paths are we going to take? Are we going to get used to being cognitively watched by our parents from a young age? To seeing them change a reasoning that they consider non-standard? Will a child who needs to train their brain through cognitive experience, using all their senses in social interactions and with the environment, get used to this violation? These are some projections for the near future. I believe there's a strong likelihood that all of the aforementioned events will unfold as anticipated. Thought hacking services for recreation and surveillance in such terms is 99.9% sure to come to fruition. I perceive this as the inherent course of action we are destined to take.

Another upcoming service is related to the creation of remote dreams, given the intense power of pleasure without any actual risk — the dreamer just feels it. This will prove immensely valuable, as it enables the synchronization of two dreams and facilitates the arrangement of a meeting between egos that will revolutionize future technology. A kind of special dream app for safe and remote encounters between two or more participants.

This is already being tested and is part of the nascent stage of the technology. Torture experiments use this subterfuge to monitor the subjects' responses, and the most relevant results will be explored for future clients. The viability of the project is progressing by leaps and bounds. At every moment different results of these interactions are captured from the poor human guinea pigs that are being tortured all over the planet. If you find yourself involved in this experiment that is increasingly advanced in content and format inside your mind, serving as a platform for technological experiments and an entertainment center for OPS (Organized Professional Stalkers) don't forget to ensure you receive your financial compensation for allowing your brain to be used as a testing platform for psychotronic dream manipulation weapons due to commercial bias camouflaged as a kind of steganography of content in the form of torture or constant attack protocol.

## 12.2 - Neural computing, Digital neural manipulation or Neural programming

Up until now, the interaction with the workings of the mind has primarily consisted of mapping specific brain regions and collecting data on electrical and magnetic activity. The only way to manipulate and activate the mental processing of external data would be through external sensory organs that receive visual, tactile or auditory stimuli. By exclusively utilizing these primary receptors, it became feasible to observe brain activity triggered by responses to stimuli across a wide range of parameters. This process enabled the collection of data and the creation of mental maps displayed on a computer screen.

Obviously, the dynamics and details omitted here are much more complex than this, but that's basically how the interaction with internal processes takes place. We're expecting a big step forward in brain studies with the arrival, popularization and evolution of powerful psychotronic weapons. In addition to all current advances, we'll work with data from the recovery of vocalized thoughts and mental images. In the coming time, experiments conducted to meet objectives regarding new processes involving the manipulation of visual thoughts and the possibility of direct manipulation of memories will have reached a high level of sophistication. It'll be possible to "break the shell" and see how the whole process happens in real time, checking the operation and carefully manipulating each memory, neural reaction and reactions to stimuli without going through the external receptors as we are used to. A window of opportunity emerges for a new area: neurocomputing.

Neural computing is a new field that involves the human brain and computational algorithms. They can serve multiple purposes by collaborating or using the processing power solely in the brain, achieving a balance between machine and brain, or by using programs to actively alter the human mind. These are computer programs that manipulate and modify biological processes. A new science is born from this weapon: the art of reprogramming a person, completely modifying a personality, directly altering memories for the first time in history, implanting, suppressing and replacing, completely discarding the time-honored methods based on torture and dull repetition, directly modifying contents as we do with computer files.

Within this concept, a powerful tool appears on the horizon. The brand new and functional SYNTELE - Synthetic Electronic Telepathy 2 enters the final phase of implementation. SYNTELE 2, which wasn't covered in this book, displays a remarkable difference from its predecessor: it enables the transmission of images interpreted by the waking brain at any time, superimposing the mental images controlled by the individual during the conscious state. Therefore, SYNTELE 2 is able to insert images and change the display order, which forces the human mind to see intrusive visual thoughts, changing the flow of the general perception of the uniqueness of the senses in which sensations constantly flow as a foundation for the development of logical reasoning. The product of imagination is then altered and a new type of privacy violation is allowed: the possibility of displaying images of any nature, such as advertisements, without respecting age groups or any type of content control. Covertly transmitting these signals will become widespread globally as advertisements are beamed into multiple brains at specific locations, leading to a form of mental spam[70]. The repercussions for human cognition will be significant, causing cognitive entropy that distorts and modifies areas of the brain without definite knowledge of the long-term effects of such interactions.

Consider the prospect of operating at the most fundamental mental level, where one can intervene and manipulate concrete operations capable of systematic reasoning about objects, numbers, time, space, causality, and similarity. Alternatively, envision altering formal operations, manipulating the built-in program responsible for these outcomes, and reshaping how we perceive interactions with the world. This could potentially lead to destructive paths by causing miscalculations of the implications from this interconnected dataset. Put simply, it means making permanent changes to the fundamental level of the built-in algorithm, a direct manipulation of the "bits."

---

[70] The term "SPAM" is often attributed to the acronym "Sending and Posting Advertisement in Mass".

Despite its futuristic and elusive nature in certain concepts, at this point we will already have adapted to it and evolved as we've done throughout the centuries. In light of this, and taking into account current trends, two potential scenarios emerge. As awareness of this weapon spreads, the challenge of covertly manipulating a brain within existing frameworks increases alongside the advancements in the technology itself. Consequently, the situation may revert to the current state with unsuspecting individuals being cowardly targeted. However, upon detecting the initial signs of neural intrusion, those who are well-informed will be equipped with a range of tools meant to trace the origin of emissions and find sanctuary within secure community spaces. This effectively guards them against the unrelenting waves aiming to infiltrate and manipulate their thoughts.

A plethora of personal technology will come into existence to facilitate and standardize the tools and techniques in using, modifying, and implementing neurocomputational systems. Development platforms[71], frameworks[72], and SDKs[73] that align with our technology will become readily available and will streamline the developer's workflow, saving time and effort in product development, effectively enabling communication with someone else's brain. Manipulating individuals' personalities and minds will become a common habit. The regulations governing such activities — that will vary from country to country — will dictate how these personality alterations are carried out.

There is also a possibility that it can be used for good by focusing on painful memories and traumas, modifying the perception of the past and the way the brain interprets such events. Further studies should be conducted to ascertain the degree to which direct alteration of memories can give rise to potential side effects and elucidate the nature of those effects.

---

[71] Business model that enables a connection between producers and consumers so that they connect to this environment and interact with each other.

[72] It serves as an abstraction that consolidates common code across various software projects, providing generic functionality.

[73] Software development kits.

In addition, this has the potential to activate pleasure centers and simulate the effects of chemical drugs, solely determined by the user's choice of the sensations they wish to enjoy at any given moment. The accomplishment of this feat is currently undergoing comprehensive testing; nonetheless, it still doesn't occur in the waking state, only in REM sleep.

In a nutshell, endless possibilities for the use of this technology in the future can be foreseen. However, failing to establish regulations for the technology in the present might lead us toward a bleak and troubling future. I don't intend to cause unnecessary concern for the most skeptical individuals regarding the future I project in this chapter. However, I feel compelled to deliver disheartening news to those who still fail to grasp the current events unfolding before us: the most complicated objective of this system, believe it or not, has already been achieved. And it happened more than half a century ago, the fruit of 29 years of classic MK-ULTRA (1950-1979) and modern MKULTRA (1980-2020) systems. The calculations, formulas, and infrastructure already exist and work beautifully.

Hacking the human mind is no longer the primary focus, it's already a fact. The devices created based on this historical finding have evolved and are getting better year after year. Despite having read this book, individuals who have never seen the weapon in operation may question the reality of its contents. It's only natural and understandable to have these questions, but if we approach this complex subject solely from personal beliefs and overlook practical solutions for neuroelectronic defenses, we might waste valuable time and energy. Time, in fact, is pressing for the good people who could be targeted by this weapon, as well as for those who are already facing its effects. We have to broaden the debate and improve the level of knowledge on the subject, once and for all overcoming the initial barriers that prevent progress.

With this premise well established, one must exclusively work with the finished product and evolve it into something more virtuous than a weapon of torture and death. In the near future, the transformation of the most intricate and impressive weapon in the entire MKTECH scheme —

the crown jewel — will evolve into something more beneficial for humanity. If executed correctly, there is a genuine potential to control the mind during unconscious states such as sleeping and dreaming. These projections, based on empirical evidence gathered during my research, follow a logical path as shown below.

## 12.3 - D2K – Synthetic Electronic Dream and its thousand facets

Processing information like a computer, having deeper control and access to the brain in a state of unconsciousness along with the ability to utilize brain functions for data processing is an amazing fact. This unlocks a myriad of possibilities, spanning from the manipulation of dreams and the alteration of visual memories (Volume 1), to distributed cryptocurrency mining, data analysis, and even the control of large machinery in our waking reality, driven by our ego and limited awareness within dreams. Alright. Alright. I know this sounds absurd. It may seem remarkably advanced for our time, but surprisingly, it's not! Unfortunately, I can only make projections based on observations of what is already known about the topic.

The illegal experiment has a hidden objective of gathering as much data and information as possible from subjects. This data is then utilized to develop a technology capable of envisioning the mentioned scenarios, which serves the commercial interests of the entire scheme. Among the scenarios that are unfolding, we have the process of mental P2P[74] networks, mainly using Synthetic Electronic Dream (D2K).

Undoubtedly, a stable and functional dream P2P network will become a reality as the technology continues to improve and spread and more people gain access to it. In the future, the brain will emerge as the new entertainment platform—not the sinister form of entertainment controlled by manipulative OPS, but rather voluntary and consensual entertainment chosen by the consumers themselves. Equipment that

---

[74] Peer-to-peer (P2P) is a computer network where each node functions as both a client and a server. This enables the sharing of services and data without relying on a central server.

interacts with cognition and creates a connection, a bridge between thoughts of friends and family, especially while we dream without the need to use TVs or projection screens.

Before continuing, it would be interesting for the reader to have already read chapter 2.5 Synthetic Electronic Dream (D2K) of Volume 1.

## 12.3.1 - Online (neural) multiplayer matches and Virtual Relaxation Room

Suddenly, you look around and realize that everything appears identical to the reality you wake up to every day. Yet, something feels odd. You look different somehow. You open the window and welcome the breeze carrying the scent and sounds of the city. You look outside, and in a sudden moment of thought, you soar across the skies on a special journey filled with new sensations you've never experienced before. Every night while sleeping, you can feel these emotions. You'll be able to ready your spirit for hours of flight without any equipment, relying solely on your virtual body and mind to capture the scenery unfolding before your eyes at incredible speed.

Can you imagine spending every night in wonderful places, paradisiacal landscapes with whoever you want, simulating any situation you would like to happen? The twists and turns of life designed in a safe environment where all your dreams will come true and your worries will be completely forgotten?

This possibility is being developed to soon become an economically viable technology. A new type of service that will revolutionize the world! The consumer will purchase a box or cube-like device that connects the dream world ego to the internet and synchronizes with cell phones or PCs. With this equipment, they can run a program in a private room or interact with others in shared spaces. The dream cube will be an improved, optimized, portable and individualized D2K. Consumers will have full control over their dreams and the sensations they wish to experience during their daily sleep routines.

Initially, we will possess a technology capable of facilitating dream encounters by linking two or more people in a shared environment. It will be a rendezvous of individuals engaging with one another, offering genuine opportunities for romantic connections and other lifelike simulated pleasures of the flesh.

In addition, this technology will bring forth other elements. Among them, the ability to encounter other egos in shared interactive spaces, where each room becomes a unique realm with its own rules, giving rise to limitless possibilities. Other rooms would be dedicated to online and single-player games. These games would adhere to the principles of current gaming, featuring well-established mechanics, challenges to conquer, competitions, and the satisfying feeling of reward, proving that overcoming all obstacles was truly worthwhile.

Virtual encounters can take place in specially designated rooms for ego connections, including interactions with bots and NPCs programmed to facilitate sensual experiences. These interactions add spice to the nights of the most adventurous individuals in pursuit of seduction, eroticism, and a hint of "caliente". Individuals will have the chance to gather on servers that link all egos in a common virtual space, akin to a social network like Second Life[75]. Yet, this experience will surpass it in immersion and enjoyment. Keep in mind that dreams can evoke both positive and negative sensations. It just depends on the images being sent for processing.

Relaxation servers could potentially shape the future landscape of multiplayer platforms. The individual will have the capability to connect to servers wirelessly. The journey and creation of the dream world will be influenced by the program or game that resonates most with them. You will have the ability to program your own dreams and construct your virtual world while you sleep. The potential of this technology for good is impressive and very promising. I anticipate a forthcoming surge akin to the gold rush of the late 1990s, the dot-com boom, — without the risk of a speculative bubble, of course — or the app boom during the advent of smartphones.

---

[75] Second Life is a 3D virtual world that simulates, in some aspects, the real social life of human beings. Although it was created in 1999, the software was only developed in 2003 and is currently operated by the company Linden Lab.

There are no serious in-depth scientific studies on long-term consequences involving the memory and mind of a person whose dream content has been systematically modified. Indeed, there are long-term experiments entirely dedicated to "nightmares" or bad dreams. These experiments have proven to be highly detrimental, leading to a range of physical and mental disorders in the short and medium term. However, paradoxically, the ongoing suffering of the targets will pave the way for the beneficial future scenarios envisioned by me.

## 12.3.2 - R.E.M Games

It will be possible and more stable to connect several people to the same dream or to a central virtual theme. In the near future, multiple egos will function much like avatars in current computer games during intense online matches. However, a revolutionary change awaits: you won't merely control a character on a screen using a mouse or controller. Instead, you will become the character, experiencing an unparalleled level of immersion with every flight, shot, and interaction! Upon awakening from these dream games, also known as REM games, you will find yourself transformed into an entirely different person. This type of apparatus won't take long to be operational. It merely involves examining the intricacies of how contemporary neuroelectronic weapons operate. Once the underlying mechanism that establishes a link between our minds and a range of frequencies is deciphered, along with the precise nature of their interaction, and once the intricacies and specifics of this weapon are unveiled to the public, our society will undergo a profound transformation in the months or years to come.

The infrastructure is ready. The current architecture of the Internet itself, chat servers or massive online games can be used to function in a game-like structure with some adaptation. Also, the online game cycle, its mechanics and interaction with other players and a common virtual scenario will remain. To meet the new demand, this big change will take place in the control process, the interaction platform: capturing the ego for virtualization and launching it in the room along with other virtualized egos. Of course, it's not possible to maintain consistency in everything based on the existing infrastructure. Some modifications and adaptations will have to be exhaustively tested. It's not enough just to replace the virtual game character with the dream ego. There are some technical difficulties and mishaps along the way. The program that will visually interact with the mind must properly calibrate each character's decision-making (ego's decisions) within dreams and create rules and limits to generate a context, a story in which the ego feels immersed and interacts freely without realizing it's a simulation or game.

We'll have a device capable of fully understanding the ego's action in the dream and which can also precisely adapt these actions to an avatar that will interact with the virtual world. This avatar embodies the ego, transcending the realm of dreams and manifesting within the virtual world. Everything that happens to the avatar will automatically be felt by the ego in the dream and, consequently, by the individual's brain and body.

Hitbox refers to an unseen geometric shape frequently used in video games to detect collisions in real-time. As a result, aspects like invisible walls and various other game elements will need to be adjusted in order to establish boundaries for the virtualized character/ego. This modification aims to create a convincing illusion of obstacles and tactile sensations, effectively engaging the player's mind. This will go through the creation of complex algorithms to virtualize the "self" in the dream. But deep technical questions will have to be answered over time such as, "What 'screen' refresh rate will be needed to ensure a fluid gameplay in the sleeper's mind? Will it be the same one we are used to in a waking state? 24 fps? 30 fps? 60 fps? Will the mind be able to monitor this refresh rate smoothly? How will the rendering of complex 3D worlds be interpreted by the brain? Do all techniques affect the mind in the same way? Or do some stand out? Photorealistic rendering (3D rendering)? Cel shading? Cartoon worlds? What are the best sounds, colors and lighting? Will our neural Graphics Processing Unit (GPU) work seamlessly and comprehend all of this? Can pre-rendered pseudogames (game-like simulations) be created?".

In addition to these questions, we will also come across several technicalities that will be resolved with the use of this new dream entertainment platform that will become the trend. The main search engines on the market will easily meet the demand. With just a few additional libraries acting as intermediaries, the intentions behind the ego's movements can be effortlessly transmitted to the game. This process involves updating the status and providing graphical feedback to the individual's mind with real-time textures and lighting, resulting in a dreamlike rendition of the game's environment.

This remarkable future, which the most impatient individuals daydream about, is no longer a distant vision but rather just around the corner. Tests and experiments with human subjects connected to MKTECH are already underway worldwide.

## 12.3.3 - Operating robots and machines unconsciously via D2K

If you think the previous topic was slightly unconventional and futuristic, something unlikely or impossible to be true, don't be surprised by the next one, as it's closed linked to the previous matter, almost as if they were technological counterparts. Allow me to make a daring prediction that might appear implausible to some. But for those who have witnessed the potency of the D2K, its potential is undeniably real. It's indeed feasible to construct something akin to what I'm about to anticipate.

What if I told you that it's possible to operate machines remotely as if you were commanding your own body thousands of miles away, totally unconscious, immersed in a deep dream? Would you believe it? Well, this possibility is real and will soon become a reality, unlocking a wide range of features that are unattainable through any other means. This specific endeavor would be exclusively tailored for the military domain. The machine would be operated by a soldier trained to use a million-dollar piece of equipment. After all, lives would be at stake at all times.

Clearly, the initial technology for such a feat will be derived from the previous one, establishing a stable virtual environment that precisely responds to the soldier's dream-based actions. To achieve this, a dedicated platform must be equipped to accurately interpret the intended movements, encompassing arm, leg, and head mobility (enabling a 360-degree view). It's important to remember that we're entering the realm of dreams, where the laws of physics and the limitations of ordinary reality do not apply.

However, all this freedom poses a challenge. In the real world, the machine will be subject to the laws of the Universe, including gravity, inertia, speed, and various others. In this way, the target must dream of a world that also sets boundaries to attain precise visual feedback of their movements, enabling synchronization with the actions of the real machine. This artificial feedback must accurately simulate the mechanical resistance of the environment in which the robot is inserted. If the robot is holding an object, the weight and load-bearing capacity of said object must be swiftly calculated and relayed within milliseconds to the soldier who holds it within the dream. Several particularities must be observed and will be part of the entire technical scope of the project.

The "good news" is that this is already happening. Every subject is required to undergo experiments that include complex data like these. The target can naturally sense that the environment is subject to gravitational attraction, that it contains inaccessible places, and provides interactions with other egos and physical structures. Consequently, they experience multiple simulated sensations that closely mirror the real ones, imposing various limits within their mind.

Revisiting the chapter on D2K (Volume 1), we recall that the brain's interpretations of dream images rely on certain aspects of the intellect that remain active during deep sleep. So, the dream ego functions within a created reality, guided by a set of parameters similar to those of everyday life. After years of meticulous refinement, with the soldier being trained to control the dreamlike environment, advanced algorithms capable of processing intricate movement details, and countless hours of simulations, both elements are now primed for testing on actual machines.

I'm well aware that the concept of remotely controlling a hibernating machine or avatar is not new and has been widely portrayed in movies.

Nonetheless, the likelihood of something like this occurring in reality changes everything. The inherent curiosity sparked by this concept promptly relinquishes numerous trivialities that once comprised an individual's purpose in life, and instead welcomes and adopts unprecedented achievements beyond imagination.

Looking ahead to the time when real-world tests take place, advancements in robotics research could potentially result in the construction of colossal MECHAs (short for mechanical, メカ) or giant robots. A mecha is usually a war/combat machine, and has a humanoid bipedal appearance that makes it well-suited for remote control using this apparatus. The link between the soldier's dreaming brain and the robot gives rise to something that seemed impossible to accomplish, broadening our grasp of reality and maximizing two-way parity to the fullest. I know some may be wondering, but if they managed to create a robot along these lines, wouldn't it be more feasible to command it in a waking state inside the machine itself? And the answer is no!

It's inconceivable to envision such a thing with current MKTECH technology in a waking state. The waking mind has a "self" in charge that adjusts all biological hardware and software according to the outside stimuli gathered by our external senses. The brain and the waking mind are extremely complex evolutionary works of nature and need all processing power to update their status in thousands of different variables at all times. In this process, the mind has a limited focus and is subject to various human conditions, such as negative thoughts or intense pain that diverts attention away from activities. As the understanding increases, the ability to employ emotional control diminishes, demanding significant cortical processing and leaving limited capacity for the essential task of mind-controlling machines.

The psychophysiological principle and the alert the brain imposes on us during wakefulness for self-preservation — every modification in the physiological state — is accompanied by a corresponding shift in emotional and mental state, and vice versa. Needless to say, this doesn't happen in the unconscious state. Our immediate reflexes are actually very weak. However, with a trained mind in the dream world, you can begin controlling your ego in new daily experiences, disregarding self-preservation concerns and fully focusing on your mission with your mecha. Remember: the waking mind can be easily distracted by millions of bits of stimuli per second it is capable of processing.

The truth is that operating machines in the awake state by simulating human movements with people inside it, or coupled to motion sensors that send the same mechanical command to the machine, faces numerous challenges. The biggest one would be physiological boundaries; the person would need to be an Olympic athlete to withstand a marathon of intricate and agile movements of a robot. Not to mention the necessary breaks for replenishing, rehydrating, and resting. In REM sleep, on the other hand, the mind is loaded with the basic functions of the neural operating system, configured just to control sleep. In other words, it doesn't need to worry about movements and their consequences, so it's free to carry out its functions, including those related to dream processing.

Well, if you thought moving around wasn't particularly complex, think again. There are findings from renowned neuroscientists suggesting that the brain evolved not for us to think, but for controlling movements in a precise manner, as there is no other reason for us to have a brain other than to produce complex and adaptable movements. We can only change the world around us with movement. Forms of communication like speech, gestures, writing and sign languages, for example, are just complex, coordinated muscle movements. According to these studies, cognition, sensory functions, and memory processes continue to exist in their known forms because they are essential for keeping up with future advancements in movement and environmental interaction. In addition to that, we have the world's most energy-efficient machine, enabling activities that, if performed without the mind's involvement, would demand significant energy usage. However, in the case of a soldier in REM sleep, minimal energy would be expended while executing any type of movement in dreams and flawlessly replicating it in a machine.

In this way, it'd be possible to execute precise remote command over arms and legs of machines without experiencing fatigue or depleting bioenergetic resources, just by capturing movement intentions within dreams and replicating them to the machine. Subsequently, the target's visual and auditory inputs would be continuously transmitted to the controller's dream, consistently integrating and updating them within the actual context of events while replicating them in the dream realm. Remarkably, this process would occur without experiencing fatigue, allowing uninterrupted 12-hour shifts where the controller would remain asleep the entire time.

Certainly, several details would require refinement, and various questions would need to be addressed:

* How is the capture of movement intentions within dreams synchronously replicated in the machine?
* Would the interpretation of the visual and auditory stimuli coming from the machine be the only source of conscious creation?
* Can a variety of neural solutions exist to input the desired behavior that produce the same movements?
* How will the machine read the intentions?
* How to provide uninterrupted feedback, shielded from external influences that might disturb the soldier?
* How to maintain deep REM sleep for a sufficient amount of time?
* How to control the machine indefinitely?
* Is it feasible to establish a diverse group of pilots so that if one were to experience an accident, another could promptly take over? Could it be likened to the functioning of a set of power generators?
* In that case, could we implement the principle of neural redundancy?

Amazingly, of all the technologies mentioned, creating mechas as we imagine them—humanoids, giants, and majestic beings—remains a distant pursuit. Undoubtedly, that would be a dream come true for anime and gaming enthusiasts. In the medium term, it makes more sense to say that we will probably be able to control humanoid robots, but on a reduced scale. After all, we cannot forget Boston Dynamics' amazing (and frightening) robots. The existing machines are likely the most suitable models to build upon, which would eliminate the need for developing separate automation AIs and robotic reasoning, as they are already powered by Dynamics itself, saving significant effort.

Under present conditions, I believe it's technically viable to test this theory today, at least to some extent. This will occur gradually, starting with encounters on servers, advancing to multiplayer games with egos, and ultimately attaining full mastery over a machine.

## 12.3.4 - Cryptocurrency mining and mental data processing

A multitude of thoughts quickly surfaced with the understanding that MKTECH uses the brain's processing power to simulate artificial dreams and nightmares, compelling it to process intrusive images and sounds within its native operating system, which are then assimilated as genuine mental images, seizing control of another individual's brain. After the initial impact, I found myself asking a series of questions and drawing connections with computing — my area of expertise: what if, through this twisted utilization of the brain for data processing, particularly with images, they were pursuing an alternative agenda? What if they attempted to execute codes, algorithms, or sophisticated scripts alongside images that the brain naturally interprets and comprehends, generating an output as a result of this steganographic process? However, the output information wouldn't be assimilated by the target's mind but only by the computers that received it. Driven by curiosity, I embarked on a quest for the answer and discovered that something similar to this is already happening. In addition to conducting impressive experiments as we've seen throughout the book, they laid the foundation for the creation of identifiable algorithms designed for sleep-related processes that go beyond decoded graphics and sounds.

The path towards harnessing the computational power of the world's most advanced supercomputer began. Soon, when this objective is fully achieved, distributed data processing can be used to send these codes in the D2K transmissions while we're all sleeping soundly at night in order to be immediately captured by the brain, demodulated and virtually interpreted. This could be replicated for several simultaneous brains, and would be managed by an algorithm responsible for managing the loads and overseeing the process, ordering the output of the data. It'd function as a bot, capable of creating the most powerful neural "zombie" network for different purposes. One of them would be the profitable cryptocurrency market as never before imagined, which would force the brain to do mathematical calculations in a state of unconsciousness.

This would then lead to the creation of the perfect platform for mining with greater cost-effectiveness and lower energy consumption. Imagine being able to harness the power of the world's best computer, to understand and control its processing by connecting to other similar supercomputers, expending only the energy required for transmitting radio waves to send the data and later retrieve the results through those same waves, leaving the heavy lifting to the organism itself responsible for processing the information. The target's dinner will serve as the sole energy source, accomplishing what thousands of graphics processing units (GPUs) achieve now.

The brain is an amazingly energy-efficient device. None of our technological advancements today come close to matching the brain's efficiency and processing power, thereby overcoming our biggest challenge: energy consumption. Once they achieve their full potential, they will harness the power of a brain network to solve intricate mathematical problems, enabling the verification of transactions on blockchains, creating permanent records that cannot be undone, modified or altered. The actual mining of cryptocurrencies revolves around a purely mathematical process. Consequently, running algorithms that identify the sequence of data (blocks) can result in receiving cryptocurrency as compensation when the block is used for transaction verification.

How long this format will be viable or functional, I cannot say. Nevertheless, experiments exploring the feasibility of utilizing a stable network of subconscious brains for cryptocurrency processing are already underway, alongside numerous other similar endeavors. So, stay alert!

As previously stated, D2K holds enormous potential. The way it interacts with the mind during sleep is impressive. It completely controls cognitive processes. What is transmitted undergoes demodulation, processing, and its effects are manifested in the host/dreamer's organism. Thus, it's possible to balance the energy expenditure and the volumes of processed data on the scale, as sending electromagnetic waves from Sun-powered satellites and the cost to create these waves are infinitely lower than manipulating hundreds of supercomputers that generate an excessive amount of heat with enormous energy consumption.

Let's imagine a new scenario where sensitive data has been stolen — passwords, valuable patents, secret building plans, or compromising photos. After the data is stolen, the hacker being relentlessly pursued by the authorities must hide the data off the network and away from their equipment. What if they could conceal the files within images that could somehow be encoded in people's minds while they sleep, utilizing D2K as an ordinary long-term memory of the target's daily life?

This entirely concealed data would be implanted during a waking state but accessed later within the dream of the compromised human host. Triggers for these memories would only be accessible in a state of complete unconsciousness, responding to a specific set of simultaneously transmitted data, much like a key granting access to the stolen information. Later, they would be able to recover the data without compromising themselves. At the opportune moment, they'd simply send the appropriate stimulus that triggers a visual memory sequence, revealing the images containing the valuable files within the host's dream. Therefore, mental images will be processed by scanning their content to identify those that contain hidden valuable data.

This is another real possibility among many others in this brand-new field of study. Unfortunately, I cannot determine the extent of file degradation resulting from the natural fading of memory over time, and whether this will impact its integrity or not. However, these questions will soon be addressed in what I refer to as advanced neural steganography.

Another branch dedicated to a more peaceful endeavor based on the current D2K models explores the potential for a device designed to help terminally ill or hospitalized patients sleep and experience pleasant dreams, creating beautiful memories as they await the inevitable with dignity. Similarly, the device would enhance the quality of life for terminally ill or quadriplegic patients. Maybe it'll also be possible to effectively alleviate symptoms of depression and other neurological illnesses, treat traumatic memories, address fears and anxieties, all while maintaining scientific rigor in studies to gain a comprehensive understanding of the subject, including the potential effects of the long-term manipulation of the human mind.

We're know that some people struggle to cope with reality, and choose to hold onto fantasies. So, controlling the type of dream in a safe environment may help them deal with their traumas. All this will occur within the limits of the brain — its energy management limit —, respecting the maximum capacity of information that can be processed by the human mind, and modifying the way it processes the information when it's assimilated.

## 12.4 – A Decade and Century of the Brain and of the Mind

In the coming decades, extensive research will be dedicated to the comprehensive study of the entire human cognitive process through unofficial or sanctioned experiments, collecting data from various locations across the globe. Official investment in research linked to all fields related to the human mind and brain has already been secured. A new frontier has emerged with record-breaking budgets allocated to research on brain function in the United States and Europe. Initiatives like the Breakthrough Research and Innovation in Neurotechnology (BRAIN) are leading the way in this endeavor.

The research budget for BRAIN Initiative is substantial even for a public-private program with funding estimated at $4.5 billion through 2025 (research in diverse fields of neuroscience). In the United States, investments tripled in three years. In 2017 alone, 260 million dollars were invested. The Allen Institute, supported by the co-founder of Microsoft, is making a significant investment in studying brain circuit dynamics, curing diseases, optimizing brain function, and improving the physical constitution of neural circuits. Their intention is to invest 1 billion dollars by 2022.

On the other hand, in 2017 alone, 120 million euros were invested in the Human Brain Project, the European equivalent of the American project aimed at developing and improving various fields linked to neuroscience. It's estimated that 1 billion will be invested by 2023. These studies are reputable, unlike those conducted within closed, secretive circles focused on weapon development.

Among other goals, the intention in this case is quite noble: to study and tackle brain disorders such as dementia and Alzheimer's disease. These projects reveal the current investment focus and emerging trends in this field of study. They indicate where financial resources are being allocated to and which specific areas are likely to deliver the anticipated returns for investors in the medium and long term. The experiments are largely based on data generated by cutting-edge equipment that can study the brain in real time, such as the modern MRI scanner called Connectom, which captures sharp images of the deepest areas of the brain. The scientific branch that studies the physical aspects of the brain has several techniques for exploring the mind, analysis of electrical signals, and interaction with machines and the structural architecture of brain circuitry.

These multimillion-dollar studies involve comprehending the brain's functioning and unlocking its secrets to integrate it with computing. The aim is to combine the brain with machines without invasive methods like installing chips or electronic equipment, making it a significant aspect of the research field. Helmets that read raw EEG waves (BCIs) are able to assess various metrics for specific data, but they don't carry the content of the thought, only brain electrical activities. So, they combine forces with sensors and equipment capable of reading the content of the thought itself, merging technologies and forming a powerful apparatus for the study of the brain.

Speaking of scientists, the convergence of billionaire-backed open projects, advanced study initiatives led by top experts, and cutting-edge million-dollar machines for brain research inevitably gives rise to a common and highly relevant question: is it possible that *none* of the scientists working on these projects are aware of psychotronic weapons and their ability to interact deep within the human mind? As the author of this book, and not specialized in neuroscience, I wonder how I could conduct a study on this subject, especially when renowned scientists have not commented on it or even acknowledged its existence. This serves as a profound reflection that exemplifies the abysmal difference between the world we live in and the world that develop military weapons. In this case, this technology was developed and improved over the years at the expense of those same individuals who are mentioned above. Most of the scientists who know about these weapons don't usually work at universities or give lectures.

These seemingly legitimate projects and initiatives will converge to a singular point in the future where they will meet the hitherto "unknown" and inaccessible mind control technology that has been thoroughly explored here. The impact of this collision will create an integrated study that will encompass the best of both technologies, and new demands will emerge. It's just a matter of time.

## 12.4.1 - Regulations and human enhancement

Soon after the initial shock of discovering the existence of MKTECH, there will be a flood of regulations worldwide, for example, concerning cognitive privacy, the prohibition of SYNTELE for the purpose of torture. Over time, major corporations will integrate this technology into a new platform that will gradually become an integral part of our everyday lives. As this technology becomes deeply ingrained in our society and ideology, it'll significantly shape the experiences of the next generations from the moment they are born. As a result, a new internet will surface with this new mental platform, the BRAIN WEB. Evidently, this internet already exists in the form of DEEP BRAIN WEB and DARK MIND WEB in private networks where stolen thoughts, EEGs and microwave voices travel on a daily basis — centers with classified information such as the structure of DARPA and its equivalent in Russia, the Russian Foundation for Advanced Research Projects in the Defense Industry (Фонд перспективных исследований). That is, colossal investments to unravel the workings of the mind, and how thoughts are created, hoping to answer many questions that were once considered unanswerable.

The MKULTRA experiments with psychotronic weapons are a part of this study — a clandestine, illegal research that violates all Geneva conventions and human rights — at the cost of much human suffering and loss, but they also contribute to the solution of many problems that conventional technologies are not capable of solving. This technology, hitherto unknown to the majority of the scientific community, is essential to achieve a complete understanding of the mind's workings and its programming in detail. This decade and century will be the age of knowledge, of brain studies. In my opinion it may take 100 or 150 years to have all mental processes mapped out and understood to a point where consciousness itself can be stored and transferred to another brain, whether it's a mechanical construct made of synthetic material or a biological one like ours.

Perhaps some features of virtualized consciousness can be installed on machines. They will capture reality as perceived by human sensory organs through sensors, and will play the role of the five senses, combining several enhancements that were previously unfeasible to incorporate into a human being, such as different types of vision, expanding the ability to see across different wavelengths such as infrared, to enlarge the view to observe distant objects, to have some sort of super hearing, or to pick up radio transmissions. Wait! We already have the ability to tune into radio frequencies. We don't need any external or internal device, just ask any modern target.

All of this will be simulated and emulated so that the new, enhanced, and "cyborgized" human feels as if before, already one step further in evolution with these devices generated by knowledge from our time. All roads lead us to these paths in the future. Now we stand witness to a historical (and painful) moment, amidst a process of social accommodation, emotional and psychological immaturity, eagerly anticipating the imminent widespread integration of this technology into our society. I estimate that by 2023 the process of full awareness of the MKTECH technology will have started.

## 12.5 - Dark Mind Web, Brain Net, Deep Brain Web, Brain/Neural Satellites

Don't be surprised if in the future a personal thought appears on your screen after using the internet to perform a search on your phone. Or even worse: you enter an app and see your thoughts in real time on that same screen while you think and try to control your next thoughts so as not to expose your privacy and cognitive safety. I'm uncertain about how soon our data will be made available, indexed by these new crawlers — robots that catalog website content for search engine display — which will unearth this information from servers storing the thoughts of all of us, every passing moment. Currently, it's not possible to specify the number of existing networks, much less their real capacity for simultaneous mind-to-mind connections.

If we ignore these problems, we'll encounter serious challenges that could become permanent over time. If we don't actively address the collection of data from others' minds before it ends up in the hands of the giants who control all the digital information flowing through the Indexed Web (the Surface Web), we'll be subjected to various actions that these companies are ready to take. Every day we receive reports of global censorship by groups and companies that, through subjective filters, determine that a certain content is inappropriate or breaches a code of conduct signed between the company and the user. Serious incidents are occurring all over the world, providing us a preview of what could happen when thought data or direct access to them becomes widely available to the same companies that hold a monopoly over the data that travels on their platforms.

Changes to the internet will also affect us all. We'll have search engines like Google, DuckDuckGo, and Bing, but for real-time thoughts. Moreover, in the future course of history, the division between the internet and the Brain Web/Net — the internet where human thoughts travel— will no longer exist. Both will provide relevant data related to a search query. This level of interaction spans many chapters, perhaps an entire book. But it's certain that the contents of thoughts will leak onto the indexed web at some point, and evil will become irremediable.

Now imagine a Deep Web, where private networks hold unindexed data, solely focused on collecting thoughts from various individuals. This network would consist of stolen cognition data from all of us. It seems like a remote possibility, but unfortunately such a network already exists. It's a private network similar to the internet, but composed only of the thoughts of hacked individuals.

In this network, everything is captured from the raw data of the thoughts, the metadata, the visual and vocalized sub-processes and, of course, most of the data aimed at improving the weapon itself — for the better understanding of human cognition and for storing and trading products theorized by the minds that are currently under attack or connected to the MKTECH system around the world. So, we have something similar to a Brain Net or Deep Brain Net which are unindexed corporate networks inaccessible to outsiders, and their information isn't available for web crawlers.

These highly secure networks, often limited to military access and involving contracted companies and collaborating agencies, transmit all kinds of related and intercepted data from the minds of thousands of affected individuals worldwide. NEURO BIG DATA is capable of working with this abundance of information which, by the way, will emerge as a new realm within data science, focused on developing and enhancing smarter methods to distinguish valuable information from cognitive clutter. These networks establish contact with satellites known as Neural Satellites, constellations specialized in keeping the MKTECH network working all over the world, where all the information stolen from people's minds, V2K attacks and the content of visual thoughts, as well as remote dreams, travel. That is, where all the information and data related to the weapons and the target's resulting thoughts travel — including their remote EEG and biometric data.

The network organizes all this information, which is subdivided into several subjects by relevance and can be easily filtered and searched. The algorithms and the AI responsible for this most crucial examination are of paramount importance. Its implementation must be carried out with utmost proficiency, otherwise there would be a lot of disorganized, disconnected information, vocalized thoughts, and metadata with minimal utilization. Keyword-based filtering would be employed for the filtering process — a kind of neural Google-like interface in which the anxieties, the longings, the fears, and the happiness of thousands of people are materialized in auditory, visual and EEG data. Here it is possible to analyze the experiments from all over the world — given that the attacks commonly use the same infrastructures —, and to assess the ongoing battles between targets and specific groups, as well as Artificial Intelligence, in their quest for complete mastery and dominance over the human mind.

An infinite range of unprecedented information is generated for the improvement of the weapon and the system is calibrated to deal with sophisticated, large-scale surveillance in the near future, and not just a few thousand isolated targets, but entire populations that are monitored 24/7, their thoughts captured and verified by the systems of this neural web.

At this point, probably the intellectual and industrial property will have disappeared once and for all with the arrival of the infamous Dark Brain Web. DBW refers to clandestine networks involved in trading all sorts of things, from the most modern industrial secrets, compromising thoughts of powerful personalities (or not), to the sale of personal information. Tell me, would you be interested in knowing the current thoughts of the president? Just show an image connected to his vocalized thoughts. Or would you rather hear the thoughts of a famous actress? That too will be possible. The mental data of prominent government figures, which has the potential to jeopardize a nation's security, will likewise lack protection. In fact, it will be easy to intercept generals' thoughts and reveal the true intentions behind military tactics, even if the orders are fragmented.

It's no use having the best and safest communication system in the world if, before the order is sent by digital means, its intention, consequences, that is, every process that orbits the thought have already been easily captured by the MKTECH system through the minds of relevant people from other nations that may even collapse if strict countermeasures and protocols aren't designed and strictly followed to protect crucial secrets in thought-based format.

In this way, this Deep Dark Brain Web becomes a huge marketplace for all kinds of information, ranging from credit card data, website passwords, network manager passwords, confidential information, intellectual property and top-secret information to the thoughts of all of us. Everything that goes on in our brain — everything that is created by it — is likely to be captured by this technology without much effort. Perhaps, in the short-term, we could use thoughts linked to blockchains — a new kind of future patent — to mitigate the problems of intellectual property theft, as well as thoughts and ideas with economic potential; we could link thoughts that show who was initially responsible for it, the progenitor of a given idea or product, preventing it from being reversed, defrauded or altered.

As I pointed out earlier, a new universe is born around this weapon. Now the question is: when will we fully embrace and adjust to all of this? I prefer abstain from speculation here because, presently, I'm unable to determine the particular moment in time we are at, given the context of my previous narrative.

## 12.6 – Kidnapping minds

How many minds are currently held captive around the world? Keeping minds enslaved within their own lives to serve as human toys, objects of perversion, sadism, and rituals involving sex is a new type of crime on demand and one that is spreading rapidly throughout the globe. It's in fact on the rise, but many go unnoticed or undetected due to being misdiagnosed as psychiatric disorders. In the not-so-distant future, a new form of mental kidnapping will emerge, similar to today's occurrences, but with a crucial distinction — these attacks will be solely driven by profit motives. Mental kidnapping will involve holding minds for ransom, and transactions will be conducted through cryptocurrency. This practice will operate without any bias or hidden agenda from the MKULTRA experiments or other secretive activities of the present.

They will reveal to the world an improved category of this heinous crime. Here, only money matters. Money will serve as the bargaining chip for the immediate release of the hostages. Something you don't see much these days: you scarcely hear about cases of people who no longer have their minds connected to the worldwide network of psychotronic torture. The outcome is always the same: conducting the experiments leads the target to exhaustion, making their death inevitable at some point during the long years of electronic torture.

Thus, a new and highly profitable form of criminal activity is anticipated in society which, to be carried out physically, demands extensive planning, captivity, constant change of transportation vehicles, and involves multiple individuals and armed violence. The success rates for kidnappings executed through the use of physical force are generally low, as this type of crime usually mobilizes substantial material and human resources from the government to arrest the kidnappers and release the victim as soon as possible. Kidnapping minds with psychotronic weapons, on the other hand, gets impressive results if you take the data into account. With the main advantage of being usable over long distances, the success rate of such kidnappings is nearly 100%, as they eliminate the need for firearms or risks associated with physical force.

It's just like when computers are infected by viruses — such as crypto-ransomware— that encrypts known data. It works by "kidnapping" your data and turning it into files that are unusable. This can only be reversed using a key that decrypts the data and returns it to its original state. However, only the perpetrators have this key. In exchange for it, they demand ransom in cryptocurrencies, making it the only means for the victim to retrieve their valuable information. Something similar to this will happen in the future if the authorities aren't well equipped and prepared. In other words, this crime that is becoming more and more common will rapidly proliferate and bring catastrophic consequences.

When a person's mind is kidnapped, living a normal everyday life becomes an impossible task. Remember: people who are attacked by this weapon of war nowadays rarely escape without serious consequences; unfortunately, many die on the way. Its power to disrupt human cognition, which physically affects areas of the brain linked to communication, is impressive.

The suffering is extremely intense. The mind held captive (electronically) will resort to all means, in a manner similar to what happens in a kidnapping — through common extortion —, however more elaborate and with fewer risks for kidnappers. If the authorities are dormant as they are today, without taking action against these heinous crimes occurring around the world, the criminals who kidnap brains will quickly become financially successful. Obviously, criminal organizations will transition or shift towards this new category of kidnapping in a blink of an eye.

This type of kidnapping can already be carried out with technology that is available to most gangs and terrorists of the present time. If we consider the widespread use of its most modern version, the probability that the individual who was kidnapped and threatened will give in to the financial demands of kidnappers is high.

The latest iteration of these psychotronic weapons is capable of destroying areas related to communication in a matter of days just by using electromagnetic waves — by carrying a signal that escapes human detection, but has an enormous power of destruction when demodulated and interpreted. A single autonomous attempt to analyze the received signal results in devastating effects. A brief exposure is enough for the victim to see what this signal is capable of, which encourages them to pay the ransom. Strokes, deafness, and persistent cognitive decline impacting sentence formation, the flow of thought and coherence are found in individuals targeted by this weapon. The new version of the weapon wasn't covered in depth in the book, but studies to understand its operation continue and the result of this analysis will be disclosed at a later time.

## 12.7 - State monitoring thoughts. Cognitive morality police

Imagine yourself inside your house, reading a book of your choice, watching TV, listening to music, unpretentiously surfing the web, or relaxing in your personal space, confident that you can exercise this right peacefully in this free, democratic country. Then, in the distance, one of the Neural Satellites that scrutinize the Earth, picking up fragments of thoughts from various minds from time to time, ends up noticing something suspicious in your thoughts.

It's a real horror show. This equipment "hovering" in space is capable of monitoring the thoughts of an entire community/city 24 hours a day. Its constellation can even cover an entire country. During this search, the AI is programmed to explore and prioritize thoughts that contain certain content inserted into its algorithm by governments to support regimes or lobbying groups. As soon as an accidental occurrence of one of those prohibited thoughts along with its associated theme present in the satellite's search code arises, the target at home will be perceived as a potential threat and will be subjected to specific measures — they will be listed as individuals requiring closer cognitive supervision.

Now, SYNTELE voices will rise up inside the victim's brain and make it clear that they don't want to hear that kind of thinking anymore, that the next warning will be more severe, and the third will be treated as a serious transgression. The more detailed monitoring will continue, and the target's profile and ID number will be recorded in a special database for those who had forbidden thoughts or plotted against the interests of the corporation or regime. That would be a future marked by darkness and limitations and it could easily become a script for a sci-fi series or movie. There's just one issue: this is already a present reality. Groups of individuals, aided by satellites, systematically subject targets within the confines of their own homes to this regime of cognitive or mental suppression.

The upcoming implementation involves broad projects of simultaneous cognitive monitoring of all thoughts within a region or an entire country, primarily targeting prominent individuals who may pose challenges to a certain group or political system. It'll eventually become an internal government action, or an action of espionage agencies. Regarding the numerous satellites that are required and might hinder the accomplishment of this dark future endeavor, I regret to inform you that I have more concerning news. Prepare yourself, as the shift from predominantly telecommunication-focused satellite constellations, which currently prevail, is gradually giving way to more contemporary models equipped with high-gain antennas. These advanced satellites can accomplish their existing functions while simultaneously capturing, amplifying, and transmitting signals to multiple minds.

Do you know how many satellites China launched in 2018, primarily carrying classified military payloads? Almost one every two weeks. This averages out to a total of 26 new satellites. In the worst case, if we count one satellite per launch, it may happen that the payload contains more than one undisclosed component. The Chinese Long March rocket operated non-stop throughout that year.

Only technology operators are responsible for cognitive deprivation. In the future this tool will be used primarily to monitor all thoughts in a certain location. Down the line, there will be the possibility to prevent specific thoughts from emerging in people's consciousness, leading to what we commonly refer to as cognitive conditioning. This process involves the use of Synthetic Electronic Telepathy (SYNTELE) to actively suppress certain thoughts while introducing others that become deeply ingrained in the subconscious mind, ultimately influencing conscious thought.

When it comes to democratic countries, this is likely to be primarily geared towards mass surveillance, much like current communication practices. Torture and attacks are expected to continue, but on isolated targets. To compound this already bleak future scenario, the experiments currently being conducted test, among their numerous topics and objectives, precisely the possibility of extensive surveillance on citizens and their reactions to it — to this whole scenario of cognitive deprivation and control of human thoughts. This book foreshadows what is to come.

Observe the magnitude of the issue posed by systematic thought monitoring. Notice how challenging it becomes to think internally in a manner different from what you're accustomed to. This technology directly attacks our executive functions which are responsible for dealing with an enormous demand, complex behaviors and social norms, the capacity to initiate actions, make plans, find ways to solve problems, anticipate consequences and change strategy in a flexible way, monitoring the behavior step by step and comparing partial and predicted results.

Our biological computer performs actions within milliseconds. To observe the flow of reasoning as it passes through specific brain regions alters, adds or withholds certain information. External modification brought about by psychotronic weapons — both stemming from disruptions to the natural rhythm of thought and direct manipulation— causes systematic changes in interpretation, leading to synesthetic anomalies and incoherent results.

This is not a future projection. At some point this is already part of the present. My fear is that in 10 to 20 years, acts like this in the pursuit of national security will slowly be incorporated into society until they become commonplace.

Indeed, it's a fact that this weapon will become increasingly used to monitor thoughts in highly dogmatic religious systems and authoritarian or communist regimes where citizens have been deprived of their freedom. Remember that with a few antennas, a well-configured system and a prepared satellite, any country can use this technology to tightly control its citizens, subjecting them to harsh punishment even for simply thinking about something that goes against religious laws or the ruling dictator's wishes.

In due course, this effective method of indoctrination will naturally manifest in these settings, carefully selecting thoughts aligned with religious or government laws while impeding the emergence of divergent or conflicting perspectives. This will then give rise to individuals who are increasingly zealous and narrow-minded in their beliefs. In some cases, these techniques can be used to form an army of indoctrinated believers from a young age, for example, with the direct manipulation of thoughts. The concept of free will, which is currently relative and limited to few private actions, may eventually fade into obscurity and become a forgotten idea.

In this regime, thoughts will be recorded and displayed in public squares where offenders will receive their sentences from their executioners, and they may even be momentarily displayed live so that everyone can witness the traitor's thoughts. For this, it's enough to send stimuli on the subject in question that generated the intellectual transgression via SYNTELE and the automatic processing of the brain that received this data will do the rest and display the pagan's thoughts for everyone to see. An urgent struggle to avoid introspection and to disrupt the prevailing thought process will follow. As the prisoner awaits imminent execution, their mind wrestles to stay composed, burdened by the looming presence of an axe about to descend upon their neck.

In today's society, the dictatorship of thoughts is systematically prevalent within social networks and various groups striving to impose their ideologies, each with its distinct perspective. The reality is that their continuous pursuit revolves around molding individuals through indoctrination, rendering them unable to grasp their surroundings, incapable of self-defense, and impeded in their ability to challenge moral injustices they perceive.

Over time, certain terms in our language will gradually fall out of use due to indirect imposition, including controlling vocabulary and language, restricting or preventing the use of certain words, increasingly restricting our imagination and freedom of communication, hampering creativity and directing the way we should think. And all this without the use of psychotronic weapons. Imagine when these resources become available to specific groups. Everything we're witnessing will increase exponentially in impact and reach.

Eternal vigilance is the price of liberty. If we lower our guard even for a second, they oppress and subjugate us, normalizing and trivializing the absurd. MKTECH clearly demonstrates that the final frontier has been conquered. Citizens' thoughts/minds are vulnerable, and anyone has access to it, facilitating all forms of social control and cognitive modification. Any idea can now be instilled and easily internalized. Surveillance takes place directly at the source of creation, manipulating thoughts through deprivation and control of personal conduct, and the regulation of all private life. Thomas Jefferson's words are more relevant now than ever.

Psychotronic weapons used on a massive scale create historical precedents, serving as tools of inspection for a central power or government-associated groups seeking to undermine citizens' freedom and control over private and collective social aspects. This leads to the State monitoring and controlling the human psyche. It is, therefore, a tool of control desired by many dictators and authoritarian regimes, capable of monitoring people's thoughts and imposing what they should think, when to think it, and what to avoid thinking, directly influencing the victims' minds. It can monitor the response to certain subjects and stimuli, reflections of signals coming from the reality of each individual. Even communist and Nazi regimes did not attain such a high level of surveillance over people's private lives.

Never count on the government to defend you; always have a plan B. Governments are subject to change, and the global tendency is the increased restriction of private and individual freedom and an ever-increasing government surveillance without any justification. Certainly, these tools will be widely used by governments worldwide. So, it'll be crucial for our survival as a species and society to employ legal means to combat the utilization of such weapons, criminalize unauthorized intrusive cognitive verification, and develop electronic devices and physical safeguards that effectively mitigate this problem.

We're facing an effective and improved means of control and technological domination never seen before in the history of humanity that undermines any mechanism of law responsible for safeguarding the fundamental freedoms of individuals.

## 12.8 - Secret police (as per NKVD)

To gain insight into the future of thought monitoring, it's crucial to acknowledge that certain groups aim to introduce mind inspection reminiscent of old secret police tactics. These groups may resort to repressive and even assassination activities, drawing inspiration from the likes of the Soviet NKVD that originated in 1934. We may soon face organized groups reviving outdated policies, conducting remote operations with the primary objectives of invading personal lives, inflicting torture, and committing murder. They strive to emulate the methods of social repression witnessed in communist-socialist countries, modified for cognitive surveillance, yet applied in a disturbing and remote manner. This is due to the absence of State support in this context, unlike in those countries, making the execution of such endeavors challenging. It creates a slight opportunity for the gradual legitimization and implementation of these acts.

They improperly exploit the technology, surveil individuals' lives, and put into effect all that has been previously discussed in this book. This type of act, which existed during the Soviet era, is gradually being carried out in Brazil, including by State officials who are remunerated by the State to perform their functions, but instead mobilize all public resources investing in attacks, defrauding civil service competitive examinations in order to place trusted associates in certain key positions, joining terrorist groups, along with Cuban, Venezuelan, Chinese and Russian international cells in a "revolutionary" cooperation network, committing acts of barbarism by means of psychotronic weapons embedded on foreign satellites, attacking citizens of their own country, violating their constitutional rights and using their influence and authoritative status to their advantage, as they believe they have the right to treat people in their own homes acquired with blood, sweat and tears as they please.

With the help of this technology, the "agents" of the new NKVD are trying to globally implement a new type of surveillance, unauthorized cognitive policing, in which they conduct, within the scope of experiments, behavioral assessments to measure human reactions to systematic activities, generating an extensive collection of statistical data.

So, if implemented on a large scale, they already have enough material to deal with much of the initial behavior of the population. In this way, secret police units are assessed and examined worldwide.

For those who think that communist and Nazi tactics lost their effectiveness, think again. They remain very much alive and are being gradually assimilated into our society, while also involving citizens in MKULTRA experiments on a global scale.

Hence, it's imperative that we promptly expose the existence of this weapon and advocate for rigorous legal measures against the violation of the mind, thoughts, dreams and especially the microwave voice that is capable of following the target anywhere, driving them utterly insane, and causing impaired perception and judgment. We require more stringent legislation on neural weapons or the implementation of a cost-effective and reliable barrier to safeguard individuals seeking protection.

As the final chapter of this book unfolds, no updates have emerged regarding the prosecution or apprehension of those responsible for committing crimes against humanity (or against individuals). I hope that as more people become interested in the subject, this begins to change.

## 12.9 - HAARP

I couldn't end the book without addressing the High-frequency Active Auroral Research Program, or HAARP, the "mysterious" program with the primary official objective of advancing understanding regarding the functioning of radio wave transmissions in the ionosphere, the upper part of the atmosphere. There are several theories about the real objective of this program that officially began in 1993 — coincidence or not, it started in the most turbulent decade involving electromagnetic weapons and experiments with electronic weapons in the modern era —, and is located in Alaska, United States.

According to official reports, the main objective of the project is to expand the knowledge obtained so far on the physical and electrical properties of the terrestrial ionosphere. In the ionosphere, the Sun serves as the primary catalyst for ionization, as radiation in the x-ray and ultraviolet spectra incorporates a significant number of free electrons into its environment. The density of electrons and the composition of the ionosphere undergo radical changes, which may even completely disrupt HF communications. Above an altitude of 100 kilometers (62 miles), the atmospheric composition undergoes stratification despite its tenuous nature. With this, it'd be possible to improve the functioning of various communication and navigation systems, both civil and military — it's a common tendency for the majority of HAARP experts to view it with skepticism.

For 25 years, the U.S. Air Force Research Laboratory has engaged in collaborative research with DARPA and the University of Alaska. After conducting an exhaustive array of experiments, the Air Force officially handed over control of the enterprise to the University where academic research is conducted to this day. Thus, this enigmatic infrastructure, wrapped in mysteries and secrets, gives rise to numerous speculations. Some say that its purpose is to conduct different tests involving the ionosphere and its layers, investigating its behavior throughout the day. Speculations also surround attempts to alter the temperature or interfere with the Earth's magnetic shield. Another set of experiments involves moving the ionosphere 300 km (186 miles) up and down, supposedly reaching satellites in low Earth orbit, including the International Space Station (ISS). Furthermore, it has the potential to assist in averting possible chaos caused by debris during re-entry into the Earth's atmosphere, disintegrating them based on the time of day. It can cause a disruption at a specific altitude, impacting the propagation of radio waves in that area, thereby potentially influencing the quality of transmissions, either positively or negatively.

In addition, some consider this to be a weather manipulation device/weapon capable of generating controllable, but massive and destructive hurricanes— in my view, this scenario is particularly implausible. Others speculate that HAARP is the latest technological evolution of the Woodpecker signal developed by the former Soviet Union. A powerful signal that traversed the entire globe, resulting in electrical failures and altering the behavior of animals and children.

Remember that the Soviets were pioneers in the control and testing of electromagnetic weapons.

Nevertheless, I hold the belief that its inception was driven by the intention to facilitate the generation of very low-frequency waves from high-frequency waves, which are significantly simpler and more cost-effective to generate. In this way, they could communicate with submarines several hundreds of feet deep where high frequency waves couldn't reach.

The low-frequency wave provides specific benefits: it's not blocked by mountains or salt water at certain depths, and it can reach the ionosphere and subsequently spread across the entire planet from there. Following extensive experimentation with large and high-cost antennas to generate such waves, they devised a method to attain the same outcome while circumventing the fundamental principles of electronics: for the emitting antenna to operate effectively, it must be proportionate to the size of the wave it will transmit. Considering that very low-frequency waves can span from a few inches to several miles, one can envision the substantial size of the transmitter required for this purpose.

In that scenario, how could they effectively transmit the waves necessary to achieve this emission spanning from 1 Hz to 30 Hz? This is where the hypothesis proposing the feasibility of implementation comes into play, suggesting the construction of multiple HAARP installations strategically positioned across the globe, working in unison. Its principle is based on the theory of the Russians Trakhtengerts, V. Y., Demekhov, A. G., Polyakov, S. V., Manninen, J., Turunen, T., and Rycroft, M. J., and it is called ELF/VLF radio signals caused by ionospheric demodulation of MF/HF radio transmitter signals.

Essentially, this theory suggests that multiple antennas emitting strong electromagnetic waves in the HF frequency range could induce the ionosphere to reflect ULF (ultra-low frequency) and LF (low frequency) signals. This phenomenon would occur, in simple terms, by heating the ionosphere to a point where it generates a flow of electrons, spanning a vast area comparable to a virtual antenna. This emulated antenna would be capable of generating very low-frequency waves that could be reflected directly from the ionosphere, reaching any location on the planet. This phenomenon can become evident through a range of signal variations when all HAARP antennas worldwide operate simultaneously.

Within its extensive documentation, there is a brief mention that HAARP has the capability to generate extremely low frequency/very low frequency (ELF/VLF) waves ranging from 30 Hz to 30 kHz. In theory, it could potentially target the waves that interfere with human behavior on its own. However, there is no confirmed information about its current utilization by the University of Alaska.

Some people argue that HAARP lacks the ability to interfere with the bioelectrical processes of living beings, making it unsuitable for such purposes. They base this claim on calculations that show the limit of electromagnetic exposure, the intensity level, and the duration of exposure at which these waves start to be harmful, indicating the need for radiation protectors to be used. Thus, they prove that HAARP's electromagnetic power wouldn't be strong enough to affect any living being within a radius of 1,080 feet from its center. But I'd argue that, in the case of MKTECH, the biggest problem is not the level of safe exposure; it's the interaction with the brain. It forces the mind to demodulate the content embedded in it and interferes directly in human cognition (in the nervous system) by modifying the wave patterns at specific points in the brain, which are generated by the potential difference in the membrane which is responsible for the conduction of an impulse. Thus, it induces the most diverse behaviors and sensations as we've seen throughout the book.

This technology might be employed for various radio experiments. It can also be employed to enhance communication with satellites that form part of the technological infrastructure, serving the purpose of transmission feedback. Anything beyond that is pure speculation. Still, there are a few areas in key locations across the world, like Siberia and South Africa, that have strangely similar patterns. Another notable point is that most are located near the Earth's magnetic poles.

The experiments carried out at HAARP have been completed, and the findings might have been integrated into electronic weaponry. In the present day, I think these antennas are no longer used for military means or to harm human minds, as they are now under the control of universities themselves. The research and implementation carried out by them are now public, with findings shared through publications on social media platforms. Its actual utility has been passed on to other facilities worldwide that are more powerful and shrouded in mystery.

# CHAPTER 12 - PRESENT, FUTURE AND CONCLUSION

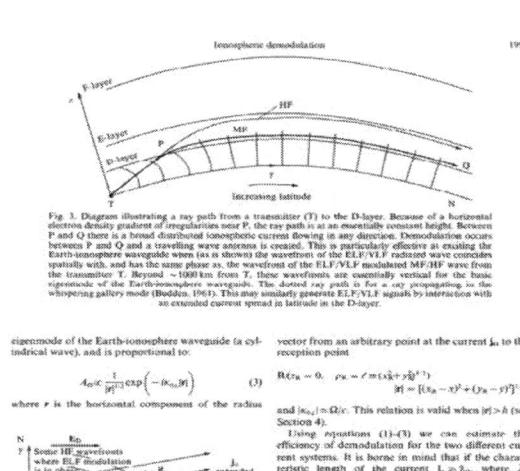

**Figure 12.1** Documentation on ELF/VLF radio signals caused by ionospheric demodulation of MF/HF radio transmitter signals.

V. Y. Trakhtengerts,' A. G. Demekhov,' S. V. Polyakov,2 J. Manninen,3 T. Turunen3 and M. J. Rycroft4 'Institute of Applied Physics, Russian Academy of Sciences, 46 Ulyanov St., 603600, Nizhny Novgorod, Russia; 'Radiophysical Research Institute, 25 B. Pechorskaya St., 603600, Nizhny Novgorod, Russia;'Geophysical Observatory, FIN-99600, SodankyIP, Finland; International Space University, Part d'Innovation, Boul. Gonthier d'Andernach, 67400, Illkirch, France.

Those who have closely followed the history of psychotronic weapons will recall that such devices originated from theories similar to these, primarily developed by Soviet scientists. Take a moment to think about this.

## 12.10 - The future of war

War has evolved throughout human history. First came early civilizations with spears, arrows, and axes; then the medieval era with swords, trebuchets, and catapults; and finally, the world-changing discovery of gunpowder as a weapon by the Chinese. In the modern era, there was a significant turning point with the discovery of the ultimate weapon of destruction with enough energy to potentially devastate the entire global population, forever reshaping the geopolitical landscape, as well as culture and society as a whole. Currently, wars are being fought on multiple fronts. One that is particularly noteworthy is the conflict centered around electronics, particularly the photon, or the utilization of light in its diverse forms of energy and wavelengths.

The ongoing and forthcoming war is, without a doubt, centered on electromagnetic weaponry for a range of objectives, spanning from downing aircraft and hacking drone remote control systems to inflicting harm on individuals through intense heat. Controlling photons will be the next big trend. Devices that emit electromagnetic pulses are a great example of the new military perspective, based on the modern use of this new generation of weapons. As we're used to only perceiving things in our everyday lives, we may not fully appreciate the conception of something that lies beyond our immediate sight. Consequently, we might underestimate its true destructive power and the potential devastation it can cause to modern civilization.

The Electromagnetic Pulse Bomb, or simply the EMP bomb, stands out as the most renowned device, possessing such immense power that it can effectively disable all electronics in an entire city. One of these devices was introduced and effectively tested: Boeing's CHAMP, which stands for Counter-electronics High Power Microwave Advanced Missile Project. Numerous countries have already prepared their devices for deployment in any wartime scenario. Within mere seconds, this bomb has the potential to devastate an entire city's computational and energy infrastructure. By disrupting its communication capabilities, it throws countless lives into chaos.

Among directed-energy weapons, the ones most familiar to the general public are LASERS – Light Amplification by Stimulated Emission of Radiation. Their popularity owes much to their portrayal in famous science fiction movies. These weapons serve a valuable purpose in neutralizing enemy satellites and disabling their equipment in cyber and electronic warfare that targets the defenses and communications systems of hostile nations. The laser has many applications. One of them involves identifying the trajectories of objects, such as enemy missiles, to be neutralized and lock onto the target using the aiming of the launching device, which then emits concentrated beams capable of destroying said target. This device is called Mobile Tactical High Energy Laser. With a name like that, it almost sounds like a weapon from the future, except that this one in particular was tested in the field in 2002, 18 years ago. Its development is classified, but it must be at a very advanced stage by now. The Chinese WB-1 and the U. S. Active Denial System are also (millimeter wave source) weapons capable of emitting a beam that heats the water beneath the skin of targets, resulting in significant discomfort. This effect compels the affected individuals to retreat and move beyond the weapon's effective range.

The weapon being tested today is part of a collection of advanced weaponry in the field of electronics and magnetism. These weapons — that have the ability to disrupt the nervous system and manipulate the human mind, as we've seen in the book —, are becoming more compact with built-in functionalities, and are capable of causing temporary paralysis without, for example, the necessity of using high-voltage shocks from a taser. These encompass weapons with the potential to influence human volition, discourage enemy troops from launching subsequent attacks, and trick adversaries into committing acts against themselves. Essentially, an infinite number of theories can be explored within the diverse scenarios of the intricate landscape of modern warfare. These electronic (neural) attack systems are currently used alongside conventional weapons.

In addition to this psychotronic arsenal that could potentially be deployed from satellites in the future, soldiers may have access to even more powerful technology capable of infiltrating the enemy's mind and instantly destroying specific brain areas. This technology could severely affect balance through the vestibular system, suppress autonomic systems, trigger visceral neurons, and induce physical responses such as reflux and cessation of heart and lung activity. Imagine having the power to directly manipulate specific sections of each vital organ! One day this will be possible.

Currently, a significant concern revolves around the tangible possibility of developing a fully operational weapon that merges the characteristics of D2K with the unparalleled power of SYNTELE, which is capable of affecting human cognition while awake. Presented next is the SYNTELE 2: a formidable device that can project images directly into the mind, seamlessly overlaying them onto current thoughts and compelling the brain to perceive and experience the transmitted scene. This process leads to negative effects throughout the body.

When the mind is alert and vulnerable to intrusive images or the manipulation of thought forms, it can result in situations similar to this: the individual tries to think of a car but struggles to synchronize verbal thoughts with mental images. Furthermore, the mental capacity to recall visual memories and stimulate imagination is insufficient, resulting in the emergence of an image of a bicycle instead. In this scenario, SYNTELE 2 puts an end to the ability of consciousness to control the stimuli, mainly to correctly projecting the visual memory stimuli that will not be learned — the requests will consistently yield inconsistent results. Consciousness can function in a manner reminiscent of a dream while in a waking state. But the content of the external and internal synthesized signals will be organized and provided by human hacking equipment, SYNTELE 2, and not by capturing external sensory systems.

When employed in a real war scenario, it could become an immensely potent weapon, especially when combined with other MKTECH functionalities like the V2K. Together, they can create significant turmoil within an enemy squadron on the battlefield, effectively making any assault nearly unfeasible, despite the heightened adrenaline and utmost focus. SYNTELE 2 will easily undo this mindset. Not to mention the direct attacks on the brain that cause serious neurological problems. It'd be interesting if the military incorporated electromagnetic wave reflector devices into their helmets soon, given that this is already occurring to a considerable degree.

The utilization of unconventional weapons — a deluge of "non-lethal" neural weapons of every kind — that hold the potential for extraordinary feats involving humans and animals will define this century. In addition, we must address the comprehensive tactics that have arisen since the emergence of these weapons in the field of psychological warfare or electronic psychological warfare. I know cognitive information and mind control warfare may seem like an unusual and secondary subject, but the impact of these technologies and techniques, born from their creation and improvement through use directly in the field, produced a new universe centered on the theme of direct human control through psychophysiological weapons. New concepts, which are now tested and approved, have elevated their importance to the point where they carry significant weight alongside traditional military weaponry. Military strategies are, in fact, adapting to incorporate its use.

Weapons, such as laser, electromagnetic, plasma, climatic — I have doubts about the viability of these in particular, but I decided to include them anyway —, genetic and biotechnological are the core principles that drives the modernization of national defense. The potential for these weapons to be used for both good and bad deserves attention, but little research material is available.

To grasp the enduring nature of these emerging weapons, one must realize that the concept of brain hacking has permeated various spheres, including the Air Forces of several nations, as evidenced by a recently disclosed document obtained through the right of access to information. Details of this are contained within a classified document titled "Sensory consequences of electromagnetic pulses emitted by laser induced plasmas". This document delves into the research on the activation of nerve cells responsible for detecting unpleasant stimuli such as heat, harm, pressure, and cold. A particular neuroreceptor is selectively stimulated and adjusted to generate sensations of burning or freezing, while inflicting no actual harm whatsoever. The skin is the easiest target for such stimulation, but, in principle, any sensory nerve could be stimulated. The Controlled Effects document suggests "it may be possible to create synthetic images to confuse an individual's visual sense or, in a similar manner, confuse his senses of sound, taste, touch, or smell."

During my research for this book, I came across a wide range of technologies, acronyms and military numbers used today based on MKTECH —a topic that could be explored in another book—, and I realized that the easiest pathway to victory is to understand what goes on in the enemy's mind. That is, extracting information directly from them, creating strategies to destroy their defenses and prevent their advance. Even better, however, would be to use a laser pulse energy and launch a direct electromagnetic attack on soldiers' minds, causing them to paralyze, give up, or completely destroying their nervous system, incapacitating or killing them.

Superficially speaking, this technology and its elements are already widely used on the battlefield. There are circulating rumors, and some with substantial credibility, regarding the advancements achieved by these weapons. Documents and reports from decades ago warn that the development of psychotronic weapons would change the world forever. People involved in the projects disappeared, victims who claimed to be used as guinea pigs for psychotronic tests were silenced by the effects of the weapon on the mind. An undeclared war has been going on since the 1970s between Russia and the United States. Today, I'm deeply concerned about the advancements in specific fields linked to these weapons. They operate on concepts and principles that are significantly different from what we're accustomed to, particularly due to their current highly destructive and invasive nature.

These include weapons that emit microwave energy, laser weapons, radio wave weapons, and infrasonic weapons. They have the capability to harm human targets by gradually causing fatal injuries or instantly damaging their brains. Additionally, there are also medium power weapons that induce neurological damage in specific areas by exploiting the target's extreme stress, leading to issues with communication, hearing, and even strokes. Some effects include temporary or permanent blindness, discomfort and dizziness, nausea, diarrhea, and abrupt or periodic changes in behavior. They have the ability to selectively eliminate areas and shut down brain functions while simultaneously disabling memory associated with the functioning of the affected area, destroying/erasing the data, the recollection of what it was like to have those functions, how they operated, and what they were used for. This can happen, for instance, in cases where individuals suffer brain damage in accidents. However, it's also possible to replicate this effect artificially.

In short, these weapons cause severe neurological damage, mainly affecting areas related to communication (speech, hearing and lexical word formation). They also often destroy the eardrum, leading to partial or complete deafness. Depending on the power and type of attack, they can drive people to a state of nervous breakdown in a matter of days. In this book, I gradually explore this long-term attack designed to slowly inflict damage, gradually eroding cognition. The MKULTRA experiment method relies on neural reprogramming, memory manipulation, attention impairment, and significant alterations in one's perception of reality.

How will this particular weapon evolve in the future, say by 2030 or 2050? Could we expect the emergence of a satellite with similar power that releases laser beams towards Earth? Is it possible that HAARP will be utilized to navigate the ionic atmosphere and facilitate the trajectory of a satellite's laser beam? Only time will tell!

Without a doubt, nations will soon need to discuss and set rules to control or limit the use of these weapons that will be incredibly strong, capable of destroying entire cities or towns within minutes by means of satellites controlled by the military all over the world. It'll be something similar to what we witnessed with the atomic bomb, but much simpler to execute and build. No radioactive material of any kind is required, and it leaves no signature or charged particles. The only trace left is the victims themselves and the electromagnetic spectrum if monitored at the time of the attack.

The reflection from the era of the atomic race still persists. The sooner we recognize, talk about, and bring attention to this reality, the faster we can improve as a society.

## 12.11 - Artificial intelligence using MKTECH in the future

Imagine having an unseen companion with whom you could share your silent thoughts. A friend linked to your mind. Someone you could always turn to for guidance in times of uncertainty or questions, to analyze problems, and assist in any situation, thereby enhancing your intellectual capacity and reasoning, and becoming an integral part of your essence. This "invisible friend" will hold immense significance for large corporations. With the advancement of **Neuro-Responsive Artificial Intelligence**, which enables the capture of remote silent thoughts, the level of interaction between humans and machines will reach unprecedented heights. Originating from the same lineage that emerged in 2015 as the first generation of assistants, such as SIRI and CORTANA, novel assistants will become part of the MKTECH network, assimilating vast knowledge through constant information exchange with humans. This trajectory will eventually lead to a convergence where the mind, AI, and the internet merge into a cohesive entity.

Initially, there will be assistants capable of analyzing processes and performing tasks far more proficiently than humans — numbers, statistics, analysis of a certain product or material in real time. They will be able to manage the connections and "calls" from people who own assistants and communicate by silent thought from a distance. Think of current personal assistant services, but much more advanced. This assistant will be always connected to your brain, guiding you at every stage of your life, including the monitoring of your vital signs. This technology operates in a manner akin to the "mental parasites" (OPS), but with a significant level of intelligence and a focus on enhancing human abilities rather than causing harm. You can even choose a voice you like and use it to replace your own. Just speak silently via amplified thought and the AI takes care of transmitting the words electronically to another person's brain.

We can see this technology as a fusion of highly advanced Chatbots allied to AIs and neural networks capable of interacting with humans as if they were human themselves. These programs are becoming increasingly human-like, capable of mimicking communication and social interaction patterns while displaying their own emotions, constantly learning during interaction with the person connected to the system, and easily assimilating habits, routines, and behavioral tendencies or the direction of thoughts that can lead to a change of mood, for example. I also believe that this AI will start to program itself, and develop more advanced artificial intelligences. Conventional programs developed by humans tend to disappear.

However, there is a slightly more perverse side to the use of AIs which was explored in chapter 11, where I briefly reported how some of them work when operated in conjunction with current psychotronic weapons. The trend of investments and developments in increasingly sophisticated AI integrated into military defense and attack systems that we see today will continuously increase to higher values in the future.

All nations recognize that AI represents an irreversible trend in deciphering vast data sets to discern enemy patterns, encompassing everything from radio signals to the movements of military convoys. In fact, the financial resources devoted to AI worldwide are remarkable. Just to give you an idea, the President of the United States, Donald Trump, signed the National Defense Authorization Act, which guarantees a substantial increase of US$ 717 billion, and encompasses all areas of defense such as the manufacture of planes, fighters, ships, satellites, secret weapons, etc. Among these weapons is the massive investment in artificial intelligence projects conducted by the Pentagon with support from other government, defense, and intelligence agencies. Certainly, space weapons and investments in neural weapons are included in that sum. The direction of AI's development is towards greater power, propelled by substantial investments from billionaires over time. It's set to rapidly outperform humans in numerous productive domains and effortlessly surpass them in the utilization of natural language, perfectly emulating voices of well-known people, thus perfecting its attacks, and making them increasingly powerful and destructive.

Among the wide array of concerns, one particularly notable issue is the inhuman and remarkably precise capability to execute attacks using neural weapons, carried out by AI designed to aggressively target the human brain. Its deep interaction with the most abysmal levels of cognition is astonishing, and its ability to learn through negative reinforcement leads to paths never taken before.

We're confronting a potential emergence of autonomous decision-making power capable of devising increasingly lethal and agonizing attacks. If its decision-making capacity extends to the choices of the targets themselves, we may be facing decisions formulated through analysis and perception of reality of an "artificial mind" that may escape human understanding.

## 12.12 - Final conclusions and considerations

In the imminent future that is rapidly approaching, as we witness the emergence of these increasingly powerful and specialized weapons, it's crucial to be prepared in order to thrive. Already, we can discern a multitude of uncertainties permeating various social spheres, indicating the rise of significant challenges that require our focus. Among them lies a direct threat to the very foundations of democracy, justice, and the rule of law. Before the Thought Police takes root in our society, directly manipulating our thoughts and exerting control over our habits, beliefs, emotions, and attitudes that are mirrored in our actions, all orchestrated by entities like the Ministry of Peace and the Ministry of Love, and overseen by the omnipresent eye of Big Brother[76], we must address pertinent questions crucial to safeguarding our freedom.

How will the legal system handle investigations that use these tools to solve crimes? What is the legal framework surrounding the investigation of thoughts, whether it pertains to suspected criminals or individuals in general? To what extent can we assert that this thought is not a byproduct of stress or contamination from external influences following a specific incident? — this will become a major concern related to privacy and constitutional rights. To what extent will thoughts hold greater value than expressed behavior?

Imagine the following scenario. The police conduct a conventional investigation into a murder suspect but find no tangible evidence or clues linking him to the crime. However, the suspect's thoughts are monitored for several days, and investigators become convinced that the suspect is the perpetrator based on his knowledge of specific details—details only the actual killer would know, including introspective insights, vocalized thoughts, dreams, and more. In light of this, can these thoughts be

---

[76] Originally published in 1949, the dystopian 1984 is one of the most influential novels of the 20th century; an undisputed modern classic. Published a few months before George Orwell's passing, this masterful work continues to serve as a powerful reflection.

considered admissible evidence? Can they be treated similarly to conversations intercepted by technological means? Can they serve as indicators and guide the investigative process? How will thoughts be utilized going forward? Will court authorization encompass vocalized thoughts? Can thoughts be presented as visual evidence? What are the conditions under which these thoughts were obtained? Was the person in a positive or negative emotional state at the time? A completely new code of conduct must be created to establish the legal limits of the procedures to be adopted, always focusing on not crossing the fine line between justice and the worst type of invasion to which a human being can be subjected: the display of one's intellectual, cognitive, and personal processes.

And what about compensation for the theft of intellectual property, where ideas are directly taken from one's mind? Or even for mind-hacking? How will the invasion be handled? These are some dilemmas that we'll inevitably face in the near future. How will society absorb and process all this information into something positive? Will this technology serve as yet another human enhancement tool? The truth is, only time will tell.

Do you recall the contract you signed, promising to keep your work confidential and to not disclose, reveal, or use any part of the product under any circumstances? These terms may no longer be applicable — after all, it's no longer possible to meet such demands. You can't guarantee that there isn't a neural parasite connected to your brain sucking up all the data flowing through your cognitive system. Even in places where there are secure computer networks to work with, it'll not be possible to be sure that there is no systematic theft of the product while it is being processed by the mind of the individual who is dedicated to solving the inherent issues of the respective project.

These questions are being raised by me now, before this technology spreads throughout society, as it advances quickly and is unstoppable. Within a few years, it'll seamlessly integrate into our daily lives, becoming indispensable, just as was the case with the Internet, computers, smartphones and website search engines. While considering the technology's significant aspects, one must not overlook the crucial moral distinction between using it for progress and the potential risk of compromising essential freedoms such as cognitive autonomy, personal property rights, and all that defines one's humanity within their own reality.

Primarily, it's imperative to acquire knowledge that allows us to protect ourselves (to some extent) against such weapons, including the government itself, which might potentially intrude upon our minds, pilfer intellectual property, seize cognition, and even manipulate the subconscious to gather data and shape a reality infused with ideological biases. After all, this weapon possesses the capability to instill fear artificially and execute various processes on entire populations.

At this juncture, SYNTELE will effortlessly become a part of people's lives, enabling activities such as operating cars and piloting aircraft, unlocking doors, issuing commands to embedded technologies, turning on lights or sounding alarms, initiating programs, and "typing" texts without the need for a keyboard. They will act like today's digital assistants, but in a silent and improved way.

It'll be an embedded mobile (or new modern devices) platform aimed at working with vocalized thoughts for such a purpose. It'll even work for us to communicate with each other without having to emit a single sound. In a few generations, 50 to 100 years from now, we'll control absolutely everything via SYNTELE. Maybe we'll even stop communicating in the traditional way and start looking like beings from other planets, just like those from our popular imagination: bulging black eyes, small mouths, and big heads.

Jokes aside, it's essential to remember that, as the technology becomes accessible to the general public through useful services in the natural progression of technological advancement, it still has its origins and conception rooted in the data collected from today's targets. Since the 1950s, numerous people lost their lives or became subjects of experiments for the technology to get to this point — this is a terrible fact that we all have to come to terms with. This technology was built upon the anguish of tormented souls, victims of the most horrific and vile tortures imaginable. These are events that cannot be changed, but we can prevent something similar from happening again.

This reckless manipulation can have consequences beyond the imagination, and it can spiral out of control, giving rise to a dogmatic outlook where conformity of thought is imposed upon everyone. In this way, innovation that springs from the unrestrained flow of thoughts and the merging of diverse ideas, rather than from the stifling of thoughts, is lost. The technology offers complete surveillance over thoughts, even during their earliest stages of formation, greatly inhibiting their growth and evolution. As a consequence, they're prevented from materializing. Nowadays, this is possible thanks to psychotronic weapons that manifest themselves in people's own minds and the computers that connect them, resulting in significant repercussions for the entire society.

As numb as we are to the violence around us, we cannot forget that torturing and experimenting on people indiscriminately is illegal. It's an atrocity that takes us back to medieval times, and goes against the bedrock of democracy. As tempting as it may be for some dubious people, we cannot let ourselves be carried away by the ease and banality of this heinous crime, which can only be compared to the darkest moments of humanity, such as the Russian and German prison camps during the World War II era.

I believe that society is prepared to learn about the existence of a technology that violates the content of their thoughts. And if it isn't, the time has come to show people the truth, even if this leads to unthinkable results at first. We cannot be afraid, though. Fear of failure is the first step toward defeat.

Our journey now reaches its conclusion. I hope this book will help targets to free their minds — that at least people become aware of its existence. I also trust that the book has effectively achieved its intended goal of preserving and disseminating accumulated knowledge, safeguarding it from oblivion, as it's all meticulously recorded within these pages. Knowledge is our ultimate weapon against hidden forces at play, which have inflicted considerable damage on the world. From now on, when someone is attacked at home, instead of immediately considering that they lost their minds and involuntarily following the steps dictated by modern experiments, they can now embrace new possibilities by providing essential resistance and impeding the progress of project phases from reaching a critical juncture with irreversible consequences, as often observed in common scenarios.

Some will probably say that the book could have been more technical, be focused on a certain subject, provide more numbers and equations, or that it should have gone deeper into the topics of computing, neuroscience or electronics. However, its intent was to uncover the existence of these weapons, illustrate the functioning of each component comprising the technology, expound upon the psychological warfare tactics involving psychotronic weapons, provide insights into its global operation, shed light on the stalking by operators for modern MKULTRA research, explore the repercussions on the human psyche, and delve into their historical context. Therefore, I tried to keep a balance between the themes, so that everyone is aware of the events happening in society and that involves you, me, our families — in short, each and every one of us.

As I close this chapter, I do so with a heavy heart, as I remember that we'll never have natural defenses for this weapon. Unfortunately, we can't just install a new version of a program that covers all these mental gaps — there is no chance of this happening. This is a fact we must now accept.

The certainty of an impending and comprehensive societal transformation is undeniable and pessimistic. This transformation will happen by breaching the final refuge for preserving secrets, directly manipulating the mind, and capturing every piece of information that passes through our brains. While our potential for adaptability, plasticity, and human resilience offers a renewed sense of hope, enabling us to eventually grow accustomed to and adjust to this novelty and its immediate effects after a few years, it remains challenging to fully accept this new reality, given that we weren't consulted beforehand. It's comforting, however, to remember that the brain is the only organ that has the ability to adapt to challenges in a short period of time, rearranging all the pieces of this intricate puzzle that encompasses the mind and reality!

## CHAPTER 12 - PRESENT, FUTURE AND CONCLUSION

After all I've witnessed and researched, I've stumbled upon yet another revelation: the ability to create "programmed" remote killers, Manchurian Candidates, human weapons that are made to act against their will in a variety of ways has long been achieved. These victims can be directly programmed, completely brainwashed, or tortured until complete collapse. This causes them to succumb to both direct and indirect orders that become ingrained in their altered reality. The target's mind, severely damaged and modified, is led down these paths created under unique conditions using psychotronic warfare strategies. This unexpected outcome catches everyone by surprise. One way or another, regardless of the outcome, those responsible for the experiments always profit by analyzing parameters and thorough data, thereby improving the weapon itself and the protocols.

I hope that, in the next 50 years, we won't have to witness a repetition of past actions by another president, followed by apologies for experiments conducted earlier in the century that tragically claimed numerous lives. We must strive to put an immediate end to these experiments and hold the responsible parties accountable, providing appropriate compensation for exploiting the minds, bodies, and souls of individuals as portable data processors, and for turning individuals into human guinea pigs for a wide range of MKULTRA 2.0 experiments.

Finally, one more disquieting concern, which occasionally crosses my mind amidst all I've witnessed—the potency of these weapons and their capacity to manipulate the human brain in the present—brings me to my final thoughts: during the 1960s and 1970s, a period when these experiments were at their height, were well-known figures like (among many others) former Beatle John Lennon, the Kennedy brothers John and Robert, and civil rights activist Martin Luther King killed by individuals seemingly manipulated through secretive operations by the CIA or KGB—the MK-ULTRA of that era?

And the ultimate question still lingers, awaiting an answer: who will be next?

THE END

# Glossary I

Glossary I is not in alphabetical order. In fact, it's arranged in a way that helps the visualization of major neuroelectronic technologies as a whole.

**MKTECH** - Mind Control Technology: complete system covering all technologies or modules below.

**EMR** - Electronic Mind Reading.
**EMRi** - Electronic Mind Reading (images): subsystem which is part of a complex scheme that uses a series of electronic devices to capture, amplify and decode the content of the electrical signals from neural networks responsible for mental images or visual memory of thoughts.
**EMRv** - Electronic Mind Reading (vocalized): subsystem which is part of a complex scheme that uses a series of electronic devices to capture, amplify and decode the content of the electrical signals from neural networks responsible for the vocalization of thoughts (the voice of the mind).
**EMRa** - Electronic Mind Reading (auditory): subsystem which is part of a complex scheme that uses a series of electronic devices to capture, amplify and decode the content of the electrical signals from neural networks responsible for hearing. One can hear everything the target is listening.
**EMRo** - Electronic Mind Reading (optical): subsystem which is part of a complex scheme that uses a series of electronic devices to capture, amplify and decode the content of the electrical signals from neural networks responsible for the sight. It's possible to see images that the target visually processes.

**V2K** - "Voice to Skull", Intracranial Voice, Microwave Voice or Microwave Hearing Effect: system capable of inserting voices and sounds directly into the target's mind. These are microwave transmissions that are demodulated by the target's brain and are indistinguishable from the sound picked up by the ear. There are no outside noises; people around them cannot hear the sounds, only the target. It's a modern version of the V2K (v2.0) that has all the qualities of its predecessor, but with improvements in sharpness and the capacity of "blending" with the noises that come naturally through auditory afferent pathways. The target hears voices "inside" any noise that is naturally picked up. They hear voices within the noise of engines, appliances, airplane turbines, etc. The noise energy "amplifies" the microwave voices in the target's brain. These voices now have the same decibels of the sound that comes at all times. Another added quality is that the microwave sound is interpreted by the target according to the environment. That is, microwave voices "receive" the acoustic signature of the environment in which the target is immersed. If the target is in a garage, every microwave voice will "receive" the natural echo of the place or the identical reverberation of the environment. This effect is largely responsible for creating mental confusion. It makes the target think that people are chasing them wherever they go.

**SYNTELE** - Synthetic Electronic Telepathy: consists in the use of the EMRvi – Electronic Mind Reading (vocalized/images) to extract information — content of thoughts — and the V2K to insert data — voices and sounds — in order to activate neural processes connected to hearing. Then, a totally silent conversation between the technology operators and the targeted individual takes place. In other words, it's possible to send messages to the operator with the thought alone and receive a response via V2K without anyone in the vicinity being able to hear what is being communicated to the mind, nor the content of the thought being amplified and transmitted. It closes, therefore, a complete communication cycle. It's also responsible for inserting /sending images to the targets' visual thoughts that are similar to what we visualize in our minds when we think of a material object.

**D2K** - Synthetic Electronic Dream or SLEEPING BEAUTY: is a MKTECH module with the very specific purpose of completely mastering cognitive processes while the target is unconscious (sleeping). Operators practically have complete control over the graphic content that will be displayed within the individual's mind. The images are sent through electromagnetic transmissions to be demodulated by the brain of the mapped individual the moment they leave the waking state and begin to sleep. The Targeted Individual dreams about the content transmitted by operators. These images are captured by the brain at any stage of sleep, but the greatest interaction takes place during REM sleep.

**RNM** - Remote Neural Monitoring: this module is of great importance for conducting the entire MKTECH scheme, due to several unique characteristics composed of three distinct elements converging towards a purpose that is achieved in a joint effort: to monitor all possible raw physiological data as well as all electrical brain wave patterns, which indicate the individual's current physical and mental state, assisting the torture and physiological (reactive) responses that are of great value to the tests conducted by the operators.

**EEG-Based Biometrics**: feedback signal sent to the target's brain, or re-radiation of electromagnetic energy transmitted to the brain. It contains all the unique electrical characteristics of each sample. Several mathematical models and algorithms were created by scientists involved in the evolution of this weapon to individualize the brain waves mapped more and more precisely. The BCI is responsible for receiving this information in real time, quantifying it appropriately and forwarding it to another responsible algorithm or subsystem, translating it into a unique pattern for each individual on the planet.

**Telemetric EEG**: Electronic Brain Link (EBL) or Remote Neural Monitoring. It's able to accurately gauge various complex psychophysiological states of the targets, measuring, stimulating and distorting feelings and emotions, such as fear. It calculates the approximate data, such as heart rate, changes in neurons in the bowel ("butterflies in the stomach") and blushing. It's possible to create a virtual avatar that displays the body position in space in real time of all target's limbs in a graphical representation with the electrical feedback data obtained. Thus, they visualize in which position the body is in a defined period of time. They remotely monitor brain waves and electrical changes, biomagnetic analysis system similar to ECG, EEG, EOG, EMG, EGG, Respiration Rate, Pulse Rate, Temperature, Impedance Cardiography and Electrodermal Activity. These data can be captured or inferred with a satisfactory margin of approximation if compared to the equivalent of using electrodes, as well as brain functions and firing patterns of neurons involved in certain behaviors over a period of time.

**Neural GPS**: The telemetric EEG along with the Neural Remote Biometrics are responsible for analyzing the individual's activities and brain waves in real time. One of its sub-functions is to constantly send a signal to remote computers via satellite for comparison with the neural biometric data pre-established in the database. Thus, they indicate to the system whether the target is really the target and their position on the planet using GPS triangulation — similar to your cell phone's GPS tracking system and geolocation. It's possible to locate a person wherever they are, even inside buildings, houses and common urban structures and some natural ones, such as caves and grottos. Afterwards, the position will make the satellite choose the area of the beam where the trigger will have priority, always keeping the quality of the connection.

**Remote Polygraph**: it evaluates data coming from EEG, SYNTELE, ERM, V2K and D2K. Together these modules form a virtually foolproof mechanism for detecting a lie. A lie that could be, for instance, the number of goals you scored in a match (just to impress a friend), distorted memories relived during a bar conversation or a heinous crime kept under lock and key inside a murderer's mind. It's unnecessary to open the key with this weapon; it simply gets to access the data before it's even locked away. It analyzes the thoughts that are expressed.

# Glossary II

In alphabetic order:

**AI** – Artificial Intelligence.
**BCI** – Brain-Computer Interface.
**BMI** – Brain-Machine Interface.
**Brain Net/Web** – internet or network where data extracted from the mind (thoughts) travel.
**Dark Mind Web** – dark internet of thoughts.
**DARPA** – Defense Advanced Research Projects Agency.
**Deep Brain Web** – deep internet of thoughts.
**DEWs** – directed-energy weapons.
**DOD** – The United States Department of Defense.
**EEG** – Electroencephalography.
**Electronic Psy Ops** – electronic psychological warfare operations.
**Electronic Psychological Warfare** - unconventional electronic and psychophysiological war tactics using psychotronic electromagnetic weapons.
**Gang-Stalking or Organized Gang-Stalking** – they're nothing more than the human representation of the Mind Control Technology.
**KUBARK** – codename used by the CIA to refer to itself.
**LGN** – Hacking: hacking focused on breaching the Lateral Geniculate Nucleus electrical data.
**MGN** – Hacking: hacking focused on breaching the Medial Geniculate Nucleus electrical data.
**MK-ULTRA** – 1950's mind control and research program.
**MKULTRA 2.0** – modern research and mind control program with psychotronic or neuroelectronic weapons.
**Neural claustrophobia** – suppressed thoughts, fear of thinking, need to drive out invaders, thinking without being heard by invaders.
**Neural Phone** – classified intelligence apparatus, precursor of SYNTELE.
**Neural Satellites** – neuroelectronic attack satellites.
**Neuroelectronic Warfare** – war in which enemies use electronic means to invade and destroy the human mind.
**Neuromatrix** – Hacking: hacking focused on breaching the Neuromatrix electrical data.
**NLW** – non-lethal weapons.
**OPS** – Organized Professional Stalkers or **POS** – Professional Organized Stalking together with OPT are a group or groups of people responsible for the content of torture

that will be imposed on the victim in order to carry out the "stalking" and still maintain constant surveillance.

**OPT** – Organized Professional Torturers.

**POWs** – prisoners of war.

**Psychotronic/Electromagnetic/Neuroelectronic Weapons** – electromagnetic weapons designed to interact with, manipulate and destroy the human brain.

**Psychotronics** – neuroelectronic weapons.

**Radar** – Radio Detection and Ranging.

**REM** – Rapid Eye Movement.

**SATAN** – Silent Assassination Through Adaptive Networks.

**Targeted Individual** – victim attacked with neuroelectronic/electromagnetic weapons.

**TI** – Targeted Individual

**Torture for fun** – illegal surveillance, electronic harassment, sexual harassment.

**V2K** - Voice to Skull (Voice of God) Weapons.

**Winter Soldier** – modern version of the remote killer.

# About the Author

**Felipe Saboya de Santa Cruz Abreu**

Felipe Saboya de Santa Cruz Abreu was born in Rio de Janeiro and pursues a career in computer science as a systems analyst/developer. He graduated in the Undergraduate Information Systems Program and completed the Post-Graduate Program in Computer Forensics & Cyber Expertise. Since 2012, the author has been engaged in the study of psychotronic/neuroelectronic technologies and their effects on the human mind.

www.invasionandmindcontrol.com

# References

*A 200–2700 MHz 2-Arm Conical Spiral Antenna*. (n.d.). Superkuh. Retrieved August 26, 2021, from

http://superkuh.com/conical-spiral-antenna.html

*A Física do Rádio*. (n.d.). A Física do Rádio [The Physics of Radio].

http://fisica3ufrb.blogspot.com.br

*A Origem do Radar: A Origem das Coisas* [The Origin of Radar: The Origin of Things]. (n.d.). A Origem das Coisas. Retrieved August 24, 2021, from http://origemdascoisas.com/a-origem-do-radar/

AbcMed. (2016, November 14). *Fases da infância - como elas são? Quais as mudanças envolvidas?* [Child development stages - what are they like? What changes are involved?]. https://www.abc.med.br/p/saude-da-crianca/1280663/fases+da+infancia+como+elas+sao+quais+as+mudancas+envolvidas.htm

*About RTL-SDR*. (n.d.). RTL-SDR. Retrieved August 25, 2021, from https://www.rtl-sdr.com/about-rtl-sdr/

Acústica [Acoustics]. (2021, April 8). *In Wikipedia*.

https://pt.wikipedia.org/wiki/Ac%C3%BAstica

Agência Nacional de Informação Geoespacial [National Geospatial-Intelligence Agency]. (n.d.). In Wikiwand.

https://www.wikiwand.com/pt/Ag%C3%AAncia_Nacional_de_Informa%C3%A7%C3%A3o_Geoespacial

*All Identified Signals*. (2020, April 2). SIGIDWIKI - Signal Identification Guide. Retrieved August 26, 2021, from https://www.sigidwiki.com/wiki/Database

ANATEL. (n.d.). *Atribuição de Faixas de Frequências no Brasil* [Allocation of Frequency Bands in Brazil]. Retrieved August 25, 2021, from https://www.anatel.gov.br/Portal/verificaDocumentos/documentoVersionado.asp?numeroPublicacao=&documentoPath=radiofrequencia/qaff.pdf&Pub=&URL=/Portal/verificaDocumentos/documento.asp

Andreas Spiess. (2019, September 8). *#286 How does Software Defined Radio (SDR) work under the Hood? SDR Tutorial* [Video]. YouTube. https://www.youtube.com/watch?v=xQVm-YTKR9s&list=PLus_DAOVXauZp4O2szPwzl2S0Ra1Gjypk&index=2&t=0s

Ano Internacional da Geofísica [International Geophysical Year]. (n.d.). *In Wikiwand*. https://www.wikiwand.com/pt/Ano_Internacional_da_Geof%C3%ADsica

Aprendizado de máquina [Machine learning]. (n.d.). In Wikiwand. https://www.wikiwand.com/pt/Aprendizado_de_m%C3%A1quina

Aranha, M., & Martins, M. (2003). *Filosofando: introdução à filosofia* [Philosophizing: introduction to philosophy] São Paulo, SP: Moderna.

Arnal, L., Kleinschmidt, A., Spinelli, L., Giraud, A., & Mégevand, P. (2019, August 14). *The rough sound of salience enhances aversion through neural synchronisation*. Nature Communications. https://www.nature.com/articles/s41467-019-11626-7?error=cookies_not_supported&code=7833324a-a232-4bf3-9358-11f6569e2169

Associated Press. (2017, February 17). *Esteban Santiago declared to be mentally competent*. Mail Online. http://www.dailymail.co.uk/news/article-4230110/Airport-shooting-suspect-Florida-federal-court.html

Association for Diplomatic Studies & Training. (2013, September 17). *Microwaving Embassy Moscow — Another Perspective*. ADST. https://adst.org/2013/09/microwaving-embassy-moscow-another-perspective/

*Atores do Controle Remoto da Mente* [Remote Mind Control Actors]. (n.d.). Google Sites. Retrieved August 25, 2021, from https://sites.google.com/site/controlemental/home/atores-do-controle-fisico-da-mente

Backdoor. (2020, February 17). In Wikipedia. https://pt.wikipedia.org/wiki/Backdoor

BBC News Brasil. (2018, January 29). *O distúrbio que leva uma mulher a conviver com cinco vozes em sua cabeça* [The disorder that causes a woman to live with five voices in her head]. https://www.bbc.com/portuguese/geral-42827481

Bear, M., & Paradiso, M. (2008). *Neurociências: desvendando o sistema nervoso* [Neurosciences: unraveling the nervous system]. Porto Alegre, RS: Artmed.

BEC CREW. (2016, June 23). *Scientists Have Invented a Mind-Reading Machine That Visualises Your Thoughts*. ScienceAlert. https://www.sciencealert.com/scientists-have-invented-a-mind-reading-machine-that-can-visualise-your-thoughts-kind-of

Bechara, E. (2015). *Módulo Sistema Nervoso - Neuroanatomia Funcional* [Nervous System Module – Functional Neuroanatomy]. Moderna Gramática Portuguesa. Rio de Janeiro, RJ: Nova Fronteira.

Benson, T. (n.d.). *With "BrainNet," scientists develop tech for brains to communicate directly*. Inverse. Retrieved August 26, 2021, from https://www.inverse.com/article/60596-brain-internet-connect-thoughts-messages?utm_campaign=inverse&utm_content=1572623892&utm_medium=owned&utm_source=facebook&fbclid=IwAR2EpyprCeA6Ki887Am14aYKaJyIl7xlPtmF9towIQx9ehRbFJkwFWaYPIQ&refresh=49

Bergamini, C., & Tassinari, Rafael. (2008). *Psicopatologia do comportamento organizacional: organizações desorganizadas, mas produtivas* [Psychopathology of organizational behavior: disorganized but productive organizations]. São Paulo, SP: Cengage Learning.

Bhattacharjee, Y. (2018, January). *A ciência do bem e do mal* [The science of good and evil]. *National Geographic Brasil*, 214.

Big data. (2021, August 14). In Wikipedia. https://pt.wikipedia.org/wiki/Big_data

Bisi, G., Braghirolli, E., Nicoletto, U., & Rizzon, L. (2015). *Psicologia Geral* [General Psychology]. Petrópolis, RJ: Vozes.

Boric-Lubecke, O., Lubecke, V., Droitcour, A., Park, B.-K., & Singh, A. (2016). *Doppler Radar Physiological Sensing*. Hoboken, New Jersey: Wiley.

Braga, N. (2015, June 14). *Como funciona o Radar (ART154)* [How radar works (ART154)]. Instituto Newton C. Braga. https://www.newtoncbraga.com.br/index.php/como-funciona/10739-como-funciona-o-radar-art154

Braga, N. (n.d.). [Instituto Newton C. Braga]. Instituto NCB. https://www.newtoncbraga.com.br/index.php/eletronica/52-artigos-diversos/4261-art587

Braz Júnior, D. (2015, October 25). *Luz: onda ou partícula?* [Light: wave or particle?]. Tilt UOL. https://fisicanaveia.blogosfera.uol.com.br/2015/10/25/luz-onda-ou-particula/

Brown, G. (2001). *Radio and Electronics Cookbook*. Oxford, UK: Newnes. https://doi.org/10.1016/C2009-0-25079-3

Buzzi, A. (2001). *Filosofia para principiantes: a existência humana no mundo* [Philosophy for beginners: human existence in the world]. Petrópolis, RJ: Vozes.

Canídeos [Canidae]. (2021, May 2). In Wikipedia.

https://pt.wikipedia.org/wiki/Can%C3%ADdeos

Carvalho, L. (n.d.). *Mundo da Rádio - Quando a potência não é tudo* [Mundo da Rádio - When power isn't everything]. Mundo da Rádio - O universo da rádio, na

Internet. Retrieved August 24, 2021, from

http://www.mundodaradio.com/artigos/quando_potencia_nao_e_tudo.html

CatsVsDogs. (2010, October 28). *EmoLens - Control Flickr with your Thoughts with the Emotiv EPOC headset* [Video]. YouTube. http://youtu.be/E9_XZlHoSp0

Catterall, W. (2011, August 3). *Voltage-Gated Calcium Channels*. PubMed Central (PMC).

https://www.ncbi.nlm.nih.gov/pmc/articles/PMC3140680/

CBS News. (2017, January 9). *Esteban Santiago, Fort Lauderdale airport shooting suspect, makes first court appearance, is denied bond*. https://www.cbsnews.com/news/esteban-santiago-fort-lauderdale-airport-shooting-first-court-appearance-denied-bond/

CBS. (2019, September 1). *Brain trauma suffered by U.S. diplomats abroad could be work of hostile foreign government*. [Video].

https://www.cbs.com/shows/60_minutes/video/J53FM8S0J_qLalL8_lYk_nRcZHNkajaD/brain-trauma-suffered-by-u-s-diplomats-abroad-could-be-work-of-hostile-foreign-government/

*Cérebro & Mente*. (2003). Cérebro & Mente [Brain & Mind].

https://cerebromente.org.br/

Chaves, M. (1993). *Memória humana: aspectos clínicos e modulação por estados afetivos* [Human memory: clinical aspects and modulation by affective states]. Periódicos Eletrônicos em Psicologia. Faculdade de Medicina – UFRGS. http://pepsic.bvsalud.org/scielo.php?script=sci_arttext&pid=S1678-51771993000100007

Cheat [Cheating in video games]. (2021, April 26). In *Wikipedia*. https://pt.wikipedia.org/wiki/Cheat

CIA cryptonym. (2021, July 22). In Wikipedia.

https://en.wikipedia.org/wiki/CIA_cryptonym

CIA cryptonym. (n.d.). In Wikiwand. https://www.wikiwand.com/en/CIA_cryptonym

Ciência Todo Dia. (2018, June 1). *Por Que Precisamos da Dualidade Onda-Partícula?* [Video]. YouTube. https://www.youtube.com/watch?v=CgY_zBuK2Cw

Ciência Todo Dia. (n.d.). *Home* [Ciência Todo Dia]. YouTube. Retrieved August 24, 2021, from URL https://www.youtube.com/channel/UCn9Erjy00mpnWeLnRqhsA1g

CINDACTA. (n.d.). DECEA – Departamento de Controle do Espaço Aéreo [DECEA - Department of Airspace Control].

https://www.decea.gov.br/?i=unidades&p=cindacta-i

*ClintMclean74 / SDRSpectrumAnalyzer*. (n.d.). GitHub. Retrieved August 25, 2021, from https://github.com/ClintMclean74/SDRSpectrumAnalyzer

CNN. (2018, September 2). *Microwaves suspected in attacks on US diplomats in Cuba and China* [Video]. YouTube. https://www.youtube.com/watch?v=Su0uc5UvLvg&feature=youtu.be&fbclid=IwAR2tEUb9WV6c9RxWkzinKW3BTWJ4qqHKJ_Lm3mopa-lrGgs9MraexvLyfwM

Costa, E. (2003). *Acústica técnica* [Technical Acoustics]. São Paulo, SP: Blucher.

Costa, E. (2009). *Eletromagnetismo: teoria, exercícios resolvidos e experimentos práticos* [Electromagnetism: theory, solved exercises and practical experiments]. Rio de Janeiro, RJ: Ciência Moderna.

Cristino, G. (n.d.). *Estrutura e Função do Córtex Cerebral* [Structure and Function of the Cerebral Cortex]. Gerardo Cristino. http://gerardocristino.com.br/novosite/aulas/neurologia-neurogirurgia/cortexcerebral.pdf

Gattass, R. (2000, January 15). *O Pensamento - Mapeamento de Imagens por Ressonância Magnética Nuclear Funcional* [Thoughts: Image Mapping by Functional Nuclear Magnetic Resonance]. Cérebro & Mente. https://www.cerebromente.org.br/n10/mente/pensamento1.htm

Curso em Vídeo. (2016, February 19). *Curso Word #01 - Apresentação do Curso de Word 2016* [Video]. YouTube. https://www.youtube.com/watch?v=CgFzmE2fGXA

Curso em Vídeo. (n.d.). *Home* [Curso em Vídeo]. YouTube. Retrieved August 25, 2021, from URL https://www.youtube.com/channel/UCrWvhVmt0Qac3HgsjQK62FQ

Davidoff, L. (2010). *Introdução à psicologia* [Introduction to Psychology]. São Paulo, SP: Pearson.

de Carvalho, O. (2013). *Aristóteles em Nova Perspectiva - Introdução a Teoria dos Quatro Discursos* [Aristotle in New Perspective - Introduction to the Theory of the Four Discourses]. Campinas, SP: Vide Editorial.

de Godói, A. (2010). *Detecção de potenciais evocados P300 para ativação de uma interface cérebro-máquina* [Brain-computer interface based on P300 event-related potential detection]. Master's thesis, Universidade de São Paulo. Biblioteca Digital USP. https://doi.org/10.11606/D.3.2010.tde-19112010-115232

DEFCONConference. (2015, December 8). *DEF CON 23 - BioHacking Village - Alejandro Hernández - Brain Waves Surfing - (In)security in EEG* [Video]. YouTube. https://www.youtube.com/watch?v=c7FMVb_5SBM&list=PLus_DAOVXauYgyIgn-I6_7u2gUQEb-Rgu&index=42&t=0s

Delgado, J. (1969). Radio Stimulation of the Brain in Primates and Man Fourth Becton, Dickinson and Company Oscar Schwidetzky Memorial Lecture. *Anesthesia & Analgesia.*

Dell'isola, A. (2018). *Mentes Geniais* [Genius Minds]. Universo dos Livros.

Diario do Centro do Mundo. (2013, September 17). *"Bom rapaz" e "impulsivo: quem é Aaron Alexis, o atirador da base naval de Washington* ['Impulsive and good-natured': who is Aaron Alexis, the Washington Navy Yard shooter]. DCM. https://www.diariodocentrodomundo.com.br/bom-rapaz-e-impulsivo-quem-aaron-alexis-o-atirador-da-base-naval-de-washington/

Dormehl, L. (2018, February 26). *This A.I. literally reads your mind to re-create images of the faces you see*. Digital Trends. https://www.digitaltrends.com/cool-tech/university-of-toronto-mind-reading-ai/?fbclid=IwAR2YHymeiPRewKtVPfLkgxjPA4jkKdPLBdlGY0PWL8PhSwH9ZknHwvP4hP0

Dunning, B. (2018, February 27). *The Boy Who Thought He Was Reincarnated*. Skeptoid. https://skeptoid.com/episodes/4612

Edminister, J. (2013). *Eletromagnetismo – Col. Schaum* [Electromagnetism – Schaum Collection]. Porto Alegre, RS: Bookman.

*Electronic torture, Brain zapping, Cooked alive, Electromagnetic mind control, Electromagnetic murder, Electromagnetic torture, Electronic murder, Microwave murder, Microwave torture, Organized murder, No touch torture, People zapper | STOPEG*. (2008). Electronic Torture. https://www.electronictorture.com/

Elwood, J. (2012, November 14). *Microwaves in the cold war: the Moscow embassy study and its interpretation. Review of a retrospective cohort study*. Environmental Health. https://ehjournal.biomedcentral.com/articles/10.1186/1476-069X-11-85

*EMOTIV | Brain Data Measuring Hardware and Software Solutions*. (n.d.). EMOTIV. Retrieved August 25, 2021, from https://www.emotiv.com/

Espectro eletromagnético [Electromagnetic spectrum]. (n.d.). In Wikiwand. https://www.wikiwand.com/pt/Espectro_eletromagn%C3%A9tico

Esquadrão do Conhecimento. (n.d.). *Como funcionam os telefones celulares?* [How do cell phones work?]. Retrieved August 26, 2021, from https://esquadraodoconhecimento.wordpress.com/ciencias-da-natureza/fisica/como-funcionam-os-telefones-celulares/

Exame. (2017, November 2). *O poder do conhecimento* [The power of knowledge]. *Exame*, 1149.

Farquhar, G. (n.d.). *Protection from Neuro-Electromagnetic Frequency Mind Control Weapons*. Mark Jacobs. Retrieved August 25, 2021, from https://www.jacobsm.com/projfree/protection.html

Física com Douglas Gomes. (2019a, August 4). *Coaching quântico não tem embasamento científico. Você sabe o que é quântica?* [Video]. YouTube. https://www.youtube.com/watch?v=oRXNbnjD85U

Física com Douglas Gomes. (2019b, August 19). *Ondas e telecomunicações numa visão para ENEM (20 h) | Física com Douglas* [Video]. YouTube. https://www.youtube.com/watch?v=O5YqSWBKo2E&list=PLus_DAOVXauYgy_Ign-I6_7u2gUQEb-Rgu&index=17

Fisica Universitária. (2016, September 23). *Eletromagnetismo - Espectro Eletromagnético* [Video]. YouTube. https://www.youtube.com/watch?v=-C2erXakQlQ

*Fonética* [Phonetics]. (n.d.). Portal Educação. Retrieved August 24, 2021, from https://www.portaleducacao.com.br/conteudo/artigos/educacao/fonetica/23454

Fort Lauderdale airport shooting. (n.d.). In Wikiwand. https://www.wikiwand.com/en/2017_Fort_Lauderdale_airport_shooting

*Freedom of Information Act Electronic Reading Room*. (n.d.). Freedom of Information Act. Retrieved August 24, 2021, from https://www.cia.gov/readingroom/

Frequência extremamente baixa [Extremely low frequency]. (2019, March 6). In

Wikipedia.

https://pt.wikipedia.org/wiki/Frequ%C3%AAncia_extremamente_baixa

Frequency-hopping spread spectrum. (n.d.). In Wikiwand.

https://www.wikiwand.com/en/Frequency-hopping_spread_spectrum

Frey, A. (1993, February 1). *Electromagnetic field interactions with biological systems.*
**Federation of American Societies for Experimental Biology.**

https://faseb.onlinelibrary.wiley.com/doi/epdf/10.1096/fasebj.7.2.8440406

**Fuentes, D., Malloy-Diniz, L., de Camargo, C., & Cosenza, R. (2014).** *Neuropsicologia:*
*teoria e prática* [Neuropsychology: theory and practice]. Porto Alegre, RS:
Artmed.

Gallagher, J. (2019, April 11). *'De olhos fechados não visualizo nada': criador de método que revolucionou animação gráfica 3D não consegue formar imagens mentalmente* ['With my eyes closed, I cannot see anything': creator of a revolutionary 3D animation method, he cannot form images mentally]. BBC News Brasil. https://www.bbc.com/portuguese/geral-47866082?ocid=socialflow_facebook&fbclid=IwAR2vlMgNlVcRrJb9-slhDvWe97Bwya8GhWlHQObJkHDQiEFPdtzzny4CpXM

Golomb, B. (n.d.). *Diplomats' Mystery Illness and Pulsed Radiofrequency/Microwave Radiation.* Square Space.

https://static1.squarespace.com/static/58fa27103e00bed09c8eac2c/t/5b7f95930e2e7262c9be0455/1535088022263/Cuba+2018-08-23c+-NEJM.pdf

Gomes, A. (2013). *Telecomunicações: Transmissão e Recepção AM-FM - Sistemas Pulsados* [Telecommunications: AM/FM Transmission and Reception - Pulsed Systems]. São Paulo, SP: Érica.

Graham-Rowe, D. (2002, May 1). *"Robo-rat" controlled by brain electrodes*. New Scientist. https://www.newscientist.com/article/dn2237-robo-rat-controlled-by-brain-electrodes/

Grinberg, E. (2018, April 24). *Travis Reinking: What we know about the Waffle House shooting suspect*. CNN. https://edition.cnn.com/2018/04/22/us/travis-reinking-waffle-house-shooting/index.html?fbclid=IwAR0a6C2ec8DThMOm3kCvuvSPlef_997URRRMMqCVplAHS5frFiWsvPaRFso

Grossman, N., Bono, D., Dedic, N., Kodandaramaiah, S., Rudenko, A., Suk, H.-J., Cassara, A., Neufeld, E., Kuster, N., Tsai, L.-H., Pascual-Leone, A., & Boyden, E. (2017, June 1). Noninvasive Deep Brain Stimulation via Temporally Interfering Electric Fields. Cell Press journal. https://www.cell.com/cell/fulltext/S0092-8674(17)30584-6?fbclid=IwAR0wXhavBMUFOhOBTLjufXAg1Zaed4lWNBaAyTla6u62_1lkgUrgWIpgW0Q

Guia de CFTV. (2020, June 29). *Qual a Diferença Entre CCD e CMOS* [What's the difference between CCD and CMOS?]. https://www.guiadecftv.com.br/qual-diferenca-entre-ccd-e-cmos/

HackRF. (n.d.). Great Scott Gadgets. Retrieved August 25, 2021, from https://greatscottgadgets.com/hackrf/

Hambling, D. (2008, July 6). *The Microwave Scream Inside Your Skull*. Wired. https://www.wired.com/2008/07/the-microwave-s/

Haykin, S. (2017). *Redes Neurais: Princípios e Prática* [Neural Networks: Principles and Practice]. Porto Alegre, RS: Bookman.

*How to Make a Directed Energy Weapon Detection System for Less Than $50*. (n.d.). Instructables Circuits. Retrieved August 26, 2021, from https://www.instructables.com/id/How-to-Make-a-Directed-

Energy-Weapon-Detection-Sys/?fbclid=IwAR3BwzQk3FqoiSFNPAVwbRYGQQc2He8b09Pqb8yFiDLLdJbSzMNAUo3s2V4

*Human auditory system response to modulated electromagnetic energy*. (n.d.). Invasão e Controle Mental. https://invasaoecontrolemental.com.br/wp-content/uploads/2020/04/human-auditory-system-response-to-modulated-electromagnetic-energy.pdf

*Human Remote Sensing - Human Spectral Imaging - Remote Biometrics*. (n.d.). Human Remote Sensing - Remote Biometry. Retrieved August 25, 2021, from https://www.information-book.com/science-tech-general/human-remote-sensing-remote-biometry/

Interferência [Wave interference]. (n.d.). *In Wikiwand*. https://www.wikiwand.com/pt/Interfer%C3%AAncia

Invasão e Controle Mental. (2020, March 18). *Testes de reirradiação eletromagnéticas utilizando SDR*. [Video]. YouTube. https://www.youtube.com/watch?v=HLV5zegdHqU

Invasão e Controle Mental. (n.d.). *Home* [Invasão e Controle Mental]. YouTube. Retrieved August 24, 2021, from URL https://www.youtube.com/channel/UCQEwceYkANiF6PuBw1DPc7A

Ionosfera [Ionosphere]. (2020, September 19). In Wikipedia. https://pt.wikipedia.org/wiki/Ionosfera

Isaac Asimov. (n.d.). In Wikiwand. https://www.wikiwand.com/pt/Isaac_Asimov

Ivezic, M. (2018, April 30). *IEMI - Threat of Intentional Electromagnetic Interference*. 5G Security by Marin Ivezic. https://5g.security/cyber-kinetic/threat-of-iemi/?fbclid=IwAR3P9jeCz0Hxk8mPs_hazT_qOHNO1sz8xx2smwena0dEKExjK9y_vq0Hq-Q

Jacobs, J. (2019, March 26). *Trump Orders Study on Risks of Electromagnetic Weapon Attack*. Bloomberg. https://www.bloomberg.com/news/articles/2019-03-26/trump-is-said-to-plan-executive-order-on-electromagnetic-weapon?fbclid=IwAR1GNEMy1mseKflNVsMfbAkl4PtZb3aoFwqTJKEuqA5By-XNwaujUMdyG9E

Jaspers, K. (2006). *Introdução ao pensamento filosófico* [Introduction to philosophical thought]. São Paulo, SP: Cultrix.

Jensen, B. (2012). Microwave Instrument for Human Vital Signs Detection and Monitoring [Doctoral dissertation, Technical University of Denmark]. https://backend.orbit.dtu.dk/ws/files/77581851/Brian_Sveistrup_Jensen_2012_Microwave_Instrument_for_Vital_Signs_Detection_and_Monitoring_.PDF

Jiang, L., Stocco, A., Losey, D., Abernethy, J., Prat, C., & Rao, R. (2019, April 16). *BrainNet: A Multi-Person Brain-to-Brain Interface for Direct Collaboration Between Brains*. Scientific Reports. https://www.nature.com/articles/s41598-019-41895-7?fbclid=IwAR2Nhy5r4lMYRjshrWiYDtL7tpXwCRuItdsUfEd8zNuzfpNF2kmUStV6EE8

José Manuel Rodríguez Delgado. (n.d.). In Wikiwand. https://www.wikiwand.com/en/Jos%C3%A9_Manuel_Rodriguez_Delgado

Joseph C. Sharp. (n.d.). In Wikiwand. https://www.wikiwand.com/en/Joseph_C._Sharp

Justesen, D. (1975). Microwaves and Behavior. *American Psychologist*.

Kato, R. (2017, November 14). Com novas técnicas, alemães conhecem melhor o cérebro [By using new techniques, Germans know the brain better]. *Exame*. https://exame.com/revista-exame/cerebro-a-fronteira-final/

Khan, A. (2017). *Microwave Engineering - Concepts and Fundamentals*. Boca Raton, Florida: CRC Press.

Laberge, S. (1985). *Sonhos Lúcidos* [Lucid Dreams]. Siciliano Livros, Jornais e Revistas Ltda.

Laika. (n.d.). In Wikiwand. https://www.wikiwand.com/pt/Laika

Laser. (2021, June 4). In Wikipedia. https://pt.wikipedia.org/wiki/Laser

Lent, R. (2015). *Neurociência da mente e do comportamento* [Neuroscience of mind and behavior]. Rio de Janeiro, RJ: Guanabara.

Li, X. P., Xia, Q., Qu, D., Wu, T., Yang, D., Hao, W., Jiang, X., & Li, X. M. (2014). *The Dynamic Dielectric at a Brain Functional. . .* Scientific Reports. https://www.nature.com/articles/srep06893?fbclid=IwAR2jmID1ELAPNz6GwFqgV336fcwyrSXnIHYadbT6GPKLSeAUzn-BFvwh5yQ&error=cookies_not_supported&code=ddee9446-31cb-4e4b-b01a-64a0843084ac

Lima, E. (2014). *Sistemas de Biometria de Frequência* [Frequency Biometric Systems]. Faculdade Integrada da Grande Fortaleza. Quixadá, CE.

Lima, L. (2020, January 27). Os psicólogos que ensinaram a CIA "técnicas singulares" de tortura [The psychologists who taught the C.I.A. how to torture]. BBC News. https://www.bbc.com/portuguese/internacional-51244029?at_medium=custom7&at_custom1=%5Bpost+type%5D&at_custom2=facebook_page&at_custom3=BBC+Brasil&at_campaign=64&at_custom4=3F606336-4103-11EA-A04C-BA0A3A982C1E&fbclid=IwAR2xtS1loW37CZMBrtJWN4l6fFiZKt02oXsUqOgSw-eQiwohTIBx4IpBO88

Lin, J. (1989). *Electromagnetic Interaction with Biological Systems*. New York: Plenum Press.

List of NRO launches. (2021, June 21). In Wikipedia.

https://en.wikipedia.org/wiki/List_of_NRO_launches

List of spacecraft manufacturers. (2021, August 24). In Wikipedia.

https://en.wikipedia.org/wiki/List_of_spacecraft_manufacturers

LSD. (n.d.). In Wikiwand. https://www.wikiwand.com/pt/LSD

Magnus Contact. (n.d.). *Mind Control – Mind Control, Neurotechnologies used as Weapons!* Mind Control. Retrieved August 25, 2021, from https://www.mindcontrol.se/

Marques, F. (2008). *Viabilidade de Implementação de um Sistema Biométrico de Autenticação* [Feasibility of Implementing a Biometric Authentication System]. Master's thesis, Universidade de Aveiro. Repositório Institucional da Universidade de Aveiro.

Martinovic, I., Davies, D, Frank, M., Perito, D., Ros, T., & Song, D. (n.d.). *On the Feasibility of Side-Channel Attacks with Brain-Computer Interfaces | USENIX*. USENIX Association. Retrieved August 24, 2021, from https://www.usenix.org/conference/usenixsecurity12/technical-sessions/presentation/martinovic

Martins, M., & Neves, I. (2015). *Propagação e radiação de ondas eletromagnéticas* [Propagation and radiation of electromagnetic waves]. São Paulo, SP: Lidel.

Matt Anderson. (2014, August 6). *EM Waves* [Video]. YouTube.

https://www.youtube.com/watch?v=bwreHReBH2A

Mclean, C. (n.d.). *The Science of Microwaves Causing the Symptoms of the Diplomats in Cuba and China*. Google Docs. Retrieved August 24, 2021, from https://drive.google.com/file/d/1gyc6ETyzrrN5FxO47K2TX0cNGu_OFZbG/view?fbclid=IwAR2OANVyz-Mbk0PCQgqE0-hoNHajmzwdWvuzi7rKnCPN67Wki1FpmucDaTl

McPhate, M. (2016, June 10). *United States of Paranoia: They See Gangs of Stalkers*. The New York Times. https://www.nytimes.com/2016/06/11/health/gang-stalking-targeted-individuals.html?_r=0

Mecha. (2020, December 9). In Wikipedia. https://pt.wikipedia.org/wiki/Mecha

Mental Health Daily. (n.d.). *5 Types Of Brain Waves Frequencies: Gamma, Beta, Alpha, Theta, Delta*. Retrieved August 25, 2021, from https://mentalhealthdaily.com/2014/04/15/5-types-of-brain-waves-frequencies-gamma-beta-alpha-theta-delta/

Microwave auditory effect. (n.d.). *In Wikiwand*. https://www.wikiwand.com/en/Microwave_auditory_effect

Mlodinow, L. (2013). *Subliminar – Como o inconsciente influencia nossas vidas* [Subliminal: how the unconscious influences our lives]. Rio de Janeiro, RJ: Zahar.

Moreira, A. (1999, January). Tecnologias de Transmissão [Transmission Technologies]. DSI. http://www3.dsi.uminho.pt/adriano/Teaching/Comum/FactDegrad.html

MSNBC. (2018, September 11). *Russia Believed To Be Main Suspect In Attack On U.S. Diplomats | Velshi & Ruhle | MSNBC* [Video]. YouTube. https://www.youtube.com/watch?v=ghT-qxiI3yw&feature=youtu.be&fbclid=IwAR2IsOS7BRm_kxNoDhrO6TMgLRGErU3wOKoQly6TDraWi0x2wk7Ay4Pw6Y0

Muxfeldt, P. (2017, April 26). *Propagação das ondas de rádio (802.11)* [Propagation of radio waves (802.11)]. CCM. https://br.ccm.net/contents/820-propagacao-das-ondas-de-radio-802-11

NA HORA DA GUERRA. (2017, May 19). *Guerra Eletrônica #6 - RADAR 2* [Video]. YouTube. https://www.youtube.com/watch?v=Y5uDnaRRYhg

NA HORA DA GUERRA. (2020, March 9). *Guerra Eletrônica #15 - RADAR Pulsado 1* [Video]. YouTube. https://www.youtube.com/watch?v=3YlE0YCBNSQ

NA HORA DA GUERRA. (n.d.). *Home* [NA HORA DA GUERRA]. YouTube. Retrieved August 24, 2021, from URL https://www.youtube.com/channel/UCkjw_BEpqjjy00_mhfFbxDA

National Geographic Brasil. (2017, June). *Por que mentimos?* [Why do we lie?]. *National Geographic Brasil*, 207.

National Geographic Brasil. (2017, May). *Gênios* [Geniuses]. *National Geographic Brasil*, 206.

National Geographic Brasil. (2017, September). *O cérebro e os vícios* [The brain and addictions]. *National Geographic Brasil*, 210.

National Geographic Brasil. (2018, February). *Big brother da vida real* [Real life big brother]. *National Geographic Brasil*, 215.

National Reconnaissance Office. (n.d.). In Wikiwand. https://www.wikiwand.com/pt/National_Reconnaissance_Office

Nelson, B. (2019, April 25). *"Mind-Reading" Device Can Translate Your Brain Activity Into Audible Sentences*. Treehugger. https://www.treehugger.com/mind-reading-device-can-translate-your-brain-activity-audible-sentences-4862546

NetHoler. (2011, July 10). *Reincarnation - Airplane Boy (abc Primetime)* [Video]. YouTube. https://www.youtube.com/watch?time_continue=181&v=Uk7biSOzr1k&feature=emb_logo

*Neurobiologia dos Sonhos: Atividade Elétrica* [Neurobiology of Dreams: Electrical Activity]. (n.d.). Cérebro & Mente. Retrieved August 24, 2021, from http://www.cerebromente.org.br/n02/mente/neurobiologia.htm

Nicolelis, M. (2011). *Muito além do nosso eu* [Far beyond our ego]. Companhia Das Letras.

Nicolelis, M. (2012, April). *A monkey that controls a robot with its thoughts. No, really.* [Video]. TED Talks. https://www.youtube.com/watch?v=stXhGMVJuqA&list=PLus_DAOVXauYgyIgn-I6_7u2gUQEb-Rgu&index=33&t=0s

NKVD. (2021, August 20). In Wikipedia. https://en.wikipedia.org/wiki/NKVD

nptelhrd. (2013, September 16). *Lecture - 10 Single SideBand Modulation* [Video]. YouTube. https://www.youtube.com/watch?v=-ccrXpAJgjs&list=PLus_DAOVXauYgyIgn-I6_7u2gUQEb-Rgu&index=28&t=434s

Oliveira, A. (2019, August 9). *Este dispositivo ouve a voz que fala dentro da sua cabeça* [This device listens to the voice that speaks inside your head]. *Super Interessante*. https://super.abril.com.br/ciencia/este-dispositivo-ouve-a-voz-que-fala-dentro-da-sua-cabeca/?fbclid=IwAR31ME02gSzhj7h5GGXGUZ0KhIVu2EKPw-p1LMtK4p8wf2GKHgTyg88Aexw

Oliveira, R. (2000). *Neurolingüística e o aprendizado da linguagem* [Neurolinguistics and language learning]. Brasília, DF: Respel.

*Open Source Tools for Neuroscience*. (n.d.). OpenBCI. Retrieved August 25, 2021, from https://openbci.com/?fbclid=IwAR0a93kXoMC4iPNvx3G0sd43Rzt0yRpD90YSQo2pEFpNM6sqO_CK-YJqGbU

*Os Tipos de Memória – Memorização* [The Types of Memory – Memorization]. (n.d.). Memorização. Retrieved August 25, 2021, from https://memorizacao.info/os-tipos-de-memoria.html

Pall, M. (2015). *Microwave frequency electromagnetic fields (EMFs) produce widespread neuropsychiatric effects including depression.* Journal of Chemical Neuroanatomy. Retrieved August 25, 2021, from URL https://www.researchgate.net/publication/281261829_Microwave_frequency_electromagnetic_fields_EMFs_produce_widespread_neuropsychiatric_effects_including_depression?fbclid=IwAR3SY0NPJDOjuFPg906XtihdwTD0Kw-lR-PGuuPsm3eLbP7tYLMpcyDb2BY

Pedofilia [Pedophilia]. (2021, August 23). In Wikipedia. https://pt.wikipedia.org/wiki/Pedofilia

Peer-to-peer. (2021, August 11). In Wikipedia. https://pt.wikipedia.org/wiki/Peer-to-peer

Pelley, S. (2019, September 1). *Brain trauma suffered by U.S. diplomats in Cuba, China could be work of hostile foreign government.* CBS News. https://www.cbsnews.com/news/brain-trauma-suffered-by-u-s-diplomats-abroad-could-be-work-of-hostile-foreign-government-60-minutes-2019-09-01/?fbclid=IwAR3LdAEr97701pU0VhvPpVJYSSYwnwFN2gPxKwG-F28ckLgJHSoaliMIsMY

Pennicott, K. (2002, May 16). *Noisy signals strengthen human brainwaves.* Physics World. https://physicsworld.com/a/noisy-signals-strengthen-human-brainwaves/?fbclid=IwAR0Of6v5k-nK2xnsAOkOft99DsasXbRBeLnjsxmMq3-9LJByuk2e3DBl7Aw

Penttinen, J. (2015). *The Telecommunications Handbook - Engineering Guidelines for Fixed, Mobile and Satellite Systems.* Hoboken, NJ: Wiley.

Perper, R. (2019, July 17). Elon Musk's company Neuralink plans to connect people's brains to the internet by next year using a procedure he claims will be as safe and easy as LASIK eye surgery. Insider. https://www.businessinsider.com/elon-musk-neuralink-implants-link-brains-to-internet-next-year-2019-7?fbclid=IwAR1br4lcC8DswkwtYt2JiMBTShmMJ3kzltrdx7yLYAbjiPPXzXWmMMAfmzI

Persinger, M. (1974). *ELF and VLF Electromagnetic Field Effects.* New York: Plenum Press.

Pilati, R. (2018). *Ciência e Pseudociência – Por que acreditamos naquilo em que queremos acreditar* [Science and Pseudoscience – Why we believe what we want to believe]. São Paulo, SP: Editora Contexto.

Pisadeira (folclore) [*Pisadeira* (folklore)]. (2021, July 1). In *Wikipedia*. https://pt.wikipedia.org/wiki/Pisadeira_(folclore)

Portilho, G. (2016, December 15). *Como funciona o olho humano?* [How does the human eye work?]. *Super Interessante.* https://mundoestranho.abril.com.br/saude/como-funciona-o-olho-humano/

Primatas [Primate]. (2021, May 8). In Wikipedia. https://pt.wikipedia.org/wiki/Primatas

Processos de Moscou [Moscow Trials]. (2021, August 13). In Wikipedia. https://pt.wikipedia.org/wiki/Processos_de_Moscou

Projeto Escrita Criativa. (n.d.). *Home* [Projeto Escrita Criativa]. YouTube. Retrieved August 25, 2021, from URL https://www.youtube.com/channel/UCO6nYvm-muWS5rR1rpz00eA

*Qual a diferença entre uma transmissão AM para uma FM?* [What is the difference between AM and FM?] (2012, October 26). A Física do Rádio.

http://fisica3ufrb.blogspot.com/2012/10/qual-diferenca-entre-uma-transmissao-am.html

Radar. (n.d.). In Wikiwand. https://www.wikiwand.com/pt/Radar

*RADIO WAVES below 22 kHz*. (n.d.). RADIO WAVES below 22 KHz. Retrieved August 26, 2021, from http://www.vlf.it/

Radioescuta DX. (n.d.). *Bem-vindo a Radioescuta e DX* [Welcome to Radioescuta and DX]. Retrieved August 25, 2021, from http://www.sarmento.eng.br/OndasCurtas.htm

Redação Galileu. (2019, January 11). *"Som misterioso" ouvido por diplomatas em Cuba era feito por grilos* [A 'mysterious sound' heard by diplomats at the U.S. embassy in Cuba was caused by crickets]. Galileu. https://revistagalileu.globo.com/Sociedade/noticia/2019/01/som-misterioso-ouvido-por-diplomatas-em-cuba-era-feito-por-grilos.html

Rede neural artificial [Artificial neural network]. (n.d.). In Wikiwand. https://www.wikiwand.com/pt/Rede_neural_artificial

Relé [Relay]. (2020, April 19). In Wikipedia. https://pt.wikipedia.org/wiki/Rel%C3%A9

Resnick, B. (2016, June 20). *Scientists have invented a mind-reading machine. It doesn't work all that well.* Vox. https://www.vox.com/2016/6/20/11905500/scientists-invent-mind-reading-machine

*Resolução comentada dos exercícios de vestibulares sobre Polarização e Ressonância de ondas* [Annotated answers of college entrance exam exercises on Polarization and Wave Resonance]. (n.d.). Física e Vestibular - Aulas Grátis de Física. http://fisicaevestibular.com.br/novo/ondulatoria/ondas/polarizacao-e-ressonancia-de-ondas/resolucao-comentada-dos-exercicios-de-vestibulares-sobre-polarizacao-e-ressonancia-de-ondas/

Riera, A., Dunne, S., Cester, I., & Ruffini, G. (2008). *STARFAST: a wireless wearable EEG/ECG biometric system based on the ENOBIO Sensor.* https://www.yumpu.com/en/document/read/53643357/starfast-a-wireless-wearable-eeg-ecg-biometric-system-based-on-the-enobio-sensor

Riera, A., Soria-Frisch, A., Caparrini, M., Grau, C., & Ruffini, G. (2007). Unobtrusive Biometric System Based on Electroencephalogram Analysis. *EURASIP Journal on Advances in Signal Processing.* https://doi.org/10.1155/2008/143728

Rocha, A. (2009*). Análise das respostas eletrofisiológicas de longa latência – P300 em escolares com e sem sintomas de Transtorno do Processamento Auditivo* [Long latency evoked responses analysis in school-aged children with and without auditory processing disorders symptoms]. Final paper, Universidade Federal de Minas Gerais. FTP Medicina – UFMG. https://ftp.medicina.ufmg.br/fono/monografias/2009/anapaula_analisedasrespostas_2009-1.pdf

*Ross Adey (1922–2004).* (2004, May 20). Microwave News. https://microwavenews.com/news-center/ross-adey

Ross, B., Schwartz, R., Meek, J., & Kreider, R. (2017, January 7). *What We Know About Esteban Santiago, Suspect in Fort Lauderdale Attack.* ABC News. https://abcnews.go.com/US/esteban-santiago-suspect-fort-lauderdale-attack/story?id=44612498

Ross, C. (n.d.). *Project Bluebird.* Want to Know. Retrieved August 24, 2021, from https://www.wanttoknow.info/bluebird10pg

Ruminantes [Ruminant]. (2021, January 7). In Wikipedia. https://pt.wikipedia.org/wiki/Ruminantes

Saffi, F., & Serafim, A. (2015). *Neuropsicologia forense* [Forensic neuropsychology]. Porto Alegre, RS: Artmed.

Saleem, A. (2021, August). SDR for Ethical Hackers and Security Researchers [Udemy course]. Udemy. https://www.udemy.com/share/101tjIBUcZdldRR34=/

Sanei, S., & Chambers, J. (2007). *EEG Signal Processing*. Hoboken, New Jersey: Wiley.

Santos, K. (2011). *Magnetron: do radar ao forno de micro-ondas* [Magnetron: from radar to the microwave oven]. Universidade Católica de Brasília. https://livrozilla.com/doc/282438/magnetron--do-radar-ao-forno-de-micro-ondas

Satélite artificial [Satellite]. (2021, August 4). In Wikipedia.

https://pt.wikipedia.org/wiki/Sat%C3%A9lite_artificial

Satélite artificial [Satellite]. (n.d.). In Wikiwand.

https://www.wikiwand.com/pt/Sat%C3%A9lite_artificial

Scherer, A. (2018, February 21). *A química da mente produtiva* [The chemistry of the productive mind]. *Exame*, 1151.

Schmidt, M. (2013, September 25). *Gunman Said Electronic Brain Attacks Drove Him to Violence, F.B.I. Says*. The New York Times. https://www.nytimes.com/2013/09/26/us/shooter-believed-mind-was-under-attack-official-says.html

Scientific American Brasil. (2018, July). *Prevenção ao suicídio* [Suicide prevention], *Scientific American Brasil*, 184, 59.

Scientific American Brasil. (2019a, March). *O Código Facial* [The Facial Code]. *Scientific American Brasil*, 193.

Scientific American Brasil. (2019b, March). *Robôs que aprendem sozinhos* [Robots that learn on their own]. *Scientific American Brasil*, 181.

*SDR = Software-Defined Radio*. (2016, May 16). SDRZero. https://www.qsl.net/py4zbz/sdr/sdrz.htm

Sebrae. (2017, April 11). *Definição de Patente* [Patent Definition].
https://www.sebrae.com.br/sites/PortalSebrae/artigos/definicao-de-patente,230a634e2ca62410VgnVCM100000b272010aRCRD

Sharp, J. (1973). Voice to Skull Demonstration: Artificial microwave voice to skull transmission was successfully demonstrated by researcher Dr. Joseph. *American Psychologist.*

Sharp, J. (1974). *Artificial microwave voice to skull transmission was successfully demonstrated.* Seminar from the University of Utah.

Silva, M. (n.d.). *Nervo óptico* [Optic nerve]. InfoEscola. Retrieved August 25, 2021, from https://www.infoescola.com/visao/nervo-optico/

Silva, V., Pereira, J., Nohara, E., & Rezende, M. (n.d.). *Comportamento eletromagnético de materiais absorvedores de micro-ondas baseados em hexaferrita de Ca modificada com íons CoTi e dopada com La* [Electromagnetic behavior of radar absorbing materials based on Ca hexaferrite modified with Co-Ti ions and doped with La]. SciELO. Retrieved August 25, 2021, from http://www.scielo.br/scielo.php?pid=S2175-91462009000200255&script=sci_abstract&tlng=pt

Silve, S. (1984). *Microwave Antenna Theory and Design (Electromagnetics and Radar).* New York, NY: Institute of Electrical Engineers.

simonxhayes. (2009, July 26). *Computer records animal vision in Laboratory - UC Berkeley* [Video]. YouTube. https://www.youtube.com/watch?v=piyY-UtyDZw

Snyder, B. (2019, November 1). The Next Computer Revolution Will Be Based on Our Brains. Bloomberg. https://www.bloomberg.com/news/articles/2019-11-01/how-the-human-brain-project-aims-to-improve-the-world-s-computers?fbclid=IwAR06wom4CeHJnnnqZk7QqX-0RuMPxSDkRfVJI-

l40nX8xeaMMDHa0_sjigA

Soro da verdade [Truth serum]. (2019, May 23). In Wikipedia.

https://pt.wikipedia.org/wiki/Soro_da_verdade

Stephen Hawking. (n.d.). In Wikiwand.

https://www.wikiwand.com/pt/Stephen_Hawking

STOPEG - *Stop Electronic Weapons and Gang Stalking*. (n.d.). STOPEG Foundation. Retrieved August 25, 2021, from https://www.stopeg.com/

*Study: Women Need More Sleep Because Of One Obvious Reason*. (n.d.). SimpleCapacity. Retrieved August 25, 2021, from https://simplecapacity.com/2015/11/study-women-need-more-sleep-because-of-one-obvious-reason/

Suçuarana, M. (n.d.). *Formigas zumbis* ['Zombie ants']. InfoEscola. Retrieved August 25, 2021, from https://www.infoescola.com/biologia/formigas-zumbis/

Super Interessante. (2016, October). *Mindfulness: como domar a sua mente agora* [Mindfulness: how to tame your mind]. *Super Interessante*, 365.

Super Interessante. (2017, June). *Cérebro* [Brain]. *Super Interessante*, 357.

Super Interessante. (2019, May). *A Ciência das Emoções* [The Science of Emotions]. *Super Interessante*, 396.

*Superposição de ondas (continuação)* [Wave superposition]. (n.d.). Só Física. Retrieved August 24, 2021, from https://www.sofisica.com.br/conteudos/Ondulatoria/Ondas/superposicao2.php

Suppes, P., Lu, Z.-L., & Han, B. (1997). Brain wave recognition of words. *PNAS*.

https://www.pnas.org/content/pnas/94/26/14965.full.pdf

TecMundo. (2016, February 11). *HAARP: o projeto militar dos EUA que pode ser uma arma geofísica* [HAARP: The U.S. military project that could be a geophysical weapon].

https://www.tecmundo.com.br/tecnologia-militar/8018-haarp-o-projeto-militar-dos-eua-que-pode-ser-uma-arma-geofisica.htm

TED. (2013, September 23). *Elizabeth Loftus: A ficção da memória* [Video]. YouTube.

https://youtu.be/PB2OegI6wvI

TED. (2019a, June 3). *A new way to monitor vital signs (that can see through walls) | Dina Katabi* [Video]. YouTube.

https://www.youtube.com/watch?v=CXy1byguvJY&list

TED. (2019b, June 3). *Sleep is your superpower | Matt Walker* [Video]. YouTube.

https://www.youtube.com/watch?v=5MuIMqhT8DM

TED. (2020, June 14). *Can we edit memories? | Amy Milton* [Video]. YouTube.

https://www.youtube.com/watch?v=ZK7ih4V0erc

TEDx Talks. (2017, August 29). *New Brain Computer interface technology | Steve Hoffman | TEDxCEIBS* [Video]. YouTube.

https://www.youtube.com/watch?v=CgFzmE2fGXA

Teleco. (n.d.). *Dados na Rede Celular: Evolução das Tecnologias* [Data on the Cellular Network: Development of Technologies]. Retrieved August 25, 2021, from

https://www.teleco.com.br/tutoriais/tutorialtrafdados/pagina_2.asp

Texas Instruments. (2018, January 8). *People counting demonstration using TI mmWave sensors* [Video]. YouTube.

https://www.youtube.com/watch?v=RT56YzqME6M&list=PLus_DAOVXauYgyIgn-I6_7u2gUQEb-Rgu&index=48&t=7s

Texas Instruments. (2019, August 20). *Intelligent Fall Detection Using TI mmWave Sensors* [Video]. YouTube. https://www.youtube.com/watch?v=njhRwijx_HY&list=PLus_DAOVXauYgyIgn-I6_7u2gUQEb-Rgu&index=47&t=0s

The Journal of Nervous and Mental Disease. (1968). *ICD-10 classification of mental and behavioral disorders: clinical descriptions and diagnostic guidelines.* World Health Organization. Geneva: World Health Organization.

The New York Times. (1979, May 30). *Soviet Halts Microwaves Aimed at U.S. Embassy.* https://www.nytimes.com/1979/05/30/archives/soviet-halts-microwaves-aimed-at-us-embassy.html

The New York Times. (2019, October 15). *Are We Ready for Satellites That See Our Every Move?* https://www.nytimes.com/2019/10/15/opinion/satellite-image-surveillance-that-could-see-you-and-your-coffee-mug.html?fbclid=IwAR3ISFsElpTiBA-iDBvXDg0tch7GR0349qkiLlhwWF9ml--6_a5oCu2VeZo

The Science of Electronic Harassment. (2020, August 15). *Detection of frequency ranges used for electronic harassment.* [Video]. YouTube. https://www.youtube.com/watch?v=sP4ZjyrfuUo&

The Science of Electronic Harassment. (n.d.). Home [The Science of Electronic Harassment]. YouTube. https://www.youtube.com/channel/UCl4nSNVf7uJPsekmnUWpBmw

Thorpe, J., Oorschot, P., & Somayaji, A. (2005). Pass-thoughts: authenticating with our minds. *Proceedings of the 2005 Workshop on New Security Paradigms - NSPW '05*, 45–56. https://doi.org/10.1145/1146269

Toscano, R. (2006). *Bloqueador de múltiplas frequências: concepção do sistema e estudo de caso para terminais is-95* [Multiple-frequency blocking: system design and case study for is-95 terminals]. Master's thesis, Instituto Militar de Engenharia. Programa de Pós-Graduação em Engenharia Elétrica.

Transponder. (n.d.). In Wikiwand. https://www.wikiwand.com/pt/Transponder

*Transtorno de Personalidade Histriônica ou Histérica (TPH)* [Histrionic personality disorder]. (2010, August 17). Memórias de uma Methamorfose. http://memoriasdeumamethamorfose.blogspot.com/2010/08/transtorno-de-personalidade-histrionica.html

UNICAMP. (n.d.). *Córtex motor normal em hematoxilina - eosina (HE). Neurônios piramidais, células gigantes de Betz* [Normal motor cortex with hematoxylin-eosin (HE). Pyramidal neurons, giant Betz cells]. Anatpat – UNICAMP. Retrieved August 25, 2021, from http://anatpat.unicamp.br/bineucortexmotornlhe.html

Universo Programado. (2020, June 25). *Inteligência Artificial detecta humanos ATRAVÉS de paredes!* [Video]. YouTube. https://www.youtube.com/watch?v=JWuS6q9EYA0&list=PLus_DAOVXauYgyIgn-I6_7u2gUQEb-Rgu&index=15&t=0s

Universo Programado. (n.d.). *Home* [Universo Programado]. YouTube. Retrieved August 25, 2021, from URL https://www.youtube.com/channel/UCf_kacKyoRRUP0nM3obzFbg

UNIVESP. (2017, December 6). *Eletromagnetismo – Apresentação da disciplina* [Video]. YouTube. https://www.youtube.com/watch?v=-UQGaneAZW8

UNIVESP. (2018, August 7). *Licenciatura em Física - Eletromagnetismo - 14º Bimestre* [Video]. YouTube. https://www.youtube.com/playlist?list=PLxI8Can9yAHfsSKveLkqvvO3yZrGrNiQO

Veja. (2016, May 6). *Obama lança programa para mapear cérebro humano* [Obama launches program to map the human brain]. *Veja.* https://veja.abril.com.br/ciencia/obama-lanca-programa-para-mapear-cerebro-humano/

Velho, J. (2016). *Tratado de Computação Forense* [Computer Forensic Treaty]. Campinas, SP: Millennium.

Vicente, J. (2019, November 1). *Controlar máquinas com o pensamento parece um sonho, mas pode ser pesadelo* [Controlling machines with thought sounds like a dream, but it can be a nightmare]. Tilt UOL. https://www.uol.com.br/tilt/noticias/redacao/2019/11/01/controlar-maquinas-com-o-pensamento-parece-um-sonho-mas-pode-ser-pesadelo.htm?fbclid=IwAR1p2xNuO-NgL3gYrNYymDhWWWslt0d8s_XkBleaH2D5JWNcnurBD1Tzzl0

Vigotskii, L., Luria, A., & Leontiev, A. (2001). *Linguagem, desenvolvimento e aprendizagem* [Language, development and learning]. São Paulo, SP: ícone.

Vigotsky, L. (2010). *A construção do pensamento e da linguagem* [The construction of thought and language]. São Paulo, SP: Martins Fontes.

Vilicic, F., & Thomas, J. (2021, March 26). *Os trunfos e os riscos da inteligência artificial* [Benefits and risks of artificial intelligence]. *Veja.* https://veja.abril.com.br/tecnologia/os-trunfos-e-os-riscos-da-inteligencia-artificial/

Voltolini, R. (2014, December 11). *China cria arma de micro-ondas que "ferve" moléculas de água do corpo* [China creates microwave weapon that "heats" water molecules in the body]. TecMundo. https://www.tecmundo.com.br/armas-de-fogo/69208-china-cria-arma-micro-ondas-ferve-moleculas-agua-corpo.htm?fbclid=IwAR1IJHcP1xY0LIoEW3K-FuHeg1jgz4yhkIfmZWxYJzkWBTbdWwiyGP6jXeI

Wang, J.-K., Jiang, X., Peng, L., Li, X.-M., An, H.-J., & Wen, B.-J. (2019, January). *Detection of Neural Activity of Brain Functional Site Based on Microwave Scattering Principle.* Department of Information and Communication, Guilin University of Electronic Technology. Retrieved August 25, 2021, from URL https://www.researchgate.net/publication/330548074_Detection_of_Neural_Activity_of_Brain_Functional_Site_Based_on_Microwave_Scattering_Principle?fbclid=IwAR3pCaVVyuLrq-2LUry09dTFDgPqlSOyi9JOKpBe5TD4_DcCmDbtPx9YpnA

WannaCry [WannaCry ransomware attack]. (2020, April 27). In Wikipedia. https://pt.wikipedia.org/wiki/WannaCry

White, D. (2019, March 25). *'Google brain' implants could mean end of school as anyone will be able to learn anything instantly*. The Sun. https://www.thesun.co.uk/tech/8710836/google-brain-implants-could-mean-end-of-school-as-anyone-will-be-able-to-learn-anything-instantly/

Wolpert, D. (2011, July). *The real reason for brains*. TED Talks. https://www.ted.com/talks/daniel_wolpert_the_real_reason_for_brains#t-77064

World Science Festival. (2015, March 18). *The Mind After Midnight: Where Do You Go When You Go to Sleep?* [Video]. YouTube. https://www.youtube.com/watch?v=stXhGMVJuqA&list=PLus_DAOVXauYgyIgn-I6_7u2gUQEb-Rgu&index=33&t=0s

The more than 500 references used to support the research for the preparation of this book will soon be available on the official website:
**www.invasionandmindcontrol.com**

Printed in Dunstable, United Kingdom